CHEMISTRY AS ART

CHEMISTRY AS ART

LINSHU LIU AND GENNADY E. ZAIKOV
EDITORS

Nova Science Publishers, Inc.
New York

Copyright © 2006 by Nova Science Publishers, Inc.

All rights reserved. No part of this book may be reproduced, stored in a retrieval system or transmitted in any form or by any means: electronic, electrostatic, magnetic, tape, mechanical photocopying, recording or otherwise without the written permission of the Publisher.

For permission to use material from this book please contact us:
Telephone 631-231-7269; Fax 631-231-8175
Web Site: http://www.novapublishers.com

NOTICE TO THE READER

The Publisher has taken reasonable care in the preparation of this book, but makes no expressed or implied warranty of any kind and assumes no responsibility for any errors or omissions. No liability is assumed for incidental or consequential damages in connection with or arising out of information contained in this book. The Publisher shall not be liable for any special, consequential, or exemplary damages resulting, in whole or in part, from the readers' use of, or reliance upon, this material.

Independent verification should be sought for any data, advice or recommendations contained in this book. In addition, no responsibility is assumed by the publisher for any injury and/or damage to persons or property arising from any methods, products, instructions, ideas or otherwise contained in this publication.

This publication is designed to provide accurate and authoritative information with regard to the subject matter cover herein. It is sold with the clear understanding that the Publisher is not engaged in rendering legal or any other professional services. If legal, medical or any other expert assistance is required, the services of a competent person should be sought. FROM A DECLARATION OF PARTICIPANTS JOINTLY ADOPTED BY A COMMITTEE OF THE AMERICAN BAR ASSOCIATION AND A COMMITTEE OF PUBLISHERS.

Library of Congress Cataloging-in-Publication Data
Chemistry as art / LinShu Liu and Gennady E. Zaikov (editors).
    p. cm.
Includes bibliographical references and index.
ISBN 1-59454-585-5
1. Chemistry. 2. Chemical processes. I. Liu, LinShu. II. Zaikov, Gennadiæi Efremovich.
QD31.3.C425   2004
540--dc22   2005020311

Published by Nova Science Publishers, Inc. ✣ *New York*

CONTENTS

Preface		vii
Chapter 1	Construction of Gene Delivery Systems from Biodegradable Polysaccharides: Mechanism, Procession and Challenges *LinShu Liu, Wendy H. Kramer, Marshall L. Fishman and Kevin B. Hicks*	1
Chapter 2	Tests of Materials. Study of Operation Mechanism at High-Temperature *G. E. Zaikov, A. A. Donskoi and M. A. Shashkina*	25
Chapter 3	Degradation of Aromatic Co-Polyesters Derived from N-Oxybenzoic, Tere- and Isophthalic Acids, and Dioxydiphenyl *E. V. Kalugina, K. Z. Gumargalieva and V. G. Zaikov*	49
Chapter 4	Linear Free Energy Relationships in Chemistry of Solutions *R. G. Makitra, A. A. Turovsky and G. E. Zaikov*	59
Chapter 5	Copolymers with Cyclic Fragments in Dimethysiloxane Backbone *O.V. Mukbaniani and G. E. Zaikov*	81
Chapter 6	Characterization of Nitroxyl Radicals Produced from Hindered Amines During Accelerated Aging of Polymers *Leonid Yu. Smoliak, Wolf D. Habicher, Gabriele Theumer, Nikolay R. Prokopchuk, Sergey G. Mikhalyonok, Svetlana V. Nesterova and Andrey V. Evsey*	139
Chapter 7	Physical Grounds for Cluster Model of Polymer Amorphous State Structure *G. V. Kozlov and G. E. Zaikov*	147
Chapter 8	Complex-Radical Polymerization of Styrene in the Presence of Metallocene Initiating Systems *Yu. B. Monakov, N. N. Sigaeva, S. V. Kolesov, A. U. Abdulgalimova and E. M. Prokudina*	203
Chapter 9	The Structural Treatment of Fluctuation Free Volume in Amorphous State of Polymers *G. V. Kozlov, G. E. Zaikov and Yu. S.Lipatov*	213

Chapter 10	Concentration Dependence of Melt Density of Ethylene-Vinylacetate Co-Polymers Binary Blends *O. V. Stoyanov, A. E. Zaikin, R. M. Khyzakhanov, R. Ya. Deberdeev, G. E. Zaikov, Ya. V. Kapitskaya, S. Yu. Sofina*	**249**
Chapter 11	Physical Chemistry of Topological Disorder in Polymers. Some Remarks on the Polymer Memory *Yu. A. Shlyapnikov and N. N. Kolesnikova*	**259**
Chapter 12	On the Structural Memory in Polymers *N. N. Kolesnikova and Yu. A. Shlyapnikov*	**267**
Chapter 13	Description of Carbon Plastics the Heat Conductivity within the Framework of Fractal Analysis *G. V. Kozlov., A. I. Burya, I. V. Dolbin and G. E. Zaikov*	**277**
Chapter 14	Anisotropy of a Fiber Structure and the Frictional Wear of Composites on the Basis of Polyarylate *G. V. Kozlov, A. I. Burya and G. E. Zaikov*	**285**
Chapter 15	A Influence of Feedback in the Structure of Carbon Plastics on Their Properties *G. V. Kozlov, A. I. Burya and G. V. Zaikov*	**291**
Chapter 16	Factors Defining the Selectivity Polymer Membranes: A Fractal Model *G. V. Kozlov, I. V. Dolbin and G. E. Zaikov*	**299**
Index		**307**

PREFACE

In chapter 1, it is shown that gene therapy for the treatment and prevention of diseases on a genetic level is still in its infancy. One of the major challenges in using gene therapy is finding efficient, safe and stable ways to introduce genes into target cells. Although a whole panel of viral and non-viral vectors is available to introduce genetic materials into cells, a major drawback of most vectors is safety, lack of target cell specificity and low gene transfer efficiency. At present, the *in vivo* gene expression levels for polysaccharide-derived gene vectors are lower than for viral vectors and for some synthetic polymer-gene vectors. However, polysaccharide-derived gene vectors present several critical advantages over other vectors, such as safety, low immunogenicity, capability to deliver large genes and large-scale production at low-cost. Thus, polysaccharide-gene vectors are preferred and have attracted increasing attention in the development of gene delivery technology. In this review paper, research on the construction of polysaccharide-derived gene delivery vectors and the recent research trials about the controlled release of plasmid DNA are briefly discussed. The paper places emphasis on the principle of polysaccharide-DNA vector preparation and occurring technical challenges. Based on recent achievements and existing obstacles, novel strategies are suggested to improve the transfection efficiency of polysaccharide-gene delivery vectors.

Chapter 6 is about polymers which contain hindered amine stabilizers (HAS) and were studied by Electron Spin Resonance 'in situ' in order to measure the nitroxyl radical (NR) concentration in processes of thermo and photooxidative degradation. The correlation links between the type of degradation, NR concentration and efficiency of stabilizers were investigated. The role of chemical and physical aspects that could affect the efficiency in polyolefins depending on the conditions of degradation were discussed. It was observed that HAS substances in Poly(ethylene terephthalate) are inactive after thermal processing but produce NR and, therefore, participate in the stabilization process during posterior UV aging.

Chapter 8 says, that for the radical polymerization of styrene initiated by metallocene systems, kinetic nonuniformity distributions of active centers have been determined based on experimentally derived molecular mass distribution curves using the Tikhonov regularization method. The polymodal pattern of the kinetic nonuniformity distribution curves indicates the existence of two types of active centers changing their kinetic activity during polymerization.

Chapter 9 shows that fluctuation free volume in polymers has fractal structure. The microvoids of free volume may be simulated by a D_f-dimensional sphere. The microvoids' sizes are controlled by the volume necessary for accumulation of energy of thermal fluctuations, required for their formation. The absolute values of relative fluctuation free volume f_g may be considered as characteristics of the nonequilibrium state of the polymer structure. For quasi-equilibrium structures the value f_g coincides with data, following from the equation Williams-Landel-Ferry. The microvoids of fluctuation free volume form structure, which reflects the polymer structure. These microvoids' structure may be represented as a specific pulsating percolation cluster due to thermodynamically nonequilibrium state of a polymer.

The multifractal nature of semi-crystalline polyethylene structure is verified and the estimation of its multifractal characteristics is given on the basis of diffusion properties. The melt of polyethylene is the Euclidean object with monodisperse distribution of microvoid sizes. These circumstances define the different form of dependencies of a diffusivity on themolecular size of a gas-diffusant. The microvoid itself is a specific multifractal, whose fractal properties are not determined by its structure or by its environment.

Chapter 10 is concerned with the study of the melt density of binary blends of copolymers ethylene and vinyl-acetate.

Chapter 11 describes how long-term structural memory of a polymeric substance is determined by concentration, size and space distribution of the elements of topological disorder present in it. Short-term structural memory is the result of reversible reconstruction of existing elements.

Chapter 12 reviews long-term structural memory of a polymeric substance determined by concentration, size and space distribution of the elements of topological disorder present in it. Short-term structural memory is the result of reversible reconstruction of existing elements.

In chapter 13 it is shown that the heat conductivity of carbon plastics on the basis of phenylone can be described within the framework of a fractal model. Depending on the dimension of filler fibres network (system) such descriptions can be obtained by the application of two limiting cases: random network of resistors (RNR) or random superconducting network (RSN).

Chapter 14 describes the influence of the anisotropy of surface structure of short fibers on an interfacial layer structure in polymer composites is shown. In its turn, mentioned changes of structure cause an essential variation of frictional wear for these materials. In this aspect the important role plays the existence of a hydrogen bonds polymer-filler.

Chapter 15 reviews a structural sense of feedback effect for carbon plastics based on phenylone and demonstrates its influence on the strength of these materials. The decrease of feedback parameter is related to a substantial increase of carbon plastic's macroscopic strength. To control the value of this parameter is possible by the change of fibres' orientation factor, which is the controlling parameter for interfacial regions.

In chapter 16 it is shown that the selectivity of the polymer nonporous membranes by the diffusivity depends on the polymer structure by rather complex way. The most important characteristic in this case is the dimension D_t controlling the gas transport processes and the molecular mobility degree influences only at large enough values of diameter of the selected gas molecules, the difference of these diameters and the dimension D_t.

In: Chemistry as Art
Editors: L. Liu and G. E. Zaikov, pp. 1-23

ISBN 1-59454-585-5

Chapter 1

CONSTRUCTION OF GENE DELIVERY SYSTEMS FROM BIODEGRADABLE POLYSACCHARIDES: MECHANISM, PROCESSION AND CHALLENGES

LinShu Liu*◊, Wendy H. Kramer, Marshall L. Fishman and Kevin B. Hicks

Eastern Regional Research Center, ARS, U.S. Department of Agriculture
600 East Mermaid Lane, Wyndmoor, PA 19038

ABSTRACT

Gene therapy for the treatment and prevention of diseases on a genetic level is still in its infancy. One of the major challenges in using gene therapy is finding efficient, safe and stable ways to introduce genes into target cells. Although a whole panel of viral and non-viral vectors is available to introduce genetic materials into cells, a major drawback of most vectors is safety, lack of target cell specificity and low gene transfer efficiency. At present, the *in vivo* gene expression levels for polysaccharide-derived gene vectors are lower than for viral vectors and for some synthetic polymer-gene vectors. However, polysaccharide-derived gene vectors present several critical advantages over other vectors, such as safety, low immunogenicity, capability to deliver large genes and large-scale production at low-cost. Thus, polysaccharide-gene vectors are preferred and have attracted increasing attention in the development of gene delivery technology. In this review paper, research on the construction of polysaccharide-derived gene delivery vectors and the recent research trials about the controlled release of plasmid DNA are briefly discussed. The paper places emphasis on the principle of polysaccharide-DNA vector preparation and occurring technical challenges. Based on recent achievements and existing obstacles, novel strategies are suggested to improve the transfection efficiency of polysaccharide-gene delivery vectors.

---

*Correspondence to LS. Liu. Tel.:+1-215-233-6486; Fax:+1-215-233-6406. E-mail address: lsliu@errc.ars.usda.gov

◊ Mention of brand or firm name does not constitute an endorsement by the U.S. Department of Agriculture above others of a similar nature not mentioned

Keywords: Polysaccharide, Gene Delivery, Hyaluronate, Pectin, Chitosan, Polysaccharide-Oligoamine

1. INTRODUCTION

Gene transfer has attracted increasing interest and attention because it presents an advanced method for the treatment of genetic disorders whether inherited or acquired. Naked genes are easily destroyed in the circulatory system; hence, the administration of naked genes is often accompanied by low efficiency and large volume injections. Research has been focused on the development of gene delivery systems, which are designed to control the location of genes within the body by affecting the distribution and access of gene expression systems to the target cells, and/or recognition by cell-surface receptors followed by intracellular trafficking and nuclear translocation. Gene delivery systems should serve to both protect a gene expression system from premature degradation and to effect nonspecific or cell-specific delivery to target cells.

Gene delivery systems are divided into two groups: viral carriers by which DNA to be delivered is inserted into a virus; and non-viral carriers by which DNA to be delivered is inserted into a polymeric (natural, semisynthetic and synthetic) vector [1, 2]. The use of non-viral vectors elucidate some important advantages: (1) They are able to introduce DNA into non-dividing cells; (2) They do not integrate into the chromosome; (3) They do not possess infective risk and (4) They are potentially less expensive than viral vectors. In contrast, the relatively low transfection efficiency of the non-viral vectors in comparison to viral gene delivery vectors is the major disadvantage. The goal in developing non-viral vectors is to design a system that simultaneously achieves high efficiency, prolonged gene expression, and low toxicity.

Cationic polymers have been intensively investigated as non-viral vectors for the potential of gene delivery. Cationic polymers based on amino acids are reported to be immunogenic and therefore not applicable for gene delivery in vivo [3, 4]. Other polymers described in the literature include polymers with a random distribution of amino groups – cationic sites, which are part of a polymer backbone such as poly(ethyleneimine), poly(amidoamine), and poly(alkylaminoglucaramide). Most of these polymers are toxic to cells and non-biodegradable.

Efforts have been made to explore the potential of polysaccharides as new non-viral vectors for gene delivery. The use of biodegradable polysaccharide carriers is especially suitable for transfection and biological applications because polysaccharide-mediated interactions play important roles in biological processes such as receptor-mediated endocytosis, opsonization, apoptosis and metastasis. Polysaccharides have been applied to cell recognition studies and the design of biomedical materials. Furthermore, they are water soluble, biocompatible, have low immunogenicity and minimal cytotoxicity. In addition, polysaccharides can be readily transported to cells *in vivo* by known biological processes [5].

In this paper, recent research in the development of polysaccharides into non-viral gene delivery vectors are briefly reviewed. Several critical issues in the chemistry regarding the preparation of polysaccharide-DNA vectors are discussed in the next section, which is followed by a section where the progress in two vectors from hyaluronate and chitosan and their derivatives are reviewed and discussed in detail. In this section, a brief description on the potential of pectin as a gene delivery vehicle is included. Also in this section, polysaccharide-oligoamines, the most recent members of the polysaccharide-based gene delivery family, are introduced. In the last part of this paper, we list the areas, where the techniques of gene therapy have been intensively investigated and showed promise in practice.

2. CRITICAL POINTS IN VECTOR CONSTRUCTION

The use of polymer-DNA vectors assumes that, before gene expression, the DNA will be able to clear several chemical and biological barriers such as DNA/vector complex formation, cell entry, escape from endosome, dissociation from the vector and finally nuclear translocation [6]. This is illustrated in Figure 1.

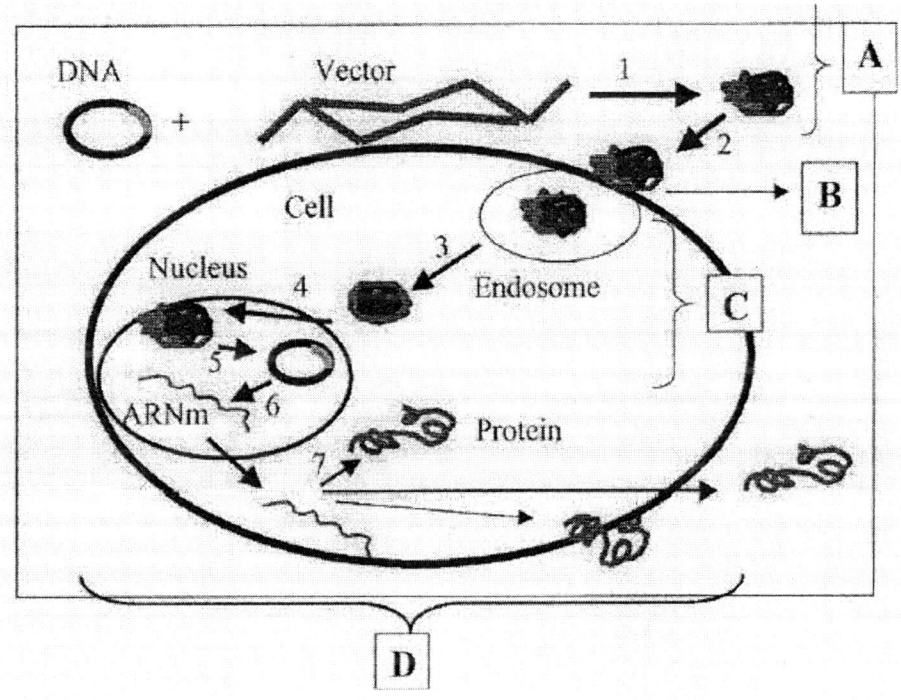

Figure 1. Simplified gene therapy mechanism. [A] Extracellulat trafficking: (1) insertion of the gene in the vector; [B] Internalization contact and crossing of the cell membrane; [C] Intracellular trafficking: (2) uptake of the vector complex into intracellular endosome; (3) comnplex release from the endosome into cytoplasma; (4) uptake of the complex in the nucleus; [D] gene expression: (5) DNA dissociation from the vector; (6) ARNm transcription from the gene; (7) protein translation from ARNm. The protein can be secreted out of the cell, be released into the cytoplasma, or fixed onto the membrane.
Adopted from Ref. [12] with permission

The delivery systems should protect plasmid DNA from DNA-degrading enzymes in the extracellular matrix. It should be able to penetrate the cell wall and to protect the DNA from degradation by lysosome and enzymes in the intracellular medium until it is internalized in the nucleus, which leads to the insertion of the genetic materials in its active form [7]. Finally, it is degraded and eliminated from the cell and tissue without causing toxicity. To realize the ultimate clinical therapeutic effect, research has focused on finding efficient and reproducible ways to construct delivery vectors, which can introduce genes into target cells [8]. Major drawbacks of most currently available vectors are: lack of stability and safety, lack of target cell specificity and low gene transfer efficiency. Strategies to control the size of vectors, to introduce chemical cross-linking on vectors' surface, and to chemically or biologically modify the surfaces of vectors have been developed to improve gene transfer efficiency.

2.1. Size of Vectors

Delivering DNA into cells require its condensation to a complex with the vector to a small size. For cationic polysaccharide polymers, this process is based on electrostatic interactions between negatively charged phosphate groups along the DNA backbone and positively charged groups on the condensing agent. For other polysaccharides, cross-linking with appropriate chemicals is required [9]. The size of gene delivery vectors influences the stability of DNA, aggregation and toxicity of vectors, cell target, and DNA transfection [10-12].

The effect of particle size on the stability of incorporated DNA in the circulatory system has been well defined [8, 12, 13]. The *in vitro* studies on vectors' stability were based on DNA degradation assays performed in tissue culture medium. It demonstrated that DNA complexed with sufficient cationic reagent or an appropriate cross-linking density to create condensed particles is protected from degradation of DNAse. However, the *in vivo* stability can not be directly predicted from the *in vitro* results.

It is generally reorganized that the small size vectors have the advantages of entering the cells and thus increase the transfection rate [12, 14-19]. Chitosan-DNA nanoparticles with the size less than 600 nm have been found effective for nasal gene delivery [20]. Typically, the particles smaller than 100 nm can be enclosed within endocytic vesicles, which permits them to enter the target cells via transferrin cytosis [17]. Endocytosis by many types of mammalian cells is limited to particles less than about 150 nm in diameter [21]. Tissues of the reticulo-endothelial system, the liver, spleen and bone marrow possess a sinusoidal endothelia structure that permits relatively free passage of materials without size restrictions, up to 100 nm [22]. There is also evidence that polymer microspheres ~1-10 μm in diameter are internalized preferentially through phagocytosis by antigen-presenting cells (APCs), such as debtritic cells, but not by other cells [23]. The ability to target APCs selectively is highly desirable in vaccine delivery, due to the pivotal role these cells play in initiating immune response against foreign antigen [24, 25].

In overview, more efforts to correlate the relationship between transfection efficiency with particles size are still desirable.

The transfection efficiency does not correlate with cell uptake. To achieve high transfection efficiency, plasmid must be ferried into the nucleus and taken from its vectors, in other words, the carrier-DNA complex should be easy to dissociate in the endosome (Figure 1). The effect of complex size on the unpack and release of DNA intracellularly is unclear and the subject has been less studied.

2.2. Surface Cross-Linking of Vectors

One strategy that has been applied to improve the *in vivo* stability of polymer-DNA vectors is to covalently cross-link the vector surface after the DNA has complexed. The cross-linking may reduce gene transfer activity of the vectors; the use of a more labile cross-linker was suggested. The blood system contains many components, such as proteins, lipids, carbohydrates and salts. These materials can bind to the vectors and cause their dissociation, degradation, decomplexation or aggregation. Furthermore, the interaction of vectors with cell surfaces plays an extremely important role. Cells surfaces contain areas of high, negative charge density. Thus, cell surfaces can promote the dissociation of cationic carbohydrate-DNA vectors, causing premature release of plasmid DNA, and escalate DNA degradation. Cross-linking may decrease the net charge density in cationic carbohydrate-DNA vectors, thus inhibiting the interaction of vectors and blood plasma components and cells.

2.3. Chemical and Biological Modification of Vectors

Surface modification with grafting oligomers or polymers is a widely accepted method to enhance the vector's stability. The method involves the use of hydrophilic segment such as polyethylene glycol, dextran, hyaluronate, poly(vinyl pyrrolidone) [9, 26-32]. The graft of these polymers dramatically suppresses particle aggregation and effectively prevents their dissociation.

The surface modification with hydrophilic polymers also inhibits the close contact between DNA and the backbone of polymers, thereby preventing the dehydration and compaction of the DNA, which leads to alternation of DNA structure. By measuring the hydrodynamic size (R_h), it was found that the ionically condensed DNA coils occupy only 10^{-3}-10^{-4} of the volume of the naked DNA coils [33], contrary to proteins, which show a unique tertiary structure. DNA coils do not condense into unique compact structures, as it can be determined by circular dichroism. As the structural DNA properties may influence the transfection efficiency, alternations in the tertiary structure of DNA by the polymers were investigated. Dextran, hyaluronate, and other hydrophilic polymers were found effective in protection of the reaction between DNA and cationic carbohydrates [34, 35].

The targeting of gene complexes to a desired cell population is another important subject in the field of gene therapy. Most studies focus on the effect of targeting ligands that are covalently attached to the DNA vectors and allow the uptake of DNA into cells via receptor-mediated endocytosis. In the other hand, inhibition of gene transfer or expression may occur by the introduction of excess ligand molecules. Different from the simple receptor-ligand systems of classical pharmacology, carbohydrate-DNA vectors are large, complex and poorly defined macromolecular assemblies. The ligand graft may also alter the local surface charge distribution, the size, shape and flexibility of the vectors; all these are strong factors in cell uptake and cytoplasmic entry. Furthermore, in most cases, large numbers of ligands are randomly immobilized on the surface of vectors, which also influences and complicates both cell binding kinetics and the mechanism of internalization [8, 36-38].

Promoting destabilization of the host cell membrane is the important attribute to allow genetic medicine to enter into the target cells. Incorporation of molecular entities, which mimic the structure of viral coat proteins, onto the vectors to increase membrane penetration has been assayed for many years. Although no significant progress has been reported in this regard due to the difficulty of the synthesis of such a material, success with this approach will be a fantastic step. Biological modification is also a strategy to improve the transit of plasmid DNA from cytoplasma into the nucleus through the nuclear envelope. The diameter of the channel for active transport is about 25 nm. A number of nuclear localization signals have been identified. These glycopeptides can be covalently attached to one terminal end of a DNA molecule; transfection enhancement was observed [39-42].

3. POLYSACCHARIDE DERIVED GENE DELIVERY SYSTEMS

3.1. Hyaluronate

Hyaluronate (HA) is one of the most important native polysaccharides in the body. It is non-toxic, biodegradable and biocompatible; thus, it has shown a wide potential in biomedical materials applications. The glycosaminoglycane consists of repeating disaccharide units of D-glucuronic acid and (β-1, 3) N-acetyl-D-glucosamine; and distributes throughout the extracellular matrix (ECM), connective tissues and organs of all high animals (Figure 2). The important functions include the organization of ECM structure, transportation of nutrients, modification of inflammation and the regulation of cell adhesion. To date, HA has been successfully used as biomaterials in the forms of hydrogels for viscosurgery and viscosupplementation [43, 44], scaffolds for wound healing [45], bone and cartilage regeneration [46-49], and hydrophobic coating for medical devices [50]. Recently, scientists explored the possibility to use native HA in construction of gene delivery systems [51, 52].

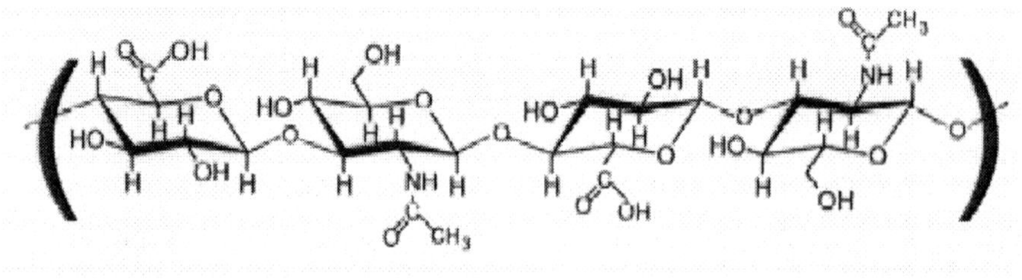

Figure 2. Chemical structure of hyaluronan. *Adopted from Ref. [51] with permission.*

3.1.1. HA-DNA Matrices

HA-DNA matrices were prepared by mixing plasmid DNA in HA solution followed by lyophilization to form a porous structure [52]. The matrices were then immersed in a N,N-dimethylformamide solution containing adipic dehydrazide and ethyl-3-[3-dimethyl amino] propyl carbodiimide for cross-linking. This method provided an uneven cross-linking pattern. Due to the greater degree of exposure of the matrix surfaces to the reagent during cross-linking, and the difficulty of diffusion of cross-linkers into the core of the structure, a higher cross-linking density was generated on the exteriors of the matrices, rather than in their core areas. This heterogeneous cross-linking may be beneficial. The HA-DNA matrices were degraded by a combined process of hyaluronidase mediated digestion of HA and the hydrolysis of the hydrazide linkages. As the matrices eroded, the lesser cross-linked layer was exposed and the rate of DNA release accelerated, compensating for the diminishing surface area.

Sustained release capability was obtained on degradation of the HA-DNA matrices. The release of DNA was gradual and continuous over an extended period and advantageous over naked DNA or liposomes, which are known to have low persistency. It was found that the release kinetics were altered by the amount of DNA loaded into the matrices and by the incubation time with the cross-linking reagents. Since HA binds to receptors on human cutaneous fibroblasts [53], the HA-DNA matrices may be a useful therapy for cutaneous wounds, such as diabetic ulcer. Studies using DNA encoding platelet-derived growth factor (PDGF) showed that the release of PDGF DNA from HA-DNA matrices had the capability to transfect cells and produce a functional protein that enhances cell proliferation. The recognition and interaction between HA and HA receptors (CD44 and RHAMM) presenting on cell membranes may be the aid in transporting DNA into cells.

3.1.2. HA-DNA Microspheres

Polymeric microspheres offer several advantages in drug and gene delivery applications, such as dosing, controlled release kinetics, and cell target via receptor interaction. Because of the high swelling behavior in aqueous solution, HA microspheres were usually prepared in organic solvent after it was converted into water insoluble esterified derivatives [54]. It puts difficulty in the process of gene encapsulation. Recently, formulation of native HA into microspheres in an aqueous environment was conducted using adipic dehydrazide mediated cross-linking chemistry [51], the method originally used to prepare HA hydrogels.

HA-DNA droplets were prepared in a water-in-oil emulsion system and the formed microspheres were isolated from the system by the addition of isopropyl alcohol to the system. This method generated HA-DNA microspheres with spherical conformation and heterogeneous size distribution. The median and mean diameters of the microspheres were 6 and 20 µm, respectively. A release kinetic study showed that the DNA release was independent from initial DNA loading in the microspheres, while controlled by hyaluronidase activity in release media. Approximately 60% of the encapsulated DNA released from the microspheres for a period of 60 days. The *in vitro* transfection of released DNA was assessed using Chinese Hamster Ovarian cells. A relatively high level of transfection was initially observed, but gradually decreased with time and corresponded to DNA amount. The *in vivo* transfection study using a rat hind limb muscle model confirmed that all injected tissues were transfected with the DNA released from HA microspheres. In addition, the native HA microspheres are adaptable for targeting cellular receptors. HA microspheres coating with a humanized mAb to E- and P-selection exhibit selective adhesion to cells expressing these receptors. The selectivity of the adhesion, as reported, is striking and significantly better than poly(caprolactone) derived microspheres coating with the same ligand. Theses results suggest that HA-DNA microspheres could be an ideal vehicle for systemic delivery of therapeutic genes.

The analytical results of gel electrophoresis suggested that (1) There was some degree of interaction between DNA and native HA; (2) Portions of the DNA released from both HA-DNA matrices and microspheres were associated with cross-linked HA fragments; and (3) Partial degradation of the released DNA as portions of uncoiled DNA could be observed. These problems could be addressed by improving the cross-linking conditions.

3.2. Chitosan

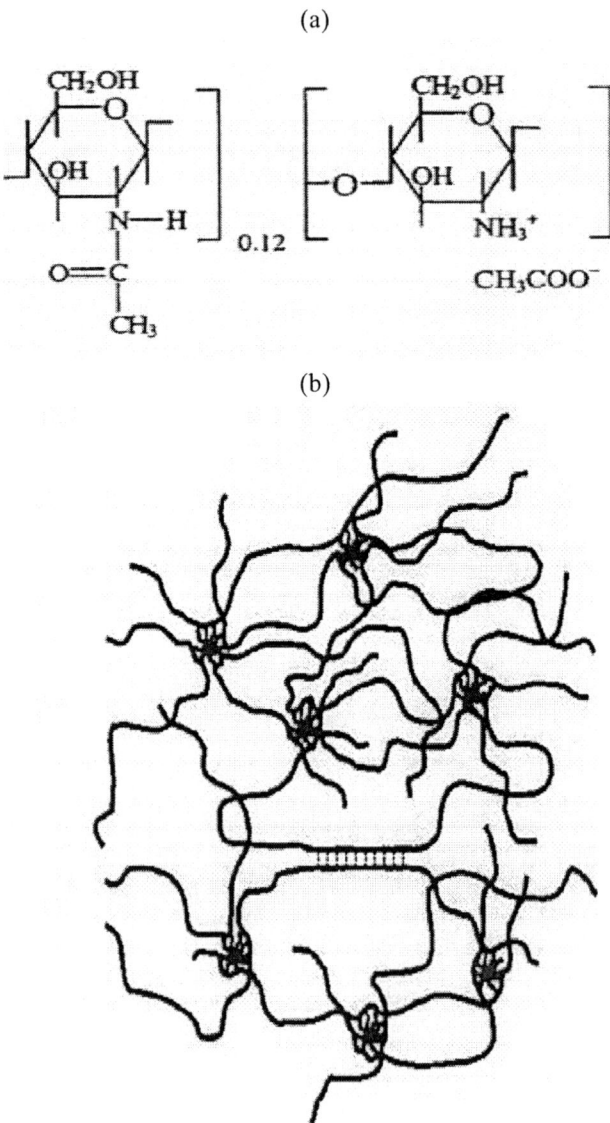

Figure 3. (a) Chemical structures of chitosan and (b) schematic representation of chitosan aggregates formed in aqueous solutions. *Adopted from Ref. [64] with permission.*

3.2.1. Chitosan as a Gene Delivery Vector: The Physical and Chemical Properties

Attention and sources have been focused on the development of chitosan into vectors for gene delivery [55]. Chitosan is a cationic polysaccharide prepared by alkaline deacetylation of chitin, which is abundant in the exoskeletons of crustaceans and insects. The molecules of chitosan consist of N-acetyl-D-glucosamine and D-glucosamine via 1,4-β linkages (Figure 3). The chitosan biopolymer is a weak base with a pKa value of about 6.2-7.0. Therefore, chitosan is insoluble in physiological solutions. In acid condition, the amine groups of chitosan are protoned, conferring to the polysaccharide a positively charged macromolecule [56]. Its cationic polyelectrolyte property provides a strong electrostatic interaction with negatively charged molecules such as DNA and some polypeptide or protein drugs. Clinical practices have demonstrated the non-toxic, low immunogenic and biodegradable nature of chitosan. Formulations from chitosan and its derivatives are used for nasal, ocular and peroral drug controlled delivery to prolong residue time of dosage forms and to improve drug absorption. In addition, chitosan excels in enhancing the transport of drugs across the cell membrane by opening the tight junctions [57-60]. These are the advantages of using chitosan as a gene delivery vehicle: (1) high gene binding efficiency, (2) protection effect of bound gene from nuclease degradation, and (3) minimizing possible DNA damage during complexation, because sonic irradiation is unnecessary and the use of organic solvents can be excluded [12]. In comparison with other cationic polymers such as gelatin, polyethyleneimine, poly(L-lysin), and poly(histidine)-grafting-poly(lysine), chitosan based gene vectors show advantages in gene transfection, biocompatibility, biodegradability and low toxicity [9, 16, 20, 34, 61-77].

Leong's group [61, 70, 78] investigated the influence of the degree of deacetylation (DD) on the level of gene transfection in the range of DD from 61.3% to 93.2%. They found that the gene expression level in serum decreased upon reducing DD, which is supposed to be from the lowered stability of the complexes. However, this instability contributed to a higher level of gene expression in muscle. Analogous to molecular weight, the degree of deacetylation is a complex factor influencing the transfection efficacy. The presence of N-acetyl groups on the chitosan backbone imparts a hydrophobic property, which renders chitosan self-aggregating in aqueous solution, probably affecting the interaction of chitosan and DNA.

The study on the relationship between the molecular weight and radius of gyration (R_g) of chitosan in aqueous solution revealed that the polyelectrolyte acted more like a Gaussian coil rather than the worm-like chain model, which was found for common polyelectrolytes [79]. The size and solution pH are two important parameters that would dictate the permeable capability of chitosan on cell membrane. Furthermore, researchers reported two types of hydrophobic aggregation in aqueous solutions of chitosan and its hydrophobically modified derivatives: one related to hydrophobic domains typical for different associating polymers with hydrophobic side chains and another inherent to chitosan itself [64]. It was considered that for chitosan the combination of electrostatic and hydrophobic forces induce destabilization of cell membrane.

3.2.2. Chitosan-DNA Nanoparticles

Chitosan-DNA nanoparticles with the mean size ranging from 100 to 600 nm can be prepared by mixing chitosan with plasmid DNA at certain ratios [10, 20, 66, 80]. The particles smaller than 100 nm can be enclosed within endocytic and/or pinocytotic vesicles. Cell viability study following incubation of cells with smaller particles also showed less toxic than with larger particles. Since the particles size is critical to blood circulation stability, cellular uptake and bioavailability, efforts have been made to develop methods to control particle sizes. Parameters influencing chitosan-DNA particle size include the molecular weight of chitosan, DNA concentration (charge ratios between chitosan and DNA), types and content of salts and pH value in media, as well as the temperature. The results have been summarized in two recent review papers by Alonso [80] and Borchard [10]. Gene transfer efficiency of chitosan-DNA nanoparticles was another subject of extensive discussion. Although several reports have indicated that chitosan with the molecular weight ranging from 15 kDa to 50 kDa were excellent as gene transfer reagents, optimal molecular weight of chitosan is dependent on cell lines. Effect of medium pH on transfection efficiency was attributed to the interaction of vectors and cells. At the pH lower than 7, amine moieties in chitosan are protoned, and the chitosan-DNA nanoparticles are positively charged, which facilitated the binding of particles to the negatively charged cell surfaces. Obviously, to do so, in-complexed amine groups should remain on the chitosan-DNA nanoparticles.

Besides the particles size, the molecular weight of chitosan influences the stability of chitosan-DNA nanoparticles. The complex particles prepared from the chitosan of larger than 5 kDa at the charge ratio of 1:1 were stable in serum [81]. Although, lower molecular weight chitosan can be administered intravenously without liver accumulation [82], the stability of nanoparticles from the chitosan of less than 5 kDa decreased in serum [69, 83].

3.2.3. Modified Chitosan Vectors

Modified chitosan derivatives were widely used to construct gene delivery vectors in order to improve chitosan solubility and gene transfection efficiency. Dextran, poly(ethylene glycol), poly(vinyl pyrrolidone) [31, 84, 85] have been grafted to chitosan; resultant chitosan derivates show an increasing water solubility. Chloroqiune and imidazole derivatives and inactivated adenovirus have been introduced to chitosan to promote the release of chitosan-DNA complex from endocytic vesicles through proton sponge mechanism [9, 86, 87].

3.2.3.1. Deoxycholic Acid Modified-Chitosan Vector

Deoxycholic acid is a main component of bile acid formed by bacterial action from cholate. In the body, deoxycholic acid acts as a detergent to solubilized fats for intestinal absorption; deoxycholic acid can self-assemble in water. Deoxycholic acid modified chitosan was prepared in aqueous solution using water soluble carbodiimide as a coupling reagent [83, 88]. The modified chitosan provides colloidally stable, self-aggregates with the mean size of 160 nm and an uneven size distribution. The transfection efficiency of the chitosan self-associate-DNA complex was enhanced in comparison with naked DNA. But, it was lower than the unmodified chitosan-DNA vector, as examined on monkey kidney COS-1 cells. However, the coupling efficiency was quite low; the hydrophobically modified chitosan only contained 5.1 deoxycholic acid groups per 100 anhydroglucose units.

3.2.3.2. Alkylated-Chitosan Vector

Alkylated-chitosan derivatives were prepared by Yao's group [34, 67, 89, 90]. Chitosan was hydrophobically modified with butyl, octyl and dodexyl halides. The resultant alkylated-chitosan assembled with DNA to form a polyelectrolyte complex. The electrostatic interaction between DNA and alkylated-chitosan results in a globular structure and thermal stability of the complex was enhanced. The complex could be dissociated by the addition of small molecular salt solution. DNA released from the complex is well protected and remained intact due to the protection from DNAse offered by the alkylated chitosan. However, no results on transfection efficacy *in vitro* or *in vivo* have been published.

3.2.3.3. Quaternized-Chitosan Vector

As chitosan is only soluble in acidic solutions, chitosan in physiological conditions can not reduce the transient opening of cell membrane, which is a critical step in gene transmembrane transport. Trimethyl chitosan (TMC) was proposed to overcome this issue, which expands chitosan applications [74, 91, 92]. When TMC was mixed with DNA, complexes were spontaneously formed. The TMC showed excellent solubility in water, regardless of the solution pH values. At both 2:100 and 2:10 DNA/oligomer ratios, the size of complex from TMC was smaller than those from unmodified chitosan. The degree of quaternization (% of amine was quaternized) of chitosan directly effect the transfection efficiency. TMC with the quaternization degree of 50% (TMC-50) markedly increased the transfection efficiency in comparison with naked DNA or lipoplexes-mediated DNA when tested on monkey kidney cell line COS-1, but the vector was not able to enhance the transfection efficiency in differentiated cells like Caco-2 cell.

3.2.4. Chitosan-Ligand Conjugated Vector

In some cases, cellular uptake of chitosan-DNA vectors appears to occur without the ligand-receptor interaction involving. The modification of chitosan with various biospecific ligands is the strategy for targeted gene to traffic into specific cells.

3.2.4.1. Galactosylated-Chitosan Vector

Galactose is a hepatocyte-binding ligand. Galactose was introduced to chitosan for liver specificity, and the resultant galactosylated-chitosan was grafted with a hydropholic Poly(ethylene glycol)(PEG) for stability in water and cell permeability [31]. DNA complexed with the galactosylated-chitosan is stable and protected from enzymatic degradation in plasma. As demonstrated by, the galactosylated-chitosan-PEG vector were efficiently transfected into Chang liver cells carrying asialoglycoprotein receptor, indicating the galactosylated-chitosan an effective hepatocyte-targeted gene carrier.

3.2.4.2. Transferrin Conjugated-Chitosan Vector

Transferrin receptor is found on the surfaces of many mammalian cells, and responsible for ion import. Hence, transferrin, as a ligand, is able to transfer small molecular weight drugs, nonbioactive macromolecules, liposomes, plasmid DNA and oligonucleotides into cells through a receptor-mediated endocytosis mechanism [76, 93, 94]. Transferrin could be conjugated to chitosan either through a disulfide bond or by reaction with the amine groups on chitosan, after creation of aldehyde groups in the polysaccharide moieties of transferrin [70]. Transfection efficiency depended on the cell type and the level of surface transferrin receptor expression, and could be enhanced by endosomalytic agents such as chloroquine. In addition, although a higher transferrin conjugation increases transfection ability, above a certain number the continuing increase of transferrin may reduce DNA transfection efficiency, which, presumably, could be attributed to the self-conjugation of transferrin. An appropriate modification degree is required.

Chitosan is a promising non-viral gene transfection vector. Through the mediation of particle size, molecular weight and surface modification, the vector stability, cell uptake and transfection efficiency can be further improved; targeted cells also can be specified. As the applications of chitosan vectors are expanded, the following subjects have attracted more attention: (1) the improvement of DNA transport efficiency, (2) understanding the mechanism of DNA unpacking from chitosan vectors, and (3) site-specific gene delivery for gene therapy using ligand modified-chitosan vectors. Moreover, the difficulty in controlling the density and distribution of cationic groups in chitosan molecule are disadvantages, which raise concerns in toxicity and relatively low transfection efficiency.

3.3. Pectin

Pectin is a cell wall structural carbohydrate present in all higher plants. Primarily, pectin contains large amounts of poly(D-galacturonic acid) bonded via a-1, 4-glycosidic linkage (Figure 4). Pectin also contains neutral sugars such as L-rhamnose which are either inserted in or attached to the main chains. In pectin from sugar beet, feruloylester substituents attach on the side chains. In pectin from all sources, the carboxyl groups are partially in the methyl ester form. The degree of esterification (DE) varies depending on the source of the pectin and enzymatic activity in the process of ripening and maturation, and the conditions under which the isolation is conducted. Some of the carboxyl groups may be converted to carboxamide groups, when ammonia is used in the process of de-esterification, producing amidated pectin. Perhaps, the most attractive property of pectin for industrial applications, both food and pharmaceutical industries, is its gelling activity. Studies on the congelative properties of pectin have led to applications of pectin in the food industry as gelling or thickening agent in the beginning, and then, as an excipient for pharmaceutical purposes.

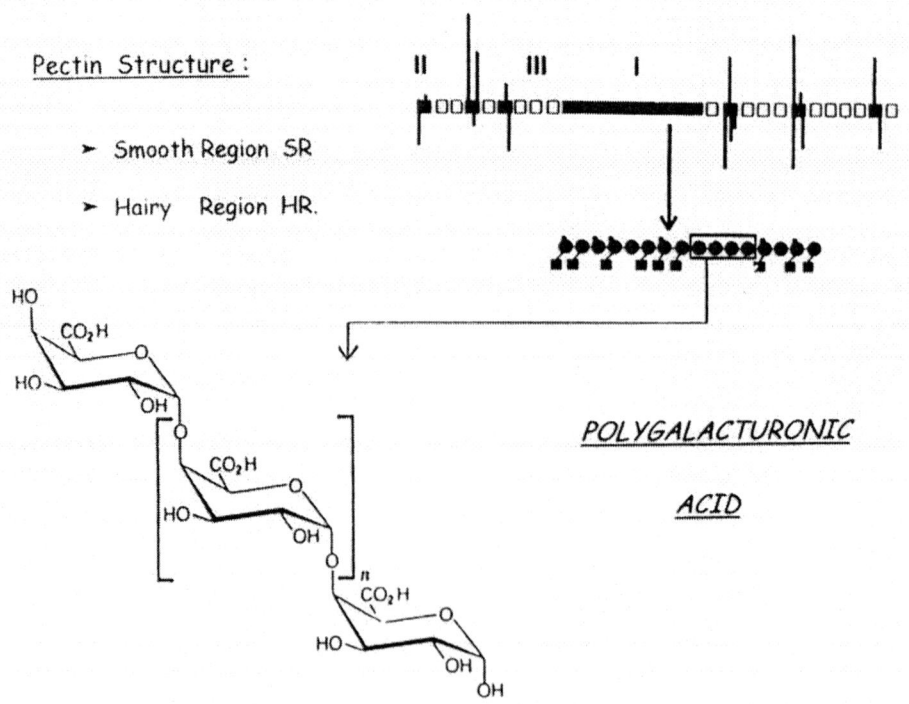

Figure 4. Structure of Pectin. *Adopted from Ref. [96]* with permission

Factors which determine whether gelation can occur and which influence gel characteristics are the types and concentrations of pectin, degree and types of modification on the carboxyl and hydroxyl groups, solution pH, temperature and the presence of cations. All these parameters are interdependent [95, 96].

Since pectins undergo extensive chemical modifications during plant growth and development, the modification of pectins also depends on the ripening process and isolation conditions. Moreover, pectin from different sources (such as citrus and apple) differs in molecular weight, composition ("hairy/smooth" regions), density and distribution of the functional groups. The advantages of these properties have motivated scientists to develop pectin into gene delivery vectors. Currently, scientists in Ben Gurion University, Israel, and scientists in the Eastern Regional Research Center, U.S. Department of Agriculture, are cooperating in a program of pectin-based gene delivery vectors. Up to the date, no *in vitro* or *in vivo* data have been published.

3.4. Oligoamine Modified Polysaccharide-DNA Vector

In contrast with the abundance of structurally different synthetic non- viral gene delivery vectors, there is only a small number of available polycations of natural origin. The recent development of polysaccharide-oligoamine conjugates added a number of new members to the polysaccharide-based gene delivery family. By this method, oligoamine of two to four amino groups were grafted onto natural polysaccharides. The resultant vectors are capable of complexing various plasmids and being administered into various cells in high yield to produce a desired protein. These polycations are expected to better meet the requirements for effective complexation and delivery of plasmid and to be biodegraded into nontoxic components at a controlled rate.

3.4.1. Chemistry of Polysaccharide-Oligoamine Conjugates

The polysaccharide-oligoamine conjugates were prepared in a multi-step reaction procedure. Firstly, aldehyde groups were created on polysaccharide by periodate oxidation; then oilgoamine molecules were grafted via Schiff-base linkages between the amine groups and the active aldehydes; the reversible Schiff-base bonds were finally converted to a stable – C-N- bond by a reduction reaction with $NaBH_4$. The concept is illustrated in Figure 5.

The difference in the degree of oxidation between the polymers relies on the polysaccharide structures. The periodate only opens the carbon-carbon bonds, on which hydroxyl groups are attached to two adjacent carbons, thus the oxidation occurred on the polymer backbone of dextran, pullulan, hyaluronate, and alginate, resulting in dialdehyde formation with some chain session; while, oxidation takes place on the branched chains of arabinogalactan. The graft of side chain oligoamines to either a linear or branched hydrophilic polysaccharides provides a two- or three-dimensional, active environment to react with anionic DNA. Supposedly, this structure is easy to be synchronized to provide an optimal complexation with various DNA [97, 98].

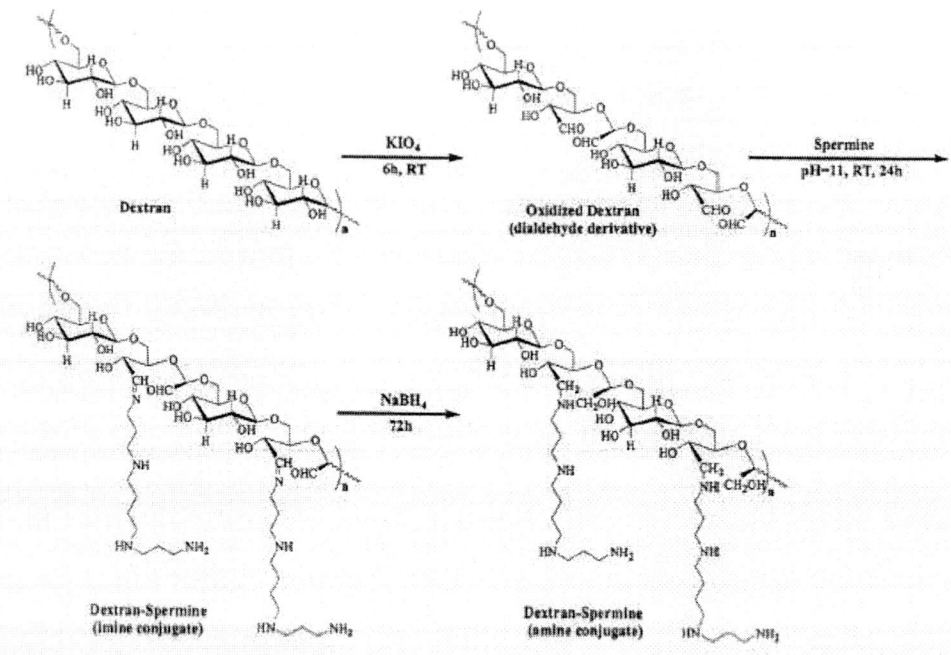

Figure 5. Grafting of spermine moities on dextran polysaccharide. *Adopted from Ref. [99] with permission*

3.4.2. Transfection Activity of Polysaccharide-Oligoamine Conjugates

As the author reported, over 300 cationic conjugates have been synthesized starting from various polysaccharides and oligoamines and tested for their transfection activity using various cell types and marker genes [36, 37, 97, 99, 100]. However, only part of the resultant complex was found to be active in transfecting cells *in vitro*. Studies on dextran derivatives were more reproducible.

Since chitosan has shown excellent in the regard of gene delivery vectors, and up to date chitosan is the only commercially available polysaccharide carrying cationic groups (if we do not count the uncertain numbers of cationic groups on the oligopeptide moieties attached to polysaccharides); the synthesis of polysaccharide-oligoamines conjugates is of important. This work indicates that the structure of polycations has a significant role in the transfection activity. Strategies to improve the transfection efficiency of the oligoamine modified polysaccharides are under investigations.

4. POTENTIAL APPLICATIONS

Adapting controlled release technologies to DNA delivery has the potential to overcome extracellular barriers that limit gene therapy. Polymeric gene vectors could be delivered intravenously, by implantation, or through the nasal and oral routes. Naturally, occurring carbohydrates vary greatly in chemical composition as well as the number of repeating units, providing a wide range of different vectors that can be easily assembled. It is becoming possible to design carbohydrate-based carriers with well-defined structural and chemical properties to meet the different requirements of "idealized" non-viral gene delivery systems. Despite the current limitations, natural carbohydrates have significant potential to serve as gene carriers for gene medicine.

4.1. Plasmid Gene Delivery and Expression in Skeletal Muscle

Muscle tissue can be used as a bioreactor for the secretion of therapeutic proteins into the surrounding tissues. Currently, plasmid DNA is delivered into muscles by the technology referred to as electroporation [101] or via an adenovirus vector [102]. Although direct injection of DNA into muscles has shown to lead to gene expression, the overall level of expression is low. Implantation of polymeric matrices or particles intramuscularly provides a more powerful set of tools since it can increase residence time within the tissue and protect against degradation, which will be broadly applicable to the development of plasmid-based gene therapies for the treatment of human diseases.

4.2. Tissue Engineering

Gene activated matrix, known as a local plasmid gene transfer technology, is the gene therapy designed specifically for tissue engineering. Animal studies in past decade, using gene activated matrix as a platform technology, have proved that plasmid genes can be delivered to acutely injured tendons, ligaments, bones, muscles, skin and nerves [30, 103-106]. The gene activated matrix consists of two ingredients: plasmid DNA and a biodegradable structural matrix carrier (Figure 6). Following implantation, plasmid DNA transfers endogenous wound healing cells, which leads to the expression of cytokines and growth factors in situ for a prolonged period. In comparison with cytokines or growth factors, plasmid DNA should not cause systemic toxicity, and is economical and relative simple to manufacture. Using this strategy, off-the-shelf tissue engineering products are expected to be available for direct placement into an acute wound bed [103].

4.3. Cancer Gene Therapy

During the past two decades, enormous research in the area of gene delivery has been conducted worldwide, in particular for cancer gene therapy applications. In the past few years, many carbohydrate-based vectors were developed and evaluated for delivery therapeutic genes that can terminate cancer cells. However, modification of current delivery systems with ligands seems an optimal strategy to enhance the transfection efficiency.

4.4. Genetic Vaccination

Genetic vaccination using DNA plasmid offers a possibility for achieving potent immune responses over many conventional vaccines of attenuated or inactivated pathogens [107, 108].

Biodegradable polymers, including biopolymers of polysaccharides and proteins, have been extensively investigated as delivery carriers for conventional vaccines and for DNA vaccines [2, 109-112]. Microparticulate encapsulation can protect DNA plasmid from physical and chemical damages. Furthermore, sustained release of DNA provides a continuing production of encoded protein antigen to generate prolonged immune responses. In comparison with some synthetic polymers such as poly(lactide-co-glycolide), naturally occurring carbohydrates are particularly useful materials for DNA delivery, because neither the staring materials nor the degradation products will alter environmental pH and DNA is pH sensitive.

In addition, recent reports indicated that DNA vaccines elicited a CD8 T-cell-mediated response, besides the stimulation of antibody-mediated and CD4 T-cell-mediated responses. The CD8 T-cell-mediated response may play a more important role in the protection from the infection from virus such as HIV-1 and cancers [107, 113].

Mucosal and pulmonary genetic immunizations have received special attention.

4.4.1. Mucosal Genetic Immunization

Since carbohydrates are one main component in mucin, which covers the mucosal membrane, several carbohydrate-based gene vectors have been tested for mucosal immunization [114, 65). The chitosan nanoparticles containing a cocktail of DNA encoding nine immunogenic syncitial virus antigens were tested against acute RSV infection by nasal administration using a mouse model. The system is able to decrease viral titers by two orders of magnitude (100-fold). Although the system is effective as single dose, dose escalation and/or the combination with prime-booster strategy might further enhance its efficiency.

4.4.2. Pulmonary Genetic Immunization

Chitosan-DNA has been tested as a typical cationic carbohydrate gene delivery vector for pulmonary gene delivery [63]. The study was conducted on mouse and the vector was constructed from low molecular weight chitosan and pCMV-Luc reporter gene. The vector showed excellent gene delivery efficiency, having a high DNA transfection rate and a high expression of the luciferase protein. The benefits of the cationic carbohydrate gene vectors for pulmonary gene delivery are expected to (1) suppress plasma DNA degradation and (2) increase the DNA activity in the lung.

REFERENCES

[1] Domb, A. J. and Levy, M. Y. Polymers in gene therapy. In *Frontiers in Biomedical Polymer Applications*. Ottenbrite, R. M., Ed. Vol. 2. Lancaster, PA: Technomic, 1999. pp. 1-16.

[2] Luo, D. and Saltzman, W. M. Synthetic DNA delivery systems. *Nat. Biotechnol.* 18(1): 33-37 (2000).

[3] Brown, M. D., Schatzlein, A. G. and Uchegbu, I. F. Gene delivery with synthetic (non viral) carriers. *Int. J. Pharm.* 229(1-2): 1-21 (2001).

[4] Tan, Y. and Huang, L. Overcoming the inflammatory toxicity of cationic gene vectors. *J. Drug Target.* 10(2): 153-160 (2002).

[5] Larsen, C., Ed. *Dextran Prodrugs*. Copenhagen: Villadsen and Christensen, 1990.

[6] Zabner, J., Fasbender, A. J., Moninger, T., Poellinger, K. A. and Welsh, M. J. Cellular and molecular barriers to gene transfer by a cationic lipid. *J. Biol. Chem.* 270(32): 18997-19007 (1995).

[7] Huang, L.and Viroonchatapan, E. In *Nonviral Vectors for Gene Therapy.* Huang, L. et al., Eds. San Diego, CA: Academic Press, 1999. pp. 3-22.

[8] Lollo, C. P., Banaszczyk, M. G. and Chiou, H. C. Obstacles and advances in non-viral gene delivery. *Curr. Opin. Mol. Ther.* 2(2): 136-142 (2000).

[9] Kim, T. H., Ihm, J. E., Choi, Y. J., Nah, J. W. and Cho, C. S. Efficient gene delivery by urocanic acid-modified chitosan. *J. Control. Release* 93(3):389-402 (2003).

[10] Borchard, G. Chitosan for gene delivery. *Adv. Drug Deliv. Rev.* 52(2): 145-150 (2001).

[11] Nagasaki, N., Hojo, M., Uno, A., Satoh, T., Koumoto, K., Mizu, M., Sakurai, K. and Shinkai, S. Long-term expression with a cationic polymer derived from a natural polysaccharide: schizophyllan. *Bioconjug. Chem.* 15(2): 249-259 (2004).

[12] Mansouri, S., Lavigne, P., Corsi, K., Benderdour, M., Beaumont, E. and Fernandes, J. C. Chitosan-DNA nanoparticles as non-viral vectors in gene therapy: strategies to improve transfection efficacy. *Eur. J. Pharm. Biopharm.* 57(1): 1-8 (2004).

[13] De Smedt, S. C., Demeester, J. and Hennink, W. E. Cationic polymer based gene delivery systems. *Pharm. Res.* 17(2): 113-126 (2000).

[14] Nakanishi, M. and Noguchi, A. Confocal and probe microscopy to study gene transfection mediated by cationic lipsomes with a cationic cholesterol derivative. *Adv. Drug Deliv. Rev* 52(3): 197-207 (2001).

[15] Leong, K. W., Mao, H. Q., Truong-Le, V. L., Roy, K., Walsh, S. M. and August, J. T. DNA-polycation nanospheres as non-viral gene delivery vehicles. *J. Control. Release* 53(1-3): 183-193 (1998).

[16] Erbacher, P., Zou, S., Bettinger, T., Steffan, A. M., and Remy, J. S. Chitosan-based vector/DNA complexes for gene delivery: biophysical characteristics and transfection ability. *Pharm. Res.* 15(9): 1332-1339 (1998).

[17] Vinogradov, S.V., Bronich, T. K. and Kabanov, A.V. Nanosized cationic hydrogels for drug delivery: preparation, properties and interactions with cells. *Adv. Drug Deliv. Rev.* 54(1): 135-147 (2002).

[18] Sato, T., Ishii, T. and Okahata, Y. In vitro gene delivery mediated by chitosan. *Proc. Controlled Release Soc..* 26: 803-804 (1999).

[19] Kong, G., Braun, R. D. and Dewhirst, M. W. Hyperthermia enables tumor-specific nanoparticle delivery: effect of particle size. *Cancer Res.* 60(16): 4440-4445 (2000).

[20] Illum, L., Jabbal-Gill, I., Hinchcliffe, M., Fisher, A. N., and Davis, S. S. Chitosan as a novel nasal delivery system for vaccines. *Adv. Drug Deliv. Rev.* 51(1-3): 81-96 (2001).

[21] Sehgal, D. and Vijay, I. K. A method for the high efficiency of water-soluble carbodiimide-mediated amidation. *Anal. Biochem.* 218(1): 87-91 (1994).

[22] Guy, J., Drabek, D. and Antoniou, M. Delevery of DNA into mammalian cells by receptor-mediated endocytosis and gene therapy. *Mol. Biotechnol* 3: 237-248 (1995).

[23] Tabata, Y. and Ikada, Y. Phagocytosis of polymer microspheres by microphages. *Adv. Polym. Sci.* 94: 107-141 (1990).

[24] Singh, M. and O'Hagan, D. Advances in vaccine adjuvants. *Nat. Biotechnol.* 17(11): 1075-1081 (1999).

[25] Donnelly, J. J., Liu, M. A. and Ulmer, J. B. Antigen presentation and DNA vaccines. *Am. J. Respir. Crit. Care Med.* 162(4, Pt. 2): S190-S193 (2000).

[26] Toncheva, V., Wolfert, M. A., Dash, P. R., Oupicky, D., Ulbrich, K., Seymour, L. W. and Schacht, E. H. Novel vectors for gene delivery formed by self-assembly of DNA with poly(L-lysine) grafted with hydrophilic polymers. *Biochim. Biophys. Acta* 1380(3): 354-368 (1998).

[27] Maruyama, A., Katoh, M., Ishihara, T. and Akaike, T. Comb-type polycations effectively stabilize DNA triplex. *Bioconjug. Chem.* 8(1): 3-6 (1997).

[28] Maruyama, A., Watanabe, H., Ferdous, A., Katoh, M., Ishihara, T. and Akaike T. Characterization of interpolyelectrolyte complexes between double-stranded DNA and polylysine comb-type copolymers having hydrophilic side chains. *Bioconjug. Chem.* 9(2): 292-299 (1998).

[29] Katayose, S. and Kataoka. K. Water-soluble polyion complex associates of DNA and poly(ethylene glycol)-poly(L-lysine) block copolymer. *Bioconjug. Chem.* 8(5): 702-707 (1997).

[30] Giannoukakis, N., Thomson, A. W. and Robbins, P. D. Gene therapy in transplantation. *Gene Ther.* 6(9): 1499-1511 (1999).

[31] Park, I. K., Kim, T. H., Park, Y. H., Shin, B. A., Choi, E. S., Chowdhury, E. H., Akaike, T., Cho, C. S. Galactosylated chitosan-graft-poly(ethylene glycol) as heptocyte-targeting DNA carrier. *J. Control. Release* 76(3): 349-362 (2001).

[32] Ogris, M., Walker, G., Blessing, T., Kircheis, R., Wolschek, M.and Wagner, E. Tumor-targeted gene therapy: strategies for the preparation of ligand-polyethylene glycol-polyethylenimine/DNA complexes. *J. Control. Release* 91(1-2): 173-181 (2003).

[33] Bloomfield, V. A. DNA condensation. *Curr. Opin. Struct. Biol.* 6(3): 334-341 (1996).

[34] Liu, W. G. and Yao, K. D. Chitosan and its derivatives--a promising non-viral vector for gene transfection. *J. Control. Release* 83(1): 1-11 (2002).

[35] Garnett, M. C. Gene-delivery systems using cationic polymers. *Crit. Rev. Ther. Drug Carrier Syst.* 16(2): 147-207 (1999).

[36] Azzam, T. and Domb, A. J. Current developments in gene transfection agents. *Cur. Drug Deliv.* 1(2): 165-193 (2004).

[37] Azzam, T., Eliyahu, H., Makovitzki, A., Linial, M., and Domb, A. J. Hydrophobized dextran-spermine conjugates as potential vector for in vitro gene transfection. *J. Control. Release* 96(2): 309-323 (2004).

[38] Kushibiki, T, and Tabata Y. A new gene delivery system based on controlled release technology. *Curr. Drug Deliv.* 1(2): 153-163 (2004).

[39] Zanta, M. A., Belguise-Valladier, P. and Behr, J. P. Gene delivery: a single nuclear localization signal peptide is sufficient to carry DNA to the cell nucleus. *Proc. Natl. Acad. Sci. U. S. A.* 96(1): 91-96 (1999).

[40] Subramanian, A., Ranganathan, P. and Diamond, S. L. Nuclear targeting peptide scaffolds for lipofection of nondividing mammalian cells. *Nat. Biotechnol.* 17(9): 873-877 (1999).

[41] Chan, C. K. and Jans, D. A. Enhancement of polylysine-mediated transferrinfection by nuclear localization sequences: polylysine does not function as a nuclear localization sequence. *Hum. Gene Ther.* 10(10): 1695-1702 (1999).

[42] Hebert, E. Improvement of exogenous DNA nuclear importation by nuclear localization signal-bearing vectors: a promising way for non-viral gene therapy? *Biol. Cell* 95(2): 59-68 (2003).

[43] Balazs, E. A. and Denlinger, J. L. Clinical uses of hyaluronan. *Ciba Found. Symp.* 143: 265-275 (1989).
[44] Davidson, J. M., Nanney, L. B., Broadley, K. N., Whitsett, J. S., Aquino, A. M., Beccaro, M. and Rastrelli, A. Hyaluronate derivatives and their application to wound healing: preliminary observations. *Clin. Mater.* 8(1-2): 171-177 (1991).
[45] Juhlin, L. Hyaluronan in skin. *J. Intern. Med.* 242(1): 61-66 (1997).
[46] Liu, L.-S., Thompson, A..Y. Heidaran, M., Poser, J. W. and Spiro, R. C. An osteoconductive collagen/hyaluronate matrix for bone regeneration. *Biomaterials* 20(12): 1097-1108 (1999).
[47] Liu, L.-S., Ng, C. K., Thompson, A. Y., Poser, J. W. and Spiro, R. C. Hyaluronate-heparin conjugate gels for the delivery of basic fibroblast growth factor (FGF-2). *J. Biomed. Mater. Res.* 62(1): 128-135 (2002).
[48] Liu, LS., Kramer, W. H, Fishman, M. L. and Hicks, K. B. Semisynthetic Composite Matrices from Collagen and Hyaluronate for Bone and Cartilage Regeneration. In *Chemical Reactions: Quantitive Level of Liquid and Solid Phase.* Chapter 16. Zaikov, G. E. and Jiménez, A., Eds. Hauppauge, NY: Nova Science Publishers, Inc., 2004. pp. 209-229.
[49] Spiro, R. C., Liu, L.-S., Heidaran, M. A., Thompson, A. Y., Ng, C. K., Pohl, J.and Poser, J. W. Inductive activity of recombinant human growth and differentiation factor-5. *Biochem. Soc. Trans.* 28(4): 362-368 (2000).
[50] Luo, Y., Kirker, K. R.and Prestwich, G. D. Modifications of Natural Polymers: Hyaluronnic Acid . In: *Methods of Tissue Engineering.* Atala, A. and Lanza, R. P., Eds. San Diego, CA: Academic Press, 2002. pp. 539-554.
[51] Yun, Y. H., Goetz, D. J., Yellen, P. and Chen, W. Hyaluronan microspheres for sustained gene delivery and site-specific targeting. *Biomaterials* 25(1): 147-157 (2004).
[52] Kim, A., Checkla, D. M., Dehazya, P. and Chen, W. Characterization of DNA-hyaluronan matrix for sustained gene transfer. *J. Control. Release* 90(1): 81-95 (2003).
[53] Bertolami, C. N., Berg, S., and Messadi, D.V. Binding and internalization of hyaluronate by human cutaneous fibroblasts. *Matrix* 12(1): 11-21 (1992).
[54] Benedetti, L. M., Topp, E.M., and Stella, V. J. Microspheres of hyaluronic acid esters-fabrications methods and in vitro hydrocortisone release. *J. Control. Release* 13(1): 33-41 (1990).
[55] Prabaharan, M. and Mano, J. F. Chitosan-based particles as controlled drug delivery systems. *Drug Deliv.* 12(1): 41-57 (2005).
[56] Muzzarelli, R. A. A. *Natural Chelating Polymers: Alginic Acid, Chitin, and Chitosan.* International Series of Monographs in Analytical Chemistry. Vol. 55. Belcher, R. and Freiser, M., Eds. Oxford: Pergamon Press, 1973. pp. 144-176.
[57] Lehr, C. M., Bouwstra, J.A., Schacht, E. H. and Junginger, H. E. In vitro evaluation of mucoadhesive properties of chitosan and some other natural polymers. *Int. J. Pharm.* 78(1-3): 43-48 (1992).
[58] Hejazi, R. and Amiji, M. Chitosan-based gastrointestinal delivery systems. *J. Control. Release* 89(2): 151-165 (2003).
[59] Fang, N., Chan, V., Mao, H. Q., and Leong, K.W. Interactions of phospholipid bilayer with chitosan: effect of molecular weight and pH. *Biomacromolecules* 2(4): 1161-1168 (2001).

[60] Romoren, K., Thu, B .J.and Evensen, O. Immersion delivery of plasmid DNA. II.. A study of the potentials of a chitosan based delivery system in rainbow trout (*Oncorhynchus mykiss*) fry. *J. Control. Release* 85(1-3): 215-225 (2002).

[61] Chew, J. L., Wolfowicz, C. B., Mao, H.Q., Leong, K.W., and Chua, K.Y. Chitosan nanoparticles containing plasmid DNA encoding house dust mite allergen, Der p 1 for oral vaccination in mice. *Vaccine* 21(21-22): 2720-2729 (2003).

[62] Cui, Z. and Mumper, R. J. Chitosan-based nanoparticles for topical genetic immunization. *J. Control. Release* 75(3): 409-419 (2001).

[63] Okamoto, H., Nishida, S., Todo, H., Sakakura, Y., Iida, K. and Danjo, K. Pulmonary gene delivery by chitosan-pDNA complex powder prepared by a supercritical carbon dioxide process. *J. Pharm. Sci.* 92(2): 371-380 (2003).

[64] Philippova, O. E., Volkov, E. V., Sitnikova, N. L., Khokhlov, A. R., Desbrieres, J. and Rinaudo, M. Two types of hydrophobic aggregates in aqueous solutions of chitosan and its hydrophobic derivative. *Biomacromolecules* 2(2): 483-490 (2001).

[65] Roy, K., Mao, H.Q., Huang, S. K. and Leong, K. W. Oral gene delivery with chitosan-DNA nanoparticles generates immunologic protection in a murine model of peanut allergy. *Nat. Med.* 5(4): 387-391 (1999).

[66] Mumper, R. J., Wang, J., Claspell, J. M.and Rolland, A.P. Novel polymeric condensing carriers for gene delivery. *Proc.Controlled Release Soc.* 22: 178-179 (1995).

[67] Liu, W.G., Zhang, X., Sun, S.J., Sun, G.J., Yao, K.D., Liang, D.C., Guo, G. and Zhang, J. Y. N-alkylated chitosan as a potential nonviral vector for gene transfection. *Bioconjug. Chem.* 14(4): 782-789 (2003).

[68] Sato, T., Ishii, T. and Okahata, Y. In vitro gene delivery mediated by chitosan. Effect of pH, serum, and molecular mass of chitosan on the transfection efficiency. *Biomaterials* 22(15): 2075-2080 (2001).

[69] MacLaughlin, F. C., Mumper, R. J., Wang, J., Tagliaferri, J. M., Gill, I., Hinchcliffe, M. and Rolland, A.P. Chitosan and depolymerized chitosan oligomers as condensing carriers for in vivo plasmid delivery. *J. Control. Release* 56(1-3): 259-272 (1998).

[70] Mao, H. Q., Roy, K., Troung-Le, V. L., Janes, K. A, Lin, K. Y., Wang ,Y., August, J.T. and Leong, K. W. Chitosan-DNA nanoparticles as gene carriers: synthesis, characterization and transfection efficiency. *J. Control. Release* 70(3): 399-421 (2001).

[71] Corsi, K., Chellat, F., Yahia, L., and Fernandes, J. C. Mesenchymal stem cells, MG63 and HEK293 transfection using chitosan-DNA nanoparticles. *Biomaterials* 24(7): 1255-1264 (2003).

[72] Fischer, D., Bieber, T., Li, Y., Elsasser, H. P., and Kissel, T. A novel non-viral vector for DNA delivery based on low molecular weight, branched polyethyleneimine: effect of molecular weight on transfection efficiency and cytotoxicity. *Pharm. Res.* 16(8): 1273-1279 (1999).

[73] Köping-Höggård, M., Mel'nikova, Y. S., Varum, K. M., Lindman, B. and Artursson, P. Relationship between the physical shape and the efficiency of oligomeric chitosan as a gene delivery system in vitro and in vivo. *J. Gene Med.* 5(2): 130-141 (2003).

[74] Thanou, M., Florea, B. I., Geldof, M., Junginger, H. E.and Borchard, G. Quaternized chitosan oligomers as novel gene delivery vectors in epithelial cell lines. *Biomaterials* 23(1): 153-159 (2002).

[75] Gao, S., Chen, J., Xu, X., Ding, Z., Yang, Y-H., Hua, Z. and Zhang, J. Galactosylated low molecular weight chitosan as DNA carrier for hepatocyte-targeting. *Int. J. Pharm.* 255(1-2): 57-68 (2003).

[76] Truong-Le, V. L., Walsh, S. M., Schweibert, E., Mao, H. Q., Guggino, W. B., August, J. T., Leong, K. W. Gene transfer by DNA-gelatin nanospheres. *Arch. Biochem. Biophys.* 361(1): 47-56 (1999).

[77] Carreño-Gómez, B. and Duncan, R. Evaluation of the biological properties of soluble chitosan and chitosan microspheres. *Int. J. Pharm.* 148(2): 231-240 (1997).

[78] Kiang, T., Wen, J., Lim, H.W. and Leong, K.W. The effect of the degree of chitosan deacetylation on the efficiency of gene transfection. *Biomaterials* 25(22): 5293-5301 (2004).

[79] Berth, G., Dautzenberg, H., and Peter, M. G. Physio-chemical characterization of chitosans varying in degree of acetylation. *Carbohydr. Polym.* 36(2-3): 205-216 (1998).

[80] Janes, K. A., Calvo, P. and Alonso, M. J. Polysaccharide colloidal particles as delivery systems for macromolecules. *Adv. Drug Deliv. Rev.* 47(1): 83-97 (2001).

[81] Richardson, S. C. W., Kolbe, H. V. J. and Duncan, R.. Potential of low molecular mass chitosan as a DNA delivery system: biocompatibility, body distribution and ability to complex and protect DNA. *Int. J. Pharm.* 178(2): 231-243 (1999).

[82] Ishii, T., Okahata, Y. and Sato, T. Mechanism of cell transfection with plasmid/chitosan complexes. *Biochem. Biophys. Acta* 1514(1): 51-64 (2001).

[83] Kim, Y. H., Gihm, S. H., Park, C. R., Lee, K. Y., and Kim, T. W., Kwon, I. C., Chung, H. and Jeong, S. Y. Structural characteristics of size-controlled self-aggregates of deoxycholic acid-modified chitosan and their application as a DNA delivery carrier. *Bioconjug. Chem.* 12(6): 932-938 (2001).

[84] Park, I. K., Park, Y. H., Shin, B. A., Choi, E. S., Kim, Y. R., Akaike, T. and Cho, C. S. Galactosylated chitosan-graft-dextran as hepatocyte-targeting DNA carrier. *J. Control. Release* 69(1): 97-108 (2000).

[85] Park, I. K., Ihm, J. E., Park, Y. H., Choi, Y. J., Kim, S. I., Kim, W. J., Akaike, T. and Cho, C.S. Galactosylated chitosan (GC)-graft-poly(vinyl pyrrolidone) (PVP) as hepatocyte-targeting DNA carrier. Preparation and physicochemical characterization of GC-graft-PVP/DNA complex (1). *J. Control. Release* 86(2-3): 349-359 (2003).

[86] Pack, D. W., Putnam, D. and Langer, R. Design of imidazole-containing endosomolytic biopolymers for gene delivery. *Biotechnol. Bioeng.* 67(2): 217-223 (2000).

[87] Benns, J. M, Choi, J. S., Mahato, R. I., Park, J. S., and Kim, S. W. pH-sensitive cationic polymer gene delivery vehicle: N-Ac-poly(L-histidine)-graft-poly(L-lysine) comb shaped polymer. *Bioconjug. Chem.* 11(5): 637-645 (2000).

[88] Lee, K. Y., Kwon, I.C., Kim, Y. H., Jo, W. H. and Jeong, S. Y. Preparation of chitosan self-aggregates as a gene delivery system. *J. Control. Release* 51(2-3): 213-220 (1998).

[89] Li, F., Liu, W.G. and Yao, K. D. Preparation of oxidized glucose-crosslinked N-alkylated chitosan membrane and in vitro studies of pH-sensitive drug delivery behaviour. *Biomaterials* 23(2): 343-347 (2002).

[90] Liu, W. G., Yao, K. D. and Liu, Q. G. Formation of a DNA/N-dodecylated chitosan complex and salt-induced gene delivery. *J. Appl. Polym. Sci.* 82(14): 3391-3395 (2001).

[91] Thanou, M. M., Kotze, A. F., Scharringhausen, T., Luessen, H. L., de Boer, A. G., Verhoef, J. C. and Junginger, H. E. Effect of degree of quaternization of N-trimethyl

chitosan chloride for enhanced transport of hydrophilic compounds across intestinal caco-2 cell monolayers. *J. Control. Release* 64(1-3): 15-25 (2000).

[92] Murata, J-I., Ohya, Y. and Ouchi, T. Design of quaternary chitosan conjugate having antennary galactose residues as a gene delivery tool. *Carbohydr. Polym.* 32(2): 105-119 (1997).

[93] Deshpande, D., Toledo-Velasquez, D., Wang, L. Y., Malanga, C. J., Ma, J. K. and Rojanasakul, Y. Receptor-mediated peptide delivery in pulmonary epithelial monolayers. *Pharm. Res.* 11(8): 1121-1126 (1994).

[94] de Lima, M. C., Simoes, S., Pires, P., Gasper, R., Slepushkin, V. and Duzgunes, N. Gene delivery mediated by cationic liposomes: from biophysical aspects to enhancement of transfection. *Mol. Membr. Biol.* 16(1): 103-109 (1999).

[95] Fishman, M. L. AND Jen, J. J., Eds. *Chemistry and Function of Pectins.* ACS Synposium Series, no. 310. Washington, DC: American Chemical Society, 1986.

[96] Liu, LS., Fishman, M. L., Kost, J., Hicks, K. B. Pectin-based systems for colon-specific drug delivery via oral route. *Biomaterials* 24(19): 3333-3343 (2003).

[97] Azzam, T., Eliyahu, H., Shapira, L., Linial, M., Barenholz, Y. and Domb, A. J. Polysaccharide-oligoamine based conjugates for gene delivery. *J. Med. Chem.*45(9): 1817-1824 (2002).

[98] Koltover, I., Salditt, T., Radler, J. O., and Safinya, C. R. An inverted hexagonal phase of cationic liposome-DNA complexes related to DNA release and delivery. *Science* 281(5373): 78-81 (1998).

[99] Azzam, T., Raskin, A., Makovitzki, A., Brem, H., Vierling, P., Lineal, M., and Domb, A. J. Cationic polysaccharides for gene delivery. *Macromolecules* 35(27): 9947-9953 (2002).

[100] Azzam, T., Eliyahu, H., Makovitzki, A. and Domb, A. J. Dextran-spermine conjugates: a efficient vector for gene delivery. *Macromol. Symp.* 195: 247-261 (2003).

[101] Smith, L.C. and Nordstrom, J. L. Advances in plasmid gene delivery and expression in skeletal muscle. *Curr. Opin. Mol. Ther.* 2(2): 150-154 (2000).

[102] Fujii, I., Suzuki, S., Igarashi, T., Matsukura, M., Miike, T. and Shimada T. Targeted and stable gene delivery into muscle cells by a two-step transfer system. *Biochem. Biophy. Res. Comm.* 275(3): 931-935 (2000).

[103] Bonadio, J. Tissue engineering via local gene delivery. *J. Mol. Med.* 78(6): 303-311 (2000).

[104] Richardson, T. P., Murphy, W. L. and Mooney, D. J. Polymeric delivery of proteins and plasmid DNA for tissue engineering and gene therapy. *Crit. Rev. Eukaryot. Gene Expr.* 11(1-3): 47-58 (2001).

[105] Xie, Y., Yang, S. T. and Kniss, D. A. Three-dimensional cell-scaffold constructs promote efficient gene transfection: implications for cell-based gene therapy. *Tissue Eng.* 7(5): 585-598 (2001).

[106] Shea, L. D., Smiley, E., Bonadio, J. and Mooney, D. J. DNA delivery from polymer matrices for tissue engineering. *Nat Biotechno.* 17(6): 551-554 (1999).

[107] Liu, M. A. DNA vaccines: a review. *J. Intern. Med.* 253(4): 402-410 (2003).

[108] Cho, H. J., Takabayashi, K., Cheng, P. M., Nguyen, M. D., Corr, M., Tuck, S. and Raz, E. Immunostimulatory DNA-based vaccines induce cytotoxic lymphocyte activity by a T-helper cell-independent mechanism. *Nat. Biotechnol.* 18(5): 509-514 (2000).

[109] Zhao, Z. and Leong, K. W. Controlled delivery of antigens and adjuvants in vaccine development. *J. Pharm. Sci.* 85(12): 1261-1270 (1996).
[110] Hanes, J., Cleland, J. L. and Langer R. New advances in microsphere-based single-dose vaccines. *Adv. Drug Deliv. Rev.* 28(1): 97-119 (1997).
[111] Mathiowitz, E., Jacob, J. S., Jong, Y. S., Carino, G. P., Chickering, D. E. Chaturvedi, P., Santos, C. A., Vijayaraghavan, K., Montgomery, S. Bassett, M. and Morrell, C. Biologically erodable microspheres as potential oral drug delivery systems. *Nature* 386(6623): 410-414 (1997).
[112] Wang, C., Ge, Q., Ting, D., Nguyen, D., Shen, H.-R., Chen, J., Eisen, H.N., Heller, J., Langer, R. and Putnam, D. Molecularly engineered poly(ortho ester) microspheres for enhanced delivery of DNA vaccines. *Nat. Materials* 3: 190-196 (2004).
[113] Barouch, D. H., et al. Control of viremia and prevention of clinical AIDS in rhesus monkeys by cytokine-augmented DNA vaccination. *Science* 290(5491): 486-492 (2000).
[114] Kumar, M., Behera, A. K., Lockey, R. F., Zhang, J., Bhullar, G., de La Cruz, C. P., Chen, L-C., Leong, K. W., Huang, S-K. and Mohapatra, S. S. Intranasal gene transfer by chitosan-DNA nanospheres protects BLAB/c mice against acute respiratory syncytial virus infection. *Hum. Gene Ther.* 13(12): 1415-1425 (2002).

Chapter 2

TESTS OF MATERIALS. STUDY OF OPERATION MECHANISM AT HIGH-TEMPERATURE

G. E. Zaikov, A. A. Donskoi[1] and M. A. Shashkina[1]

N.M.Emanuel Institute of Biochemical Physics Russian
Academy of Sciences, 4 Kosygin str., Moscow 119991, Russia
[1] Institute of Aviation Materials, 17 Radio str., Moscow, 119105, Russia

Development of fire and heat shield materials with reduced combustibility consists of several stages.

Heat energy from an external source, delivered to the material surface, can increase its temperature so that degradation of material components or chemical reaction between them begins. Combustible degradation products formed are capable of igniting and combust independently of flame spreading. The problem of creating fire and heat shield materials is closely connected with the questions of material combustibility decrease.

Under normal conditions, heat transfer proceeds from higher heated (hot) bodies to less heated (cold) ones by means of heat conductivity or convection. Under fire conditions, irradiation is the main type of heat transfer from flame or incandescent material surface. The radiant flux falling on the body surface consists of three components [1 - 3].

The radiant flux permeates through the cover to the protected object surface and heats it up, which is unwanted for fire and heat shield materials. Heat flux absorbed by the material is accumulated in it or transforms into different types of energy [4].

The level and spectral characteristics of the radiant flux, absorption properties of the material, and reflecting power of the surface in relation to emission spectrum of falling flux affect the amount of energy absorbed by polymeric composite material.

Transmission and absorption coefficients depend on the chemical composition and structure of the substance. If the absorption coefficient is low, the radiant flux transmits throughout the material. If it is high, a thin layer of the material is rapidly heated up to a critical temperature of material decomposition. Therewith, the maximum temperature is observed on the surface protecting the layers below.

From positions of increasing fire resistance, it is desirable to increase reflecting power of materials, which is low for organic compounds, polymers in particular [4]. The reflecting power can be increased by both surface treatment and injection of fillers capable of reflecting the radiant energy in the infrared spectrum part into the compounding of composite material. Oxides and salt of various metals can be applied as such materials.

Fillers are often subdivided into active and passive ones. However, this division is rather relative, because one and the same filler may be a passive diluter of the condensed phase at low temperatures under different operation conditions, and under other conditions it may be transformed physically and chemically consuming heat energy for these processes.

At decreasing combustibility of polymeric materials in high-temperature flux, it is common practice to reduce the role of mineral fillers in the polymeric matrix to dilution of the condensed phase, change of thermophysical properties of the system and physical and chemical transformations. However, one may note a fortuitous coincidence that some compounds used as combustion decelerators are polymerization and polycondensation catalysts [5]. One may suggest these compounds to be capable of catalyzing reactions of cross-linking and coke formation under thermal degradation conditions.

1. COKE FORMATION AS THE INVESTIGATION METHOD OF FIRE AND HEAT SHIELD MATERIALS AND COVERS

Covers from rubbers and rubber-like materials with the surface affected by high-temperature flux are capable of forming a coke-like solid residue with a porous structure. Such layer formed on the surface possesses low heat conductivity. This layer is gradually destroyed and substituted by a new layer of changed material formed on the surface. As underlying layers of the cover degrade thermally, volatile products are released transmitting though the "coke" and forming the border gas layer representing an additional heat resistance. It is noted that the use of rubbers has allowed a two-fold reduction in the weight of fire and heat shield covers.

There is an opinion that the ideal heat shield material is the one completely exhausted up to the end of high-temperature flux effect [6].

The importance of coke forming ability of the organic base of fire- and heat shielding materials is also outlined in ref. [7]: "Carbon properties relate directly to consideration of organic materials, because the main product of their thermal degradation under optimal conditions is carbon. It can be stated that application of plastics is just a special way of making carbon contact with incandescent gas flow". It is noted that the problem of application of protective materials at high temperatures has no unique solution: both chemical and physical aspects are of the same importance. Consideration of enthalpy of gases in the composition of solid propellant combustion products and melting temperatures of cover components are important, because if the latter is much lower than the temperature of cover heating, the phase transition heat is not used effectively enough due to rapidly carry off of the melt. Melting temperatures of components must be chosen so that melting proceeds not too rapidly, for example, refrasil melting at 2000 K is the most effective filler at solid propellant combusting at 3300 K. Fillers of the silicon dioxide type are suitable, because at this temperature their melting heat is effectively consumed in the neutral oxidative medium.

Ref. [8] shows calculation of heat transfer through the cover under the following assumptions: the cover surface obtains temperature of propellant combustion products immediately, the cover thickness does not change with time, the organic part of the material degrades at a definite temperature and known energy consumption, a stable coke-like layer is formed on the surface, the unchanged part of the cover possesses low heat conductivity, gaseous products possess no cooling capability, heat flux is perpendicular to the surface, and thermal constants are independent of temperature. The calculation formula is the following:

$$a = R\sqrt{\tau},$$

where a is the coordinate of the cover pyrolysis zone (the origin of coordinates is located on the surface contacting with the gas flow); τ is the duration of high temperature influence; R is a constant typical of the material.

Hence, the penetration depth of the thermal degradation zone into the cover is proportional to the square root of the duration of heating. Calculations show that degradation temperature increase causes an insignificant effect on the cover destruction rate, and the change of latent pyrolysis heat changes it significantly. Thermophysical constants of the unchanged part of the cover are of secondary importance, but it is useful if they provide for possibly higher temperature gradient.

Application of fillers or reinforcement complicates the mechanism of cover destruction, because several zones are formed in it; but the general character of cover operation remains practically unchanged. The coke-like layer should not increase heat conductivity of the cover, and fillers (oxides, nitrides, carbides, zirconates, and titanates) should change with absorption of maximum possible energy at high temperatures. Erosion behavior of such composites in oxygen-acetylene flame is better than of graphite, but this was not confirmed by tests in a high-rate flux.

Contrary to thermal degradation processes of natural products (mineral coal, crude oil and gas), which are used in industry for producing a great variety of materials, including carbon black production widely used in rubber industry, up to quite recent times, pyrolysis of high-polymeric materials was used only for studying their structure, kinetics and aging mechanism, stability at increased temperatures and under effect of oxidants, and identification in analyses of rubbers, in particular. Among great variety of works on the study of high polymers, investigations of their thermal degradation are few.

Hereinafter, by high temperatures we mean the ones at which "valence states, compounds formed and general properties of systems are significantly different from valence states, compounds and properties at room temperature" [9].

Initial works studying thermal degradation products of elastomers date back to 1929, when products of natural rubber refining in batches 7 kg each treated in iron vessels under atmospheric pressure and temperature of 973 K were studied in detail [10]. Among liquid products, 23 compounds possessing from 5 to 10 carbon atoms, including isoprene (10%) and dipentene (20%), were successfully identified. Later, thermal degradation of rubbers was studied mostly on smaller samples [11, 12].

Refs. [13, 14] present a detailed review of works on degradation of high polymers under the effect of various chemical and physical factors, and heat in particular. Depolymerization processes at moderate heating and thermal degradation of polybutadiene, polyisobutylene, butadiene-styrene rubber, natural rubber, polyethylene, polytetrafluoroethylene, etc. are discussed. Possible mechanisms of degradation and depolymerization are described.

Pyrolysis of polyisoprene and polyisobutylene is initiated at 300°C and terminated at 400°C; for polybutadiene, approximately at 350°C and 477°C; and for polyethylene, at 360°C and 475°C, respectively. Degradation of these polymers is almost complete: the residues after pyrolysis are 3.8% for polyisoprene, 0.3% for polybutadiene at 500°C, 0.4% for butadiene-styrene rubber at 455°C, and 1.4% for polyethylene at 475°C. These residues dissolve in benzene and cyclohexane, which is untypical of the carbon residue.

Ref. [15] also provides information about pyrolysis of butadiene-styrene rubber. It is noted that exothermal degradation is initiated at 380°C. At 430°C, it becomes so intense that ignition may occur.

If linear rubbers form no noticeable solid residue during pyrolysis, pyrolysis of phenol-formaldehyde resins gives up to 50% of it, and thermoreactive organosilicon resins give 85% [16].

In the case of pyrolysis of aliphatic compounds, the coke residue is absent or negligibly low. The presence of great amount of condensed ring structures produces a significant coke residue, because carbonaceous residues are formed at pyrolysis of cyclic compounds. Thus it becomes clear that preliminary cyclization during dehydration at high temperature is desirable for raising coking capacity of linear polymers.

Carbonization of samples of the materials studied can be held in both oxidation and reduction media. The medium composition is usually selected depending on expected operation conditions of the material. One of the investigation methods in reduction media is the following [16]. Samples of appropriate geometrical shape are placed into crucibles and covered by a coal charge. Temperature increase rate is 4°/min. As a temperature of 1,000°C is reached, the samples are exposed for 1 hour at this temperature. The amount of solid residue is calculated by the formula:

$$K = \frac{P_{fin}}{P_{ini}} \cdot 100\%,$$

where K is the amount of solid residue; P_{ini} is the initial sample weight; P_{fin} is the sample weight after test.

Proper geometrical shape allows a study of strength properties of the solid residue obtained.

2. THE ROLE OF INTUMESCENCE IN THE PROBLEM OF FIRE PROTECTION OF POLYMERS

The fire shield mechanism of polymeric material components is quite complicated and manifold. To study the protective mechanism of the entire material and its separate components, the determining sign of the process and the main operation principle of the present material should be outlined. Intumescence can be this sign, which considers simultaneously processes of gasification and coke formation in polymeric material under high-temperature influence.

The technology of intumescence is quite new in the polymer science as the method providing for polymer protection from flame effects. Intumescent systems terminate polymer combustion in its early stage, i.e. at the stage of thermal degradation of the polymer accompanied by release of combustible gas products.

The intumescent process consists in combining coke formation and puffing up of combusting polymer surface. The foamed up cellular coke-like layer formed, the density of which reduces with temperature [17], protects the protective material from heat flux and flame effects.

2.1. Intumescence Chemistry

Intumescent additives usually include three components: an acidic component necessary for acidic catalytic effect, polyalcohols (as carbonizing compounds) and foaming agent. In the initial stage ($T > 280°C$), the acidic component interacts with carbonizing agent [18].

Carbonization proceeds at about 280°C. Therewith, Friedel-Craft reactions and free-radical processes take place [19]. Later, the foaming agent decays with release of gas products, which induces foaming up of the coke layer. Such intumescent material decomposes with temperature increase.

When protecting a polymer from the effect of high temperature, carbonized material participates in two chemical processes:

- Reactions between free-radical fragments of foamed material and radical products of the gas phase, which are the products of polymer composite degradation. These free-radical fragments of the coke layer can participate in termination reactions of radical chains formed during pyrolysis of polymeric composites in the condensed phase;
- Acid-catalytic reactions with oxidation products formed during thermooxidative degradation of the material.

In composition, the intumescent coke layer is a heterogeneous system. It represents the condensed phase, phosphorus-carbonized cells of which contain gaseous products. In turn, the condensed phase consists of solid and liquid phases (acid-catalyzed resins), which contain liquid and gaseous products of polymer degradation. Carbonized fractions of the condensed phase consist of polyaromatic fragments, formed into layers typical of graphite structures.

Further on, phosphorus-carbonized material forms definite areas in the material consisting of crystalline intermolecular polyaromatic layers linked by bridge bonds of polymeric chains and phosphate (poly-, di- or ortho-) groups, crystalline particles and the amorphous phase, disseminated into crystalline zones. This amorphous phase consists of small polyaromatic molecules, obtained by hydrolysis of phosphate fragments of alkyl chains – the decomposition products of material components and fragments of polymeric chains. Phosphorus-carbonized material possesses fire shield properties under the following conditions: it should coat completely the unchanged protected polymer and possess strength providing mechanical resistance of the cover formed.

2.2. Protection through Intumescence

The protection mechanism suggested is based on the coke layer effect as a physical barrier reducing heat and mass exchange between the gas and condensed phases. The presence of polymeric chain fragments in the intumescent layer tindicates that it absorbs combustible gaseous products of polymer pyrolysis. Moreover, the intumescent layer prevents diffusion of gaseous fuel into the flame zone, as well as restricts oxidant access from the surrounding air to the polymeric material.

Stability of the intumescent material limits formation of gaseous fuel and leads to spontaneous extinguish under standard conditions.

Recently, many intumescent additives have been studied [20 - 22], but they contained three main components: catalyst, coke forming and foaming up agents.

Coke formation catalysts are usually phosphorus-containing compounds. The the fact that they must be added in relatively large amounts (up to 15-20%) does not correlate with common ideas of catalysis. That is whyit can be suggested that these compounds may participate in formation of a new structure.

Ref. [23] shows analysis of the literature data on the study of fire shielding of some polymers (polypropylene, ethylene-butyl acrylate copolymer with maleic anhydride, poly(ethylene terephthalate), cotton, and polyacrylonitrile) containing ammonia polyphosphates and ethylene diammonia phosphates as fire shield additives combined with various synergic additives providing for the intumescent effect. Pentaerythritol, thimethylol melamine, hexabromocyclodecane, and triphenyl phosphate were studied as additives. The analysis indicates comparatively low efficiency of ammonium polyphosphates, which increases in the presence of synergists. The effect of these additives depends on the chemical structure.

Injection of compounds containing nitrogen, halogens, and antimony into the composites can increase efficiency of phosphorus-containing components. Studies of various systems showed synergism of the effect of these elements during intumescence.

For example, phosphorus-nitrogen bonds can participate in forming volumetric network structures, in which phosphorus will be fixed and thus diffusion will be complicated [24]. The element analysis and IR-spectroscopy carried out by Weil et al. confirmed existence of phosphorus-nitrogen bonds on the coke surface, formed during combustion of the composite containing ethylenevinyl acetate, pyrophosphate melamine, hexamethyltrisdioxyphosphorus, and melamine trioxide [25].

Antimony-halogens synergism for reducing combustibility of polymeric systems is commonly known. The halogen-antimony synergism is displayed in the condensed and the gas phases [26].

The main effect of the presence of antimony trioxide is observed in the gas phase. Antimony halides formed in the gas phase react with atomic oxygen, water and hydroxyl radicals, giving SbO and hydrogen halide. Dispersed solid SbO and Sb are formed in the flame zone. They catalyze recombination of hydrogen radicals. Moreover, there is an opinion that antimony halides decelerate halogen release from flame promoting the decrease of combustible components content.

Data on synergic interaction between bromine and chlorine exist in the literature [27 28]. In the majority of cases, the highest effect is observed when they are in equal concentrations and at a total content of 10 – 12%.

There are several notes about existence of bromine-phosphorus synergism in the literature. Studies of the oxygen index show that bromine compounds cause no noticeable effect in the gas phase, but act as foaming up agents during coke formation [23].

Comparison of phosphorus-bromine and antimony-bromine synergic efficiencies displays higher values for the second pair.

Ref. [29] suggests that phosphorus compounds are capable of causing the synergic effect with halogens. This is confirmed for the case of polymers containing oxygen that leads to a significant decrease of additive concentration used in the polymeric system.

2.3. The Role of Ammonium Polyphosphates in Protection of Polymers from Fire

Synergic efficiency of ammonium polyphosphates in polypropylene is higher than in other systems [23]. As such polymeric composite combusts in the presence of pentaerythritol, the following processes proceed: ammonia polyphosphate decomposition with ammonia and water release, pentaerythritol phosphorylation and polypropylene thermooxidation, dehydration, dephosphorylation, network cross-linking, carbonization and coke structure formation. Foaming up agent is delivered into the gas phase and decomposes to incombustible products.

Studying pyrolysis of the composition containing pentaerythritol and ammonia polyphosphate, Brauman [30] suggested that ammonium polyphosphate not only formed a protective layer, but also participated in chemical reactions in the condensed phase. These processes proceed at very high but different rates. Predominance of one or other reaction at any time depends on the relation of ratios that shows up on the coke properties during combustion.

Comparative analysis of systems containing different phosphates [31] shows that quite low values of the oxygen index are associated with formation of phosphorus oxides during combustion, which reduces phosphorus concentration in the coke. Addition of low amounts of zeolites restricts formation of condensed polyphosphate products and increases content of acid phosphate products in the intumescent material [32, 33].

However, higher thermostable metal-ammonium polyphosphates induce a destabilizing effect shown on the example of polyamide [34].

Good fire shield properties were observed for phosphamacyclomatrix inorganic polymer with high thermal stability [35]. Phospham structure is not detected: chemical analysis indicates a structure with formula of the type $(PN_2H)_x$.

Phospham is a combustion decelerator for polyamide, which is displayed by the oxygen index change, shown in Table 1.

The highest jump-like increase of the oxygen index is observed at injection of 10% of phospham. Further increase of its content causes no sharp rise of the index. The coke formed contains phospham; however, polyamide also participates in formation of solid residue, which confirms increased coke yield compared with the amount of injected modifier (Table 1).

Studies of polyalcohol-melamine coking system enabled the role of melamine in this process to be understood. Evaporation from the solid phase and degradation in flame are accompanied by an endothermal effect. Melamine also participates in phosphorylation at simultaneous foaming up during intumescence, because the following gaseous products are formed during melamine degradation: water, CO, CO_2, ammonia, and hydrocarbons promoting foaming up of the solid residue.

Table 1. Oxygen index and coke residue yield during combustion of polyamide containing phospham

Phospham, wt.%	Oxygen index, cond. units	Coke yield, wt.%
-	25.2	-
10	29.2	17.5
20	31.6	23.5
30	34.8	32.2

Studying intumescent additive consisting of ammonium polyphosphate and pentaerythritol in different polymeric matrices (polypropylene, polyethylene, polystyrene) using IR-spectroscopy, Sevostianov and Novikov [36] showed that chemical structure of the coke was hardly influenced by the type of polymeric matrix, and contents of carbon and phosphorus atoms in the solid residue corresponded to their amount in the intumescent additive.

Studying the composite based on propylene with the same additive, Delobel et al. [31] showed that polyphosphate chains were formed under thermal influence, and the solid residue contained increased amount of orthophosphate compounds. Ammonium polyphosphate substitution by diammonium pyrophosphate caused formation of pyrophosphate fragments instead of orthophosphate ones [37].

Study of the coke pore structure and the surface shows that the ratio of phosphorus to carbon increases with temperature (up to 500°C), whereas this ratio decreases in the main mass of the system. The oxygen-carbon ratio is of the same type, which suggested to the authors migration of phosphonates to the surface accompanied by their oxidation.

In ref. [38] it is shown that injection of zeolites into the polymeric system increases efficiency of intumescent additive reducing heat release and suppressing smoke formation. Authors of ref. [38] suggest that phosphorus-carbon structures are formed in the intumescent system, which become more stable in the presence of zeolite. Zeolite promotes formation of organic phosphates and/or aluminum phosphates in clusters of polymeric chains and thereby restricts depolymerization and, consequently, the amount of combustible gaseous products delivered into the flame zone. Moreover, it is shown that zeolite promotes formation of higher "coherent" structure in the polymeric material. Occurrence of a "coherent" macromolecular network and interaction with polymeric chains increase fire shield properties of the material. Actually, formation of polyaromatic structures in the intumescent protecting barrier makes the material stronger. The material surface becomes more flexible, which reduces probability of cracks occurring on the surface under high temperature effect. Therewith, diffusion of oxygen into polymeric matrix and combustible products of polymer degradation into the combustion zone is decelerated.

Polymers can also be used as carbonizing agents, which reduces the peak of heat release and delays sample ignition in time. Such resistance to heat flux effect can be explained by formation of a new fire shield layer, more resistant to cracking, formed in reactions of material components with injected polymer. Study of heat transfer in some systems shows that such protective layer is quite effective in limiting destruction of the material surface [23].

Ref. [23] discusses in detail the intumescent mechanism of material protection from flame and high temperature effects, composed of simpler simultaneously proceeding processes, including formation of degradation gas products and solid residue on the surface of sample or cover. These processes, the main ones in the intumescence mechanism, must be adjusted in time and by rate of proceeding for every polymeric linkage and conditions of thermal effect. During polymer combustion, a part of supplied heat energy is absorbed by the material and consumed for its degradation. Entering the gas phase, volatile products sustain combustion as fuel or oxidant. Obviously, the combustion rate should be determined by the amount of supplied heat and the amount that is used for heating up, phase transformations and polymer degradation.

Estimation of flame temperature effect on the rate of polymer degradation by the linear pyrolysis method consisted in studying processes of material degradation at one-side effect of heat energy, it showed that the rate of polymer degradation increases with the flux temperature. However, the degradation rate is essentially associated with the nature of the polymer [39].

The study of temperature profiles on linear pyrolysis of a series of polymers shows that surface temperatures of degrading polystyrene, polyethylene, foam polystyrene and epoxy resin increase with heat flux power, whereas the surface temperature of poly(methyl methacrylate) remains practically constant. In the case of the latter polymer, surface structure change was observed, which led to an increase in the sample specific surface.

The rate of linear pyrolysis increases and values of oxygen index decrease with surface temperature. Dependence of oxygen index on temperature is of nonlinear type and possesses curve bendings in the area of phase transitions, which is created by heat expended in these processes.

Measurements of temperature profiles in the condensed phase during combustion of poly(methyl methacrylate) and polystyrene show that surface temperatures and temperature fields near the surface, i.e. in the degradation zone, are identical. This indicates invariability of the combustion rate under critical conditions.

Hence, existence of the oxygen index is, in fact, stipulated by minimal rates of polymer degradation, at which the necessary amount of fuel is delivered to the gas phase. The role of fuel is played by volatile products of polymer degradation [39, 40]. Decrease of the oxygen index with temperature affecting the surface should mean reduction of the heat flux delivered from flame to the polymer surface. Obviously, the heat flux decrease is associated with the flame temperature decrease with oxygen concentration. One may suggest that gasification of polymers requires comparatively low amounts of heat energy, and if combustible volatile products are released during pyrolysis of polymers at a rate exceeding the minimum one, then the atmosphere possesses enough oxygen to sustain stable combustion [40].

Analysis of gaseous products selected from various flame zones of combusting poly(methyl methacrylate) in the nitrogen-oxygen mixture possessing various concentrations of components displays the presence of oxygen in the whole volume of the flame. The highest amount of oxygen entering reactions is observed in high-temperature zones and the ones above the surface. The latter indicates the possibility of proceeding of polymer thermooxidative degradation during combustion. Estimation of the linear pyrolysis rate of a series of polymers in heated gas flows with different oxygen concentrations shows that the presence of oxygen in flame can show up differently on the rate of polymer degradation. Oxygen concentration increase does not affect pyrolysis of polystyrene, induces an insignificant increase of poly(methyl methacrylate) degradation rate, but exerts a significant effect on polyethylene, which is associated with rapid oxidation reactions in the surface layer. Studies of epoxy resin at low oxygen concentrations show that the pyrolysis rate in the presence of oxygen is higher than in inert medium. However, starting from 3 vol.%, the degradation rate decreases abruptly, and therewith carbonized residue is formed on the polymeric sample surface [40, 41].

The notion of "carbonized" or "coke" residue is widely used in works on combustion of polymers. The term includes thermoresistant products of polymer pyrolysis, which, beside carbon, can contain nitrogen, oxygen, phosphorus and other elements [42, 43].

Formation of a carbonized surface layer promotes a decrease of polymeric material combustibility [44 - 47]. The main reasons for suppression of combustion due to carbonization are the following:

- The coke formed makes penetration of the heat energy into the condensed phase difficult;
- Carbonized layer prevents oxygen diffusion from the surroundings to degrading polymer;
- The presence of carbonized layer prevents exit of gaseous and liquid products to the surface.

Clear presentation of the role of separate process mechanisms, the degree of their participation and ratio in decreasing combustibility of polymeric materials has not yet been formulated.

At combustion of carbonizing polymers it is observed that, in some cases, flame combustion continues after the end of flame influence on the coke crust formation. One may suppose that, on the one hand, coke occurrence prevents heat permeation into the condensed phase decreasing the material combustibility and, on the other hand, the presence of porous structure promotes mass transfer of liquid degradation products to heated up surface of the material, which in its turn, promotes their combustion [44].

Behavior of a carbonizing polymeric system under the effect of radiant flux was studied on the model representing polymer samples covered by plates of porous thermoresistant material. It is found that the rate of poly(methyl methacrylate) degradation increases with heat flux intensity and reduces with the thickness of foam material. The rate of linear pyrolysis is also reduced with decrease in the condensed phase heating through.

The change of condensed phase heating rate has an essential effect on the mechanism of polymer pyrolysis, which is associated with competing processes of degradation and structuring. Study of degradation of polymers possessing reactive groups shows that degree of decomposition increases with the heating rate [48].

The change of combustion and pyrolysis conditions of polymers has an essential effect on toxicity of products formed. The gas chromatography method of studying gaseous degradation products shows the presence of definite ranges of oxygen concentration, in which the maximum amount of carbon oxide is released from different polymers [48 - 50].

Coke formed during combustion possesses porous structure, and gaseous degradation products permeate through the carbonized layer and enter the combustion zone. For liquid products, coke can be a "fuse", by which the liquid rises due to capillary forces and, entering the surface, sustains combustion.

Tests on linear pyrolysis of samples from poly(methyl methacrylate) covered by plates from porous thermoresistant material show that increase of protecting plate thickness at the same heat effect induces surface temperature increase and changes the level height of liquid degradation products in the foam material. The latter is associated with distribution of temperatures in the porous materials and its heat conductivity.

Liquid motion in foam materials adheres to the Darsy law [51]. Combustion rate of carbonizing polymers is determined by the rate of coke gasification and the transmission rate of gas and liquid products of polymeric material degradation through the carbonized layer. The amount of gas and liquid degradation products decreases with increase of carbonized residue yield. In the limit, the combustion rate may be determined by the rate of oxidation pyrolysis of the carbonized layer. Simultaneously, permeability of cokes formed during pyrolysis of the basic polymeric compound varies in a wide range. Calculations show [294] that at the present permeability, the carbonaceous layer becomes the obstacle for releasing volatile degradation products into the gas phase.

Permeability data [42] show that coke possesses about one-third of its volume as through pores with small diameter, by which liquids can rise by capillary forces. Viscosity of liquid products is the important factor affecting the rate of their motion by the carbonized layer [51, 52]. Calculations and experiments show that polymer melts can also transmit through the surface carbonized layer [52]. If polymer melting temperature is low, and the melt is of low viscosity, or liquid products are easily formed during pyrolysis, then coke formed on the surface cannot be the effective protection from fire [51].

Considering combustion of carbonizing polymers, it is desirable to separate two surfaces: the carbonized coke surface contacting the gas phase and the surface of degrading polymer contacting the coke. Analysis shows that the amount of heat absorbed by the polymer decreases with temperature rise of the coke surface, because the contribution of heat energy due to convection decreases and heat release by irradiation increases due to surface temperature rise. Heat losses for heating and gasification of the condensed phase also increase. Therewith, a great contribution is made by rise of enthalpy of gas products permeating through the coke, because heat capacity of gases is much higher than that of solids.

Existence of the minimal limit rates of combustion implies that creation of incombustible polymers requires striving for a decrease in the release rate of pyrolysis volatile products capable of igniting and possessing high heating capacity. Decrease of polymer combustibility is favored by processes proceeding in the condensed phase with high heat absorption. An effective measure for decreasing combustibility may be the carbonized layer formed on the surface [39].

Analyzing the above-said, the following ways of decreasing combustibility of carbonizing polymers can be outlined:

- Increase of carbonized residue yield, which decreases the amount of volatile products released into the combustion zone;

- Increase of coke thermoresistance increasing temperature of the material surface that promotes decrease of convective heat flux, and rise of heat energy scattering by the surface by irradiation and heat losses for material heating. The presence of coke on the surface forms a barrier for heat flux from flame and, consequently, it is desirable to increase the thickness of the carbonized layer and to decrease its heat conductivity;
- Decrease of coke permeability and increase of viscosity of polymer degradation liquid products to restrict their rise by carbonized product.

An empirical approach to decrease of polymer combustibility helped in detecting the main classes of combustion inhibitors. However, many problems connected with the mechanism of these compounds are not clear yet.

Chlorine- and bromine-containing compounds are widely applied to decelerate combustion of polymeric materials; fluorine and iodine derivatives are not used. Based on the interaction scheme of hydrogen halides with active radicals in the hydrocarbon flame, thermodynamic and kinetic calculations were executed using reference data. These calculations show that practically all the reactions with participation of hydrogen halides in the temperature range of 700 – 1,700°C are thermodynamically abandoned, and that hydrogen bromides and iodides are the most reactive compounds. This was confirmed experimentally by estimating oxygen indices of several polymers filled with zeolites, preliminarily saturated with hydrogen halides. Somewhat underestimated results were obtained for estimation of hydrogen iodide inhibiting ability, which is associated with its decomposition into components with temperature [53].

For reducing combustibility of polymers, phosphorus-containing compounds of various structures are the most widely used. Under combustion conditions, they promote formation of carbonized residue. Taking into account that many phosphorus-containing combustion inhibitors decompose to acids at heating and assuming the possibility of their oxidation, it is desirable to pay attention to phosphoric acids and their ammonium salts [48 - 58].

It is commonly assumed that phosphorus-containing compounds decrease combustion, because they promote carbonization of polymeric materials during pyrolysis and combustion. On the other hand, measurements of permeability of carbonized residues obtained during pyrolysis of phenol-formaldehyde resins to which ammonium monophosphate has been added show that the presence of a phosphorus-containing compound reduces the Darsy constant by several times [52, 59]. Consequently, one of the reasons for polymer combustibility decrease in the presence of phosphorus-containing compounds is the decrease of carbonized layer permeability.

Poly(vinyl alcohol) and epoxy resin combust in air, and at dynamic heating up to 500°C, i.e. to the level of combusting polymer surface temperature, they decomposed almost completely. Therewith, addition of phosphoric acids or their ammonium salts induces formation of a thermoresistant residue, combustibility of polymers being decreased simultaneously (the oxygen index of epoxy resin increases to 30) [60 - 64].

Combustion inhibition by phosphorus-containing compounds shows up not only in formation of carbonized layer, but also in decrease of coke permeability. This fact was confirmed by analyzing the coke residue obtained from phenol-formaldehyde resin and poly(phenylene dimaleimide), thermally treated by phosphoric acid, as well as in pyrolysis of polymeric materials modified by ammonium phosphate. Introduction of phosphorus compounds into composite material leads to a decrease of Darsy constants by almost 15 times, which shows up on combustibility decrease of polymeric material. For example, oxygen indices of materials based on epoxy resin, covered by coke plates treated by phosphorus-containing compounds, equal 76 and 52. If there was no treatment by phosphoric compounds performed, the limit of oxygen concentration is 45 and 35%, respectively. Permeability lowering in porous materials accompanying treatment by phosphoric compounds is probably associated with the fact that polyphosphates formed at high temperatures possess high viscosity and fill in the coke pores [65 - 67].

Some boric compounds also cause such effect on permeability of carbonized layer [50, 59]. Combustibility of the system based on polystyrene covered by a coke layer, obtained by pyrolysis of phenol-formaldehyde resin filled with boron oxide and cross-linked by hexamethylenetetramine, is estimated in the works mentioned. However, it is noted that permeability of cokes depends on temperature in the presence of boric compounds: it decreases first with temperature rise to 450°C, and then increases, which shows up on the material combustibility characterized by oxygen index. Oxygen index reaches 50 at 450°C, and thereafter reduces to 40 – 42. This is associated with the fact that at 450°C boric oxide becomes a limpid liquid, which probably covers coke pores and promotes permeability lowering. Further heating leads to lowering viscosity of films from boron-containing compounds. As a consequence, the surface of carbonized material is uncovered and burnt off.

Oxidation processes induce formation of carbonized residue during pyrolysis and combustion of some polymeric materials. The study of potassium hypochlorite (decomposing with oxygen release) influence on epoxy resin combustion shows that at low concentrations this substance increases coke residue yield and reduces material combustibility (oxygen index of epoxy resin equals 23 at 5 wt.% concentration). Increase of potassium hypochlorite concentration leads to an abrupt decrease of the oxygen index.

The oxidation process can be affected by multi-valent metal compounds. The studying of pyrolysis of materials based on epoxy resins in the presence of compounds with copper, tin, antimony, and cobalt shows that initiation of polymeric matrix degradation in the presence of tin compounds is shifted by 60 – 80°C towards high temperatures, and the solid residue yield is increased. Electron-spectroscopic studies of polymer oxidative degradation in the temperature range of 240 – 320°C show that addition of tin compounds significantly increases the induction period of paramagnetic particle occurrence, reduces the rate of their accumulation and total amount that shows up on thermooxidative degradation [68 - 72]. The presence of tin- and cobalt-containing compounds promotes formation of carbonized layer and, consequently, decrease of the combustion rate. Chromatographic studies also show quantitative change of the composition of pyrolysis volatile products.

Carbonization processes are also promoted by the presence of sulfogroups. However, the data on application of sulfocontaining compounds as inhibitors of polymeric material combustion, shown in the literature, are contradictory. Measurements of permeability of pyrolysis products of phenol-formaldehyde resin added by sulfponilamide show that the presence of sulfocontaining compound, similar to phosphorus-containing one, leads to a decrease of the Darsy constant almost by an order of magnitude. However, in this case, permeability lowering may be associated with foaming of the mixture of phenolic resin with sulfonilamide, which provides a way for liquid degradation products to erupt into the combustion zone [45]. Sulfur derivatives can participate in the coke composition and remain stable at high temperatures. Study of pyrolysis and foam formation of aryl sulfonilamide compounds, which are capable of forming thin foam, shows that they may have potential for creating fire shield materials and their operation mechanism should be studied.

Efficiency of carbonized layer increases with formation of foamed coke layer on the cover surface. A stable foamed coke can be formed in the case of coincidence of gas formation and polymeric system viscosity increase rates with future transition of polymer into solid. If these rates are different, the foam is unstable and subsides as it forms. In this case, protective properties of the foam are not displayed. Formation of a foam stable at heating requires system transition from melt to solid with sharp increase of the melt viscosity providing fixing of gas product bubbles in the product formed. Such viscosity increase at heating is possible by using components with several types of reactive groups, interaction of which at high temperatures leads to formation of non-melting compounds with spatial cross-linking of macromolecules. As an example, let us present the system containing para-aminobenzene sulfonilamide, described in ref. [73].

Thermogravimetric study shows that under conditions of isothermal heating of the given compound, two areas of intensive degradation are observed: in the range of 280 – 350°C with the mass loss of 20 – 50% and in the range of 450 – 550°C with the mass loss of 70 – 95%. Foaming up with gas products release, both volatile and nonvolatile substances, proceeds at 280 – 350°C, which corresponds to the first area of intensive degradation. In the second stage, nonvolatile foam material degrades.

The authors of ref. [73] have found that ammona release in the first stage probably proceeds under interaction of the electrophilic sulfur atom with a nucleophilic nitrogen atom of the amino group. Consequently, redistribution of electron densities at heating forms favorable conditions for polyamination and reamination to occur. Formation of such structures was confirmed by infrared spectroscopy data of volatile products obtained at 240°C, which display an absorption band at 3250 cm–1 typical of valence oscillations of secondary –NH- amine bond. Therewith, intensity of bands at 3200 and 3300 cm–1 typical of sulfonamide grouping is significantly reduced.

Ammonia release shifts the equilibrium to the right, and thereafter the second molecule of para-aminobenzene sulfonilamide can be attached forming an oligomer, which increases viscosity. Solid products formed at 250 – 260°C represent cross-linked insoluble spatial structures. Formation of spatially cross-linked insoluble products during pyrolysis becomes possible owing to interaction of the end amino group with secondary amino group [74].

As temperatures up to 360°C, sulfonilamide groups begin interacting with one another, as well as bonds in sulfonilamide compound backbone break with formation of foam cokes. It is found that SO_3 is the main gaseous product of the reactions in the given temperature range. The amount of SO_3 released increases with temperature.

According to the element analysis data of carbonized residue, a tendency to reduce sulfur content, but concentration of nitrogen and carbon with temperature is observed. IR- and mass-spectroscopy of solid degradation products shows bands occurring in the range of 1,400 cm–1 typical of R-SO_2-OR sulfonates, and R-O-SO_2-O-R esters of sulfonic acids. Asymmetric structure of R-SO_2-S-R is confirmed by observing asymmetric oscillations of S-O bond accompanying symmetric ones in the range of 1,150 cm–1. Occurrence of an absorption band in the range of 1,040 – 1,020 cm–1 characterizes valence symmetric oscillations of RSO_3H

and RSO3H4 sulfoacids and their salts. Mass-spectrum data indicate that backbone and cross-linking breaks with formation of high-molecular compounds proceed simultaneously at para-aminobenzene sulfonilamide is heated. It is suggested that oligomer degradation proceeds as the result of S-N bond breaking, which are the weakest in the chain. This confirms the possibility of forming sulfonic acids, which as known are very unstable and transform into esters of thiosulfoacids. Being strong agents of dehydration, cyclization and cross-linking, sulfoacids formed can promote carbonization and lead to foam coke formation [74].

Based on the above-mentioned data and considering analysis of the literature, one may suggest that a significant role at thermal decomposition of para-aminobenzene sulfonilamide in the temperature range of 280 – 360°C is played by free radicals formation at the stage of the backbone rupture with their future recombination. One of the main reactions is the interaction of highly active phenyl radicals with aromatic rings, which leads to formation of a spatially cross-linked structure.

Hence, based on the mechanism suggested it can be concluded that condensation reactions with ammonia release proceed in the initial stage at thermal decomposition of para-aminobenzene sulfonilamide. Thereafter weak bonds rupture with SO3 release and formation of free radicals participating in the foam coke formation. The foam material formed is characterized by spatially cross-linked structures, which make an essential contribution to reduction of polymeric material combustibility.

Addition of dicarboxylic acids, for example, terephthalic acid, which interact with sulfonilamide or products of its decomposition under combustion conditions, increases thermal resistance of the foam. On evidence of thermogravimetric analysis under conditions of isothermal mode at 360°C, terephthalic acid is volatilized, sulfonilamide loses 55% of mass, and the sulfenamide + terephthalic acid mixture loses 25% of mass only, which indicates formation of higher thermoresistant products. As the result of chemical transformations proceeding at heating sulfonilamide mixture with terephthalic acid, the coke structure is changed, which is confirmed by X-ray patterns [74, 75], and a more regulated coke structure is observed. Traces of the crystalline phase typical of hexagonal graphite-like structure occur at 350°C that leads to a sharp increase of fire resistance of foam cokes.

Injection of fire shield additives into polymeric composites, shaped as mechanical admixtures, during the stage of component mixing is the most widespread method of increasing fire resistance of polymeric materials. However, an irregular distribution of components of the composite material is possible in this method. This leads to a significant instability of material properties.

A more effective method of conferring fire resistance or even full incombustibility on the polymeric material is polymer modification during the stage of its production by injecting combustion inhibitor into the polymeric chain structure.

As hydrogen halides inhibit combustion and bromine is one of the most active halogens, the supposition that vinyl bromide will decompose releasing hydrogen bromide is checked on a series of copolymers with vinyl bromide, synthesized specially for the purpose [319]. The study of the oxygen index of polymers obtained shows that compared with homopolymers, methyl methacrylate copolymers with butyl acrylate display a negligible decrease of combustibility, and styrene and acrylonitrile ones display a greater rise of the oxygen index. Thermogravimetric analysis shows that vinyl bromide injection reduces thermal resistance of polymers. In the case of copolymers based on styrene and acrylonitrile, a carbonized residue is formed during combustion and pyrolysis that reduces the degradation rate at one-side heating [76].

Of great interest are dehydrochlorination and aromatization of hydrocarbons in the presence of catalysts – titanium, cobalt, and aluminum oxides, and aluminum phosphates [5]. The mechanism of highly effective combustion decelerators of the intumescent type is associated with catalysis of coke formation proceeding with participation of polymer molecules or a carbon forming component, specially injected into the system [77].

Based on the thermal balance equation of polymer combustion, substances degrading or reacting with heat absorption should promote combustibility decrease. Decomposition enthalpies of a series of salt crystalline hydrates, hydroxides and other substances capable of transforming with heat energy absorption were measured on a differential scanning microcalorimeter. The study of the oxygen index of polymeric materials containing these compounds shows an increase in the presence of the latter [78 - 80].

Exothermal processes proceeding in the condensed phase are accompanied by combustion increase of polymeric materials.

The study of the influence of carbonaceous residue on combustion of polymeric material shows that a carbonized layer can serve as a powerful screen protecting from heat energy permeation into the lower layers of the material. The insulating effect increases with the layer thickness and decrease of its heat conductivity that can be achieved by forming a foamed up material under the fire effect. This principle is used in developing fire and heat shield materials. Foaming up covers increase in volume under the heat flow effect and form incombustible foam material [81 - 84].

Estimation of the temperature field in the expanding cover, temperature variations with the cover thickness, and dependencies on heat effects of the reactions proceeding in the cover shows the main contribution of the height of the foam material formed and its heat conductivity into the durability of fire and heat shield covers. Processes proceeding in covers with significant endoeffects are negligible in the absence of carbonized foamed layer [85].

In the initial stage of one-side heating of fire shield foaming covers, heat is absorbed and removed to underlying layers at a rate limited by thermophysical characteristics of the initial polymeric composite. The material degrades with heating, and the upper layers begin foaming up, when reaching a definite temperature. During the initial period, temperature on the protected surface rises rapidly. Thereafter the rate of heating decreases as the consequence of foam material formation and hardly changes for some time. Note that the protected surface temperature does not increase with heat flux intensification if such materials are used. Foaming up material realizes the effect of foaming self-control, when rise of the heat flux intensity induces increase of the heat insulating foam volume, which prevents an intensive heating through of the protected surface. Temperature increase of the surroundings promotes deeper heating through of the cover and involving a high volume of the material into foam formation.

The foam formation rate and height of foam coke layer formed increase with the initial cover thickness. However, a tendency to overestimate the height of the foamed material is observed. Measurements of the temperature field in foaming covers show that bottom layers are heated up at a low rate, and they do not reach the foam formation temperature for a long time, i.e. they play the role of thermal insulation [86].

Double-layer covers are used, in which the upper layer provides the function of foaming up heat shield material, and the bottom one represents a polymeric matrix filled with substances decomposing with heat absorption. Salt crystalline hydrates, boric acid, etc. are used as "cooling down" materials. Such combination allows the weight of the initial cover to be reduced by 25 – 30% preserving fire shield properties at the same level [80]. An efficient method of increasing fire shield properties of the covers is found in application of a foaming material with higher heat insulating properties as the bottom layer. In the initial stage of cover operation, this prevents heating of the protected object and increases the heating rate of the upper layer up to the foaming initiation. The latter also increases the rate of foam coke formation and its thickness.

The mechanism of foaming up fire shield covers differs from foam production from liquids or foam materials. The necessary condition for the material to foam up under fire effect is the initial composite material transition into the rubbery state with future irreversible transition into the solid state [81, 84, 87, 88].

Considering the scheme of foaming up, one should take into account kinetics of gas products release, diffusion of gas bubbles due to expulsion force, and removal of gas products with regard to permeability of the condensed phase. The surrounding pressure, resistance of polymeric matrix to gas bubbles formation and overcoming the surface tension forces may produce a significant effect on this process.

Stable foam cannot be formed under conditions of one-side heating, when composite surface temperature increases continuously, if the system does not transform irreversibly to the solid state at high temperatures. Such process is observed at carbonized residue formation. At the initial moment of flame influence, the polymer is in the solid state. That is why formation of a porous structure requires high pressure in gas bubbles during gas products release to stretch the polymer. Further on, elasticity modulus of the polymeric material is reduced and, consequently, gas bubbles are able to leave the condensed phase pressed out by the repulsion force. Temperature in the condensed phase increases under the flame influence, viscosity is reduced, and therewith, foam subsidence is observed. As the reaction proceeds with formation of rubbery compounds accompanied by increase of the composite rigidity, the foamed layer is fixed [86].

Hence, obtaining of a foamed material under the fire effect required synchronous gas release and increase of viscosity and rigidity of the composite. At high degrees of foaming transition of the system from the viscous flow state into the solid one proceeds at a low rate, and as the rate of changing viscoelastic properties increases, thin dispersed material is formed. The transition rate of the polymeric material into the solid state is determined by reactions of rigid ladder and three-dimensional structures, the occurrence of which requires the presence of two or more reactive groups in the composite material [65, 66, 89]. Mixtures of polyatomic alcohols with phosphoric acid or compounds forming phosphoric acids during decomposition at increased temperatures may form components of such systems. Injection of compounds containing amino groups into such systems promotes increase in coke yield during pyrolysis. Addition of amines mixed with pentaerythritol and phosphoric acid shifts the initiation of coke formation towards lower temperatures. Interaction between alcohols and phosphoric acid increases system viscosity, which shows up favorably on the foam formation. Injection of nitrogen-containing compounds increases the melt viscosity and reduces the foaming up temperature, which shows up on the foam coke obtained. Further heating induces formation of double bonds and cross-linked structures [88, 90].

REFERENCES

[1] Deribere M., *Practical Application of Infrared Rays*, Gosenergoizdat, Moscow-Leningrad, 1959, 440 p. (Rus)
[2] Polezhaev Yu.V. and Yurevich F.B., *Heat Protection*, Ed. A.V. Lyikov, Energiya, Moscow, 1976, 380 p. (Rus)
[3] Margolin I.V. and Rumyantsev N.P., *Grounds of Infrared Engineering*, Voenizdat Min. Oborony, 1957, 308 p. (Rus)
[4] Lecont J., *Infrared Radiation*, Izdat. Fiz.-Mat. Literatury, Moscow, 1958, 584 p. (Rus)
[5] Dolgov B.N., *Catalysis in Organic Chemistry*, Moscow-Leningrad, Goskhimizdat, 1949, 560 p. (Rus)
[6] Ridder R.C., *Plast. Eng.*, 1977, vol. 33(1), pp. 38 – 42.
[7] Grunfest I.J., *British Plastics*, 1958, vol. 31(2), pp. 530, 531, 539.
[8] Hourt W.C., *Ind. Eng. Chemistry*, 1960, vol. 52(9), pp. 761 – 763.
[9] Bruer L. and Sirsi A.V., *Khimia Vysokikh Temperatur*, 1958, vol. 27(8), pp. 966 – 989. (Rus)
[10] Midgley T. and Henne A.L., *J. Chem. Soc.*, 1929, vol. 51(4), pp. 1215 – 1226.
[11] Burchfield H.P., *Ind. Eng. Chem. Anal. End.*, 1944, vol. 16(7), pp. 424 – 426.
[12] Wall L.A., *J. Res.Nat. Bur Stand.*, 1948, vol. 41(4), pp. 315 – 322.
[13] Grassi N., *Chemistry of Polymer Degradation Processes*, Moscow, Inostrannaya Literatura, 1959, 252 p. (Rus)
[14] Madorsky S.L., Strams S., Thompson D., and Williamson L., *J. Polym. Sci.*, 1968, vol. 4(5), p. 639.
[15] Simon V.L., Nitrile Rubbers, In Coll.: *Synthetic Rubber*, Ed. G.S. Witby, Leningrad, Goskhimizdat, 1957, pp. 777 – 822. (Rus)
[16] Madorsky S.L. and Strans S., *WADS Technical Report*, 1959, pp. 59 – 64.
[17] Bourbigot S., Morice L., and Leroy J., *Fire Retardancy of Polymers*, Ed. M. Le Bras, G. Camini, S. Bourbigot, and R. Delobel, 1988, No.224, Cambridge, pp. 129 – 139.
[18] Le Bras M., Bourbigot S., Le Tallec Y., and Laurens J., *Polym. and Stab.*, 1997, vol. 56, pp. 11 – 21.
[19] Le Bras M., Bourbigot S., Delporte C., Siat C., and Le Tallec J., *Fire and Materials*, 1996, vol. 20, pp. 191 – 203.
[20] Sharf D., In: *Rec. Adv. in Fr. Of Polym. Mat.*, vol. 2, Ed. M. Lewin, BCC, 1991, p. 55.
[21] Sharf D., Nalpta R., Heflin R., and Wus T., *Fire Safety J.*, 1992, vol. 8(19), p. 103.
[22] Camio G., Costa L., and Troassarelli L., *Polym. Degr. Stab.*, 1984, vol. 8, p. 243; *Polym. Degr. Stab.*, 1985, vol. 12, p. 203.
[23] Ruban L.V. and Zaikov G.E., The role of Intumescence in Fire Protection of Polymers, *Plasticheskie Massy*, 2000, No. 1, pp. 39 – 43. (Rus)
[24] Langley J., Drews M.J., and Barker R.T., *J. Appl. Polym. Sci.*, 1974, vol. 25, p. 243.
[25] Zhu W. et al., *J. Appl. Polym. Sci.*, 1996, vol. 62, p. 2267.
[26] Costa L., Goberti L., Paganetto P., Camio G., and Squarzi P., *Proc. 3rd Meeting on FR Polymers*, Torino, 1989, p. 19.
[27] Cleave R.F., *Plastics and Polymer*, 1979, vol. 20, p. 783.
[28] Markezich R.L. and Mundhenke R.F., In: *Rec. Adv. in Fire Retardance of Polym. Mat.*, Ed. M. Lewin, BCC, 1995, vol. 66, p. 177.

[29] Dombrowski R. and Huggard M., In: *4th Intern. Sumposium, Additives-95*, Clearwater Beach, FL, 1996, p. 1.
[30] Brauman S., *J. Fire Retardant Chem.*, 1980, vol. 7(2), p. 61.
[31] Delobel R., Le Bras M., Quassou N., and Alistiqsa F., *J. Fire Sci.*, 1990, vol. 8(3-4), p. 85.
[32] Bourbigot S., Le Bras M., Delobel R., and Breant P., *Polym. Degr. Stab.*, 1996, vol. 54, p. 275.
[33] Bourbigot S., Le Bras M., Breant P., Tremillon J.-M., and Delobel R., *Fire and Materials*, 1996, vol. 20, p. 145.
[34] Selevich A.F., Levchik G.F., Lesnikovich A.I., and Levchik C.V., *Belarus Patent* No. 00260-016 (Belorussian University), 1993. (Rus)
[35] Levchik S.V., Levchik G.F., Camino G., Costa L., and Lesnikovich A.I., *Fire Mater.*, 1996, vol. 20, p. 183.
[36] Sevostyanov M. and Novikov S., *Vysokomol. Soedin.*, 1991, vol. 33, p. 1568. (Rus)
[37] Bourbigot S., Le Bras M., and Delobel R., *Appl. Surf. Sci.*, 1994, vol. 81, p. 299.
[38] Bourbigot S., Le Bras M., and Le Laureyns Y., *Polym. Degr. and Stab.*, 1997, vol. 56, p. 11.
[39] Zhubanov T.B. and Gibov K.M., Minimal Border Rates of Polymer Combustion, *Collection of Reports on International Conference "Flammability of Polymers"*, Smolensk, 1985, pp. 42 – 43. (Rus)
[40] Zhubanov T.B., Gibov K.M., and Zhubanov B.A., On Connection between Oxygen Indices and Border Combustion Rates of Polymers, *Vysokomol. Soedin.*, 1986, vol. 28B, p. 250. (Rus)
[41] Zhubanov T.B. and Gibov K.M., Polymer Combustion Rate as the Estimation Measure of its Combustibility, *Thes. XXII All-Union Conference on High-molecular Compounds*, Alma-Ata, Oct., 1985, p. 42. (Rus)
[42] Zhubanov B.A., Abdicarimov M.N., and Gibov K.M., Pyrolysis Processes under Combustion Conditions, *Coll. Rep. Intern. Conference "Nehorl avost Polymernykh Materialov"*, Bratislava, 1976, pp. 30 – 31. (Rus)
[43] Zhubanov T.B. and Gibov K.M., Linear Pyrolysis of Carbonizing Polymers. Model Systems, *Materials of VIII All-Union Symposium on Combustion and Explosion*, Oct., 1986, Tashkent, Coll.: *Combustion of Heterogeneous and Phase Systems*, Chernogolovka, 1986, pp. 101 – 103. (Rus)
[44] Zhubanov T.B., Gibov K.M., and Shapovalova L.N., Capillary Events During Combustion of Carbonized Polymers, In: *Fire Shield Polymeric Materials, Problems of Estimating Properties, Meeting Thes.*, Tallinn, 1981, p. 93. (Rus)
[45] Shapovalova L.N., Study of Foam Formation in Polymeric Composites. *Proc. Scientific Conference of Young Scientists of the Institute of Chemical Sciences*, Alma-Ata, 1983, pp. 135 – 136. (Rus)
[46] Shapovalova L.N. and Karzhubaeva R.G., Pyrolysis and Combustion of Sulfur-containing Polymeric Materials, In: *Polymer Combustion and Creation of Limitedly Combusting Materials, Thes. Rep. 5th All-Union Conference*, Volgograd, 1983, p. 40. (Rus)
[47] *Patent* No. 1,215,343. Acrylonitrile (Styrene) Compolymers with Sulfonilamide Methacrylate Possessing Reduced Combustibility and Method of Their Production. (Rus)

[48] Druz N.N. and Gibov K.M., Toxicity of Pyrolysis and Phenoplast Combustion Gas Products, *Thes. Rep. Coord. Meeting on Phenoplasts*, Kemerovo, 1983, p. 158. (Rus)
[49] Druz N.N. and Gibov K.M., Influence of Polymer Pyrolysis Conditions on Carbon Oxide Formation, *Thes. Rep. All-Union Conference on Polymer Combustion and Creation of Limitedly Combustible Materials*, Volgograd, 1983, p. 91. (Rus)
[50] Druz N.N. and Gibov K.M., Toxicity of Polymeric Material Combustion Products, *Proc. IKhN AN KazSSR "Chemistry and Physics of Polymers"*, Alma-Ata, 1984, p. 145. (Rus)
[51] Gibov K.M., Zhubanov B.A., and Shapovalova L.N., Influence of Carbonized Surface Layer Porosity on Combustion of Polymers, *Vysokomol. Soedin.*, 1984, vol. 26B(2), pp. 108 – 110. (Rus)
[52] Gibov K.M., Zhubanov B.A., and Shapovalova L.N., Mass Transfer of Pyrolysis Products through Carbonized Layer during Combustion of Polymers, *Thes. Rep. First Symposium on Macroscopic Kinetics and Chemical Gas Dynamics*, Chernogolovka, 1984, vol. 1, Part 2, p. 4. (Rus)
[53] Gibov K.M., Nazarova S.A., and Zhubanov B.A., Deceleration of Polymer Combustion by Hydrogen Halides, *Izv. AN KazSSR, Ser. Khim.*, 1978, No. 1, p. 60. (Rus)
[54] Druz N.N. and Gibov K.M., Toxicity of Polymer Material Combustion Products. *Proc. IKhN AN KazSSR. Chemistry and Physics of Polymers*, Alma-Ata, 1984, vol. 62, p. 145. (Rus)
[55] Zhubanov Yu.A., Gibov K.M., Nikitina I.I., and Sarsembinova B.T., On Some Features of Thermal Transformations of Ammonium Phosphates in Polymeric Materials, *Thes. Rep. VI All-Union Conference "Phosphates-84"*, Alma-Ata, 1984, p. 45.
[56] Nikitina I.I., Gibov K.M., Kan A.A., and Zhubanov B.A., Study of High-temperature Reactions Between Phosphoric Compounds and Hydroxyl-containing Substances, Scientific Collection: *Theoretical and Applied Aspects of Fire Protection of Wooden Materials, Institute of Wood Chemistry AS LatSSR*, Riga, Zinatne, 1985, p. 93. (Rus)
[57] Nikitina I.I., Gibov K.M., and Kan A.A., On Phosphoric Compounds Reactions During Combustion of Polymeric Materials. *Thes. Rep. XXII All-Union Conference on High-molecular Compounds*, Alma-Ata, Oct., 1985, p. 53. (Rus)
[58] Sarsembinova B.T., Nikitina I.I., Gibov K.M., and Zhubanov B.A., On the Mechanism of Phosphorus-containing Decelerators of Polymer Combustion, Scientific Collection: *Studies of Monomers and Polymers ICS AS KazSSR*, Alma-Ata, 1986, vol. 66, p. 158. (Rus)
[59] Shapovalova L.N. and Zhubanov T.B., On the Role of Phosphorus- and Boron-Containing Compounds in Fire Shield Covers, *Thes. Rep. Republ. Scientific and Practical Conference on "Introduction of Scientific-Research and Production-Technical Works on Chemical Technology"*, Karaganda, 1985, p. 141. (Rus)
[60] Ksandopulo G.I., Chuvasheva S.P., Kononenko K.M., and Gibov K.M., Deceleration of Epoxy Resin Combustion by Compounds Containing Halogens and Phosphorus, *Collection of Reports of All-Union Meeting on the Inhibition Mechanism of Chain Gas Reactions*, Kaz. S.M. Kirov State University, il Alma-Ata, 1971, pp. 223 – 229. (Rus)
[61] Akulova D.V., Ivanov B.A., Ksandopulo G.I., Chuvasheva S.P., and Gibov K.M., Some Problems of Polystyrene Combustion Deceleration, *Thes. Rep. III All-Union Scientific-Technical Conference "Combustion Processes and Problems of Fire Extinguishing"*, VNIIPO, Moscow, 1972, p. 58. (Rus)

[62] Akulova D.V., Chuvasheva S.P., and Gibov K.M., On Pyrolysis Influence on Polystyrene Combustion, *Collection of Works on Applied and Theoretical Chemistry*, Kaz. S.M. Kirov State University, Iss. 5, Alma-Ata, 1974, pp. 133 – 139. (Rus)

[63] Gibov K.M., Kapyirina V.Ya., and Dovchilin T.Kh.. Fire Shield Composites Based on Epoxy Resin, *Plastmassy*, 1977, No. 12, p. 46. (Rus)

[64] Abdikarimov M.N., Gibov K.M., and Zhubanov B.A., Pyrolysis and Combustion of Epoxy Resin in the Presence of Amines Hydrochlorides and Sulfates, *Izvestiya AN KazSSR, Ser. Khim.*, 1980, No. 2, p. 42. (Rus)

[65] Gibov K.M., Nikitina I.I., and Gadranova Kh.S., On Condensation Reactions During Combustion of Polymers, *Proc. VIII Intern. Microsymposium on Polycondensation*, Alma-Ata, 1980, p. 44. (Rus)

[66] Gibov K.M., Zhubanov B.A., Nikitina I.I., and Gadranov Kh.S., Some Features of Polymer Carbonization, *Works of ICS AS KazSSR "Synthesis of Monomers and Polymers"*, Alma-Ata, Nauka, 1982, vol. 57, pp. 54 – 77. (Rus)

[67] Nikitina I.I., Gibov K.M., and Galyizhanov, On the Influence of Some Carbonizing Additives on Polymer Combustibility Reduction, *Izvestiya AN KazSSR, Ser. Khim.*, 1986, No. 2, p. 73. (Rus)

[68] Ksandopulo G.I., Chuvasheva S.P., and Gibov K.M., On the Influence of Some Additives on Foam Polystyrene Combustion, *Collection of Scientific Works on Chemistry*, Kaz. S.M. Kirov State University, Alma-Ata, 1971, Iss. 1, Part 2, pp. 40 – 55. (Rus)

[69] Ksandopulo G.I., Sumarokova G.I., Kononenko K.M., Gibov K.M., Chuvasheva S.P., and Surpina D., Influence of Complex Tin Compounds with Amines on Combustion of Condensed Systems, *Coll. Rep. II All-Union Scientific-Technical Conference "Combustion Processes and Problems of Fire Extinguish"*, VNIIPO, Moscow, 1972, pp. 43 – 48. (Rus)

[70] Chuvasheva S.P. and Gibov K.M., Influence of Cobalt Salts on Epoxy Resin Combustion, *Collection of Scientific Works on Applied and Theoretical Chemistry*, Kaz. S.M. Kirov State University, Alma-Ata, 1974, Iss. 5, pp. 139 – 145. (Rus)

[71] Ksandopulo G.I., Pivovarov A.P., and Gibov K.M., Study of Variable Valence Metal Salt Additives Influence on Epoxy Resin Pyrolysis by ESR Method, *Collection of Scientific Works on Applied and Theoretical Chemistry*, Kaz. S.M. Kirov State University, Alma-Ata, 1974, Iss. 5, pp. 145 – 151. (Rus)

[72] Chuvasheva S.P., Gibov K.M., and Ksandopulo G.I., Influence of Tin and Cobalt Salts on Epoxy Resin Pyrolysis Under Combustion Conditions, *Coll. Rep. IV All-Union Scientific-Technical Conference on "Problems of Combustion and Fire Extinguish"*, VNIIPO, Moscow, 1975, p. 63. (Rus)

[73] Shapovalova L.N. and Smirnova T.Ya., Features of Polycondensation and Pyrolysis Mechanism of Para-aminobenzene Sulfonilamide, *Thes. Rep. 23rd Conference on High-molecular Compounds*, Alma-Ata, 1985, p. 83. (Rus)

[74] Shapovalova L.N. and Cherdabaev A.Sh., On Foam Material Structure from Para-aminobenzene Sulfanilamine, *Thes. Rep. of Young Scientists, AS KazSSR "Youth and Scientific-Technical Progress"*, Alma-Ata, 1986, p. 35. (Rus)

[75] Karachubaeva R.G., Shapovalova L.N., and Gibov K.M., Radical Copolymerization of Methacrylic Acid Sulfonilamide With Several Vinylic Monomers, *Intern. Symp. on the*

Radical Polymerization, Kinetics and Mechanisms, Preprints, Italy, 1987, pp. 159 – 162. (Rus)

[76] Gibov K.M., Zhubanov B.A., Abdicarimov M.N., and Asodchi E.F., Linear Pyrolysis and Combustion Vinylic Polymers with Vinyl Bromide, Proc. VI All-Union Symposium on Combustion and Explosion, Collection: *Combustion of Condensed and Heterogeneous Systems*, Alma-Ata, Chernogolovka, 1980, pp. 55 – 58. (Rus)

[77] Zhubanov B.A., Dovlichin T.Kh., and Gibov K.M., Influence of Oxygen Concentration on Diffusional Combustion of Poly(methyl methacrylate), *Vysokomol. Soedin.*, 1975, vol. 17B, p. 746. (Rus)

[78] Chuvasheva S.P., Ksandopulo G.I., Kononenko K.M., and Gibov K.M., Deceleration of Polymeric Materials Combustion, *Coll. Rep. II All-Union Scientific-Technical Conference on "Combustion Processes and Problems of Fire Extinguish"*, VNIIPO, Moscow, 1972, pp. 56 – 62. (Rus)

[79] Nikitina I.I., Gibov K.M., Galyimzhanova S.A., and Paltseva N.G., On the Influence of Some Additives on Polymer Combustion Decrease, *Thes. Rep. Scientific-Technical Conference on "Polymeric Materials in Mechanical Engineering"*, Izhevsk, 1983, p. 37. (Rus)

[80] Gibov K.M. and Paltseva N.G., On the Role of Endothermal Processes in Reducing Polymer Combustibility, *Thes. Rep. 1st All-Union Symposium on Macroscopic Kinetics and Chemical Gas Dynamics*, Alma-Ata, Chernogolovka, 1984, vol. 1, Part 2, p. 5. (Rus)

[81] Gibov K.M., Zhubanov B.A., Dovlichin T.Kh., and Mamleev V.Sh., On the Mecanism of Fire Shield Effect of Foaming Polymeric Covers, *Coll. Rep. Intern. Conference "Nehorl avost Polymernych Materialov"*, Bratislava, 1976, pp. 69 – 71. (Rus)

[82] Gibov K.M., Zhubanov B.A., and Dovlichin T.Kh., Fire Shield Polymeric Covers, "Chemistry and Physical Chemistry of Polymers", *Trudy IkhN AN KazSSR*, Alma-Ata, Nauka, 1979, vol. 49, pp. 43 – 56. (Rus)

[83] Smirnova T.Ya., Ergazieva K.I., Nikitina I.I., Jadranova Zh.S., and Gibov K.M., Application of Oligomeric Systems to Creation of Foaming Covers, *Thes. Rep. II All-Union Conference on Chemistry and Physical Chemistry of Oligomers*, Alma-Ata, Chernogolovka, 1979, p. 59. (Rus)

[84] Nikitina I.I., Gibov K.M., Kan A.A., and Zhubanov B.A., Fire Shield Foaming Covers, *Coll. Rep. Intern. Conference "Flammability of Polymers"*, Smolenict, 1985, pp. 20 – 21. (Rus)

[85] Kan A.A., K.M. Gibov, and Saisembinova B.T., Study of Fire Shield Foaming Covers Behavior at One-side Heating, *Izvestiya AN KazSSR, Ser. Khim.*, 1982, No. 3, p. 80. (Rus)

[86] Gibov K.M., Zhubanov, Dovlichin T.Kh., Mamleev V.Sh., and Nikitina I.I., Heat Mode of Foaming Fire Shield Covers, *Izvestiya AN KazSSR, Ser. Khim.*, 1972, No. 5, p. 36. (Rus)

[87] Gibov K.M., Zhubanov B.A., Nikitina I.I., Kan A.A., and Krasnikov A.V., *Thes. Rep. 1st All-Union Conference on Composite Materials and Their Application in National Economy*, Tashkent, 1980, p. 87. (Rus)

[88] Kan A.A. and Gibov K.M., Foam Coke Formation at High-temperature Pyrolysis of Polymers, *Thes. Proc. All-Union Conference "Combustion of Polymers and Creation of Limitedly Combustible Materials"*, Volgograd, 1983, p. 28. (Rus)

[89] Karxhaubaeva R.G., Gibov K.M., Shapovalova L.N., Ergozhin E.E., and Chigir L.V., Synthesis and Study of Sulfur-containing Copolymers with Reduced Combustibility, *Collection of Works on Chemistry*, Kaz. S.M. Kirov State University, Alma-Ata, 1984, Iss. 8, pp. 49 – 56. (Rus)

[90] Gibov K.M., Zhubanov B.A., Nikitina I.I., and Jadranova Zh.S., Carbonization of Fire Shield Foaming Systems, *High School Collection of Scientific Works "Chemistry and Technology of Elementorganic Semi-products and Polymers*, Volgograd, Polytechnical Intitute, 1981, pp. 30 – 35. (Rus)

Chapter 3

DEGRADATION OF AROMATIC CO-POLYESTERS DERIVED FROM N-OXYBENZOIC, TERE- AND ISOPHTHALIC ACIDS, AND DIOXYDIPHENYL

E. V. Kalugina, K. Z. Gumargalieva[1] and V. G. Zaikov[2]

Polyplastic Co., 14A, General Dorokhov st., Moscow 119530, Russia
[1] N.N.Semenov Institute of Chemical Physics, 4, Kosygin str., Moscow 119991, Russia
[2] N.M.Emanuel Institute of Biochemical Physics, 4, Kosygin str., Moscow 119991, Russia

Special attention to these polymers is defined by their specific feature, which is orientation in the melt, mostly associated with the intense development in computer technologies. Owing to this property such polymers are devoted to the "family" of liquid-crystal polymers. The liquid-crystal properties are also observed for PAI with uneven number of CH_2-groups [1]. It should be noted that polyalkanimide (PA-12), discussed in [2 - 14], also displays liquid-crystal properties under definite processing modes.

Liquid-crystal aromatic copolyesters (LCP) were studied. They were derived from dioxydiphenyl diacetate, acetoxybenzoic, iso- and terephthalic acids (IPA and TPA, respectively): 100/0, 75/25, 50/50, 25/75, 0/100.

$n = 50, m = 25, f = 25$

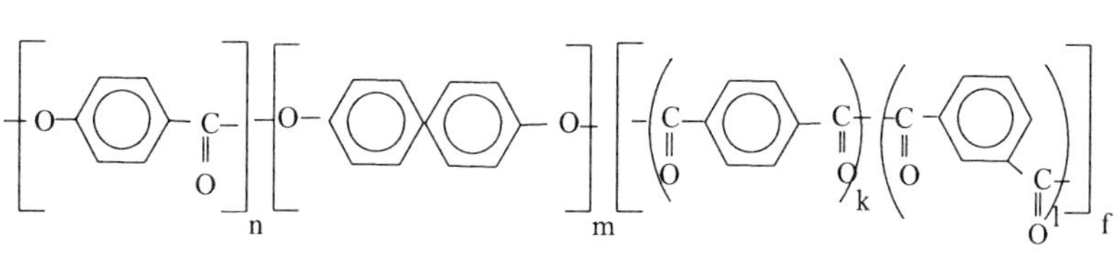

International analogue – *Xydar* (Amoco)

Thermal stability of LCP with different TPA/IPA ratio was studied by dynamic TGA/DTA techniques. Table 1 shows DTA/TGA data obtained in argon flow. Data obtained in air are shown in Figure 1.

Without air oxygen LCP degrade in one stage, forming significant amount of coke residue (up to 40wt.% at 700°C). Two endothermic heat effects were observed on DTA curves: a low one in the melting range and quite intense one in the polymer degradation zone. Calculation of the heat effect gave $\Delta H = 1 - 2.5$ kJ/mol. As decided from the studies of 4-hydroxybenzoic and 2,6-hydroxynaphthoic acid copolymer by DSC method [15], the heat effect of about 1 kJ/mol relates not to real melting, but to changes in the order strength at transition from crystal to mesophase. Apparently, due to superimposition of heat effects which accompany degradation, and transition to the isotropic melt, nobody succeeded in detecting temperature transition associated with LCP changing the isotropic degree.

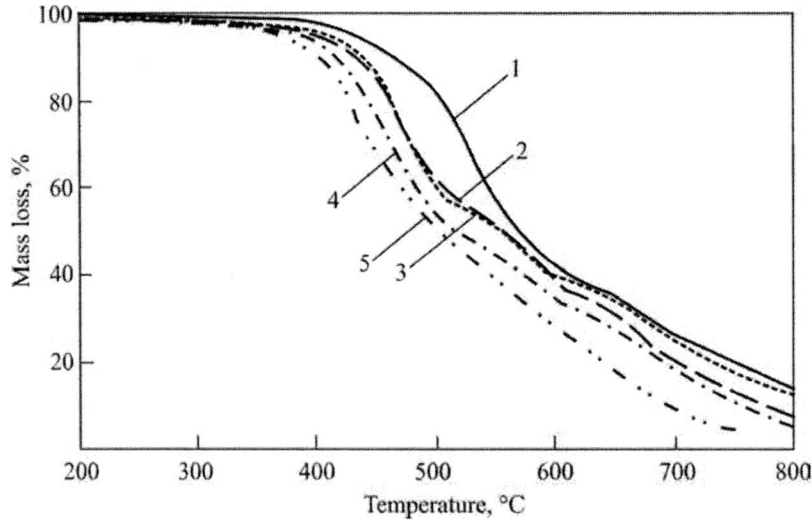

Figure 1. TGA data obtained at the heating rate of 5°/min in air for liquid-crystal polymers (LCP): KI-0 (1), KI-25 (2), KI-50 (3), KI-75 (4), and KI-100 (5)

Table 1. LCP sample characteristics

Sample	TPA/IPA ratio, mol%	TGA/DTA data in argon flow (with heating rate 10°/min)	
		Melting range, °C	Degradation initiation temperature, °C
KI-0	100/0	410 - 415	490
KI-25	75/25	360 - 365	480
KI-50	50/50	340 - 350	472
KI-75	25/75	300 - 315	455
KI-100	0/100	315 - 325	450

Thermal stability of LCP in air is significantly (by 25 - 30°C) lower than in argon. According to dynamic TGA data in air (Figure 1), mass losses of studied LCP are observed in the temperature range of 350 - 800°C. The degradation proceeds in two stages: the first stage at 350 - 550°C is accompanied by mass losses up to 40 wt.%; the second stage is slower and proceeds in the temperature range of 550 - 800°C up to full degradation of the polymer. The coke content at 750 - 800°C equals 1 – 3 wt.%. As shown by DTA data, LCP degradation stages are accompanied by exothermal heat effects. As tested in the air, a low endothermic heat effect is absent in the melting range, apparently, due to overlapping by exothermal effects of degradation reactions. LCP rating in the sequence with thermal stability decrease is the following: KI-0 > KI-25 ≥ KI-50 > KI-75 > KI-100.

The increase in IPA content shifts the melting range towards lower temperatures and reduces LCP thermal stability. The study of LCP phase transitions by X-ray analysis in the temperature range of 20 - 400°C indicates similar changes in all LCP. Annealing at 300°C causes an insignificant increase of the main crystalline reflex. As an example, Figure 2 shows the difractogram for powder-like and mold samples of KI-75 LCP. No phase transitions (reflex occurrence and elimination) were detected in the studied temperature range in LCP. This may prove the DTA results and suggestions about closeness of degradation temperatures and transitions to mesophase [15].

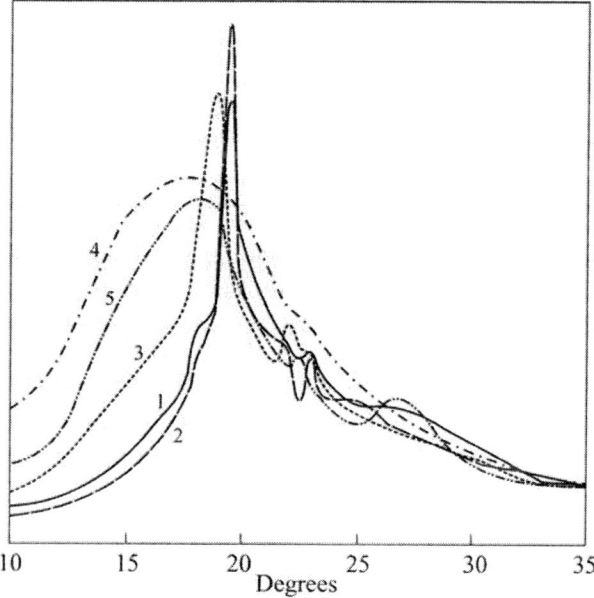

1. powder-like sample (20°C)
2. powder-like sample (300°C)
3. powder-like sample (350°C)
4. powder-like sample (400°C)
5. mold sample

Figure 2. Diffraction patterns for LCP-2 powder at heating

The ability of studied materials to transit to the so-called "liquid-crystal state" characterizes their behavior at processing temperatures. At softening temperature (as regards to the structure, this range falls within 300 - 400°C) a jump-like viscosity decrease is observed in all polymers. Hence, extremely strong fibers are formed from the melt. This effect is explained [16] by cooperative orientation of large macromolecule axes along the flow direction (viscosity anisotropy), which is realized only in LCP. Thermal stability of polymers depends upon several factors: structure, molecular-mass parameters, content of macrochain defects, labile end groups (currently, hydroxyl ones), low-molecular organic (non-reacted, residual monomers) or inorganic (increments of metal ions from the raw material and equipment) additives in the macromolecules.

The composition of inorganic additives in monomers and LCP were studied by plasma-emission spectroscopy technique. The example of KI-75 LCP, studied by TGA, shows the effect of some metal increments on thermal stability. The additive content was increased by injection of inorganic salts of appropriate metal into the polymer. Table 2 shows comparative data on Cu, Fe, Ni, Ca, and Al content and temperatures of degradation initiation by TGA.

The results obtained show different effect of metal increments on LCP thermal oxidative stability. In the studied range of concentrations, aluminum and metals of the alkaline sequence (Ca, Na, K, etc.) do not practically affect the thermal stability. Iron causes the negative effect, whereas Cu and Ni, vice versa, increase thermal stability of LCP. It should be noted that injected concentrations are quite corresponded to usual content of metal increments in industrial samples of engineering, bulk polymers, such as polycarbonate, aliphatic polyamides, polystyrene, etc.

Table 2. The effect of metal increments on the thermal oxidative stability of KI-75 LCP

Metal	Content, wt.%	Degradation initiation temperature (T_d) by TGA in air, 10°/min heating rate
Fe	1.3×10^{-3}	320
	1.3×10^{-2}	300
Al	1.4×10^{-3}	320
	1.4×10^{-2}	320
Ca	4×10^{-3}	320
	4×10^{-2}	320
Ni	1.0×10^{-3}	320
	2.0×10^{-2}	325
Cu	1.3×10^{-5}	320
	2.0×10^{-3}	330

The composition and content of organic additives to LCP were studied by the mass-spectrophotometry technique. Phenol and sioxydiphenyl (94 and 186 m/e, respectively) were identified. They represent the hydrolysis products of the initial monomer of dioxydiphenyl diacetate and heavy fragments of the following structure:

m/e=214   HO—⟨○⟩⟨○⟩—O—$\underset{\underset{O}{\|}}{C}$H

m/e=306   HO—⟨○⟩—$\underset{\underset{O}{\|}}{C}$—O—⟨○⟩⟨○⟩—OH

Total amount of organic additives in different samples has not exceeded $(1.0 - 2.0) \times 10^{-2}$ wt.%, which does not practically affect thermal stability of the polymers.

As shown by kinetics of oxygen absorption at 350°C (the processing temperature), thermal stability of LCP is reduced with increase of IPA content (Figure 3). This result confirms the TGA data. The kinetics of O_2 absorption is the two-stage process with absorption of 1 mole of O_2 per monomeric unit at the first, quick stage (2 – 3 h), and 0.2 mol/base-mol during following 7 – 8 h of thermal oxidation. Analogous dependencies are displayed by CO_2 release (Figure 4) - the main gas product of the studied LCP degradation product. Shown below are heavy, highly boiling LCP degradation products, identified by NMR and MS techniques [17 – 20].

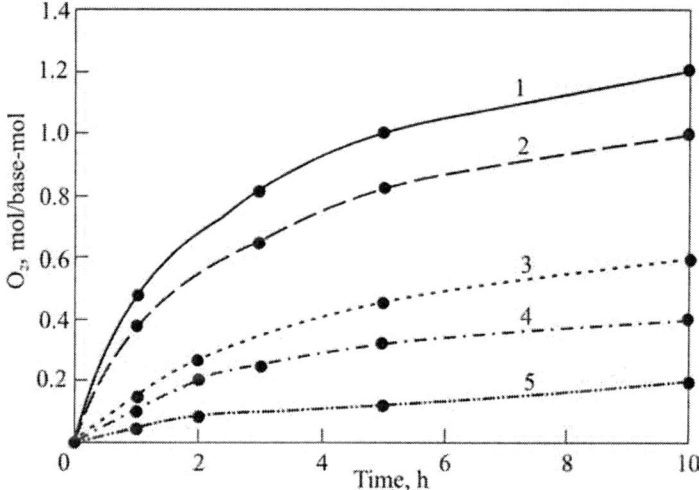

Figure 3. Kinetics of O_2 absorption by LCP KI-100 (1), KI-75(2), KI-50 (3), KI-25 (4), and KI-0 (5)

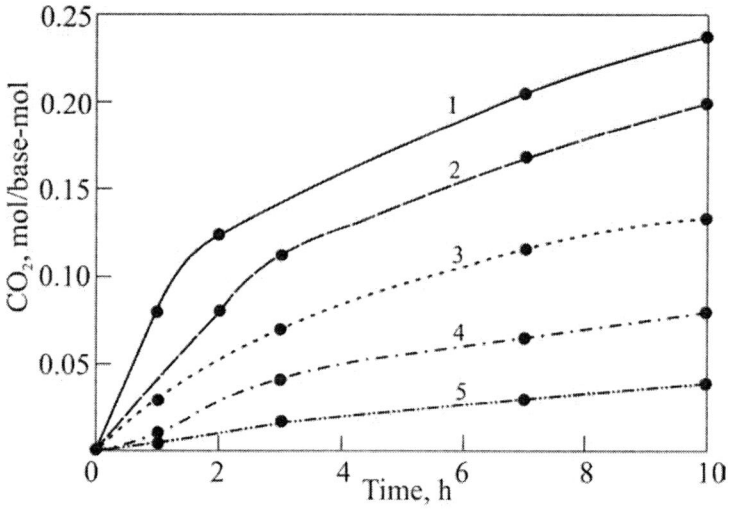

Figure 4. Kinetics of CO_2 release during LCP thermal oxidation (350°C, in air): KI-100 (1), KI-75 (2), KI-50 (3), KI-25 (4), and KI-0 (5)

HEAVY PRODUCTS OF LCP DEGRADATION

HO—⟨◯⟩—C(=O)—O—⟨◯◯⟩

HO—⟨◯⟩—C(=O)—O—⟨◯◯⟩—OH

HO—⟨◯◯⟩—OH

HOOC—⟨◯◯⟩—COOH

HO—⟨◯◯⟩—O—COOH

HO—⟨◯⟩—C(=O)—O—⟨◯⟩

As observed from the composition, heavy products of LCP degradation, precipitated on cold zones of the ampoule, represent a mixture of the initial monomer (dioxyphenyl diacetate), products of its hydrolysis (dioxydiphenyl – DODP - for example), and products of DODP and p-OBA interaction. It should be noted that the composition of the above-mentioned products is identical for all studied LCP and slightly differs just by the ratio of components. Degradation transformations were studied on the example of KI-75 LCP in the processing range. As shown (Figures 5 - 6), at the melting point or during melting (300 and 320°C) the oxidation rate is much lower than at 350°C, when according to the X-ray diffraction analysis the whole polymer transits to isotropic melt. Besides the main gas product (CO_2), hydrogen (at early oxidation stages at 0.5 – 1 h exposure) and water (at 4 h exposure) were also detected. As observed from dynamics of the elemental composition change, hydrogen content decreases and carbon content increases in KI-75 during its thermal oxidation, e.g. a graphite-like structure is formed.

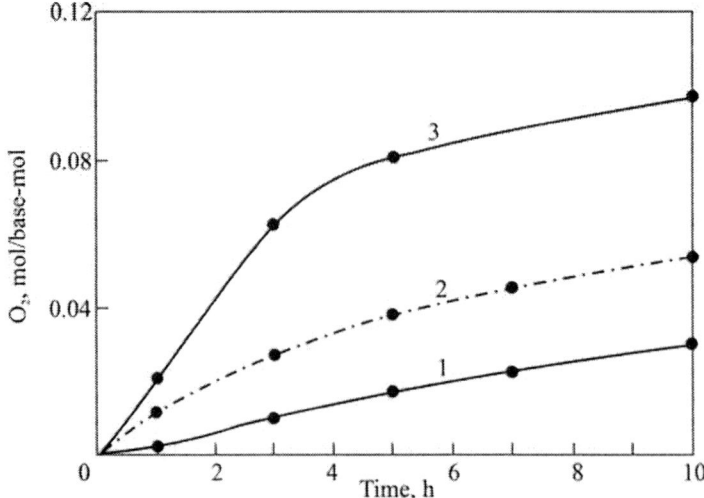

Figure 5. Oxygen absorption kinetics for KI-75 at 300 (1), 320 (2) and 350°C (3) in air

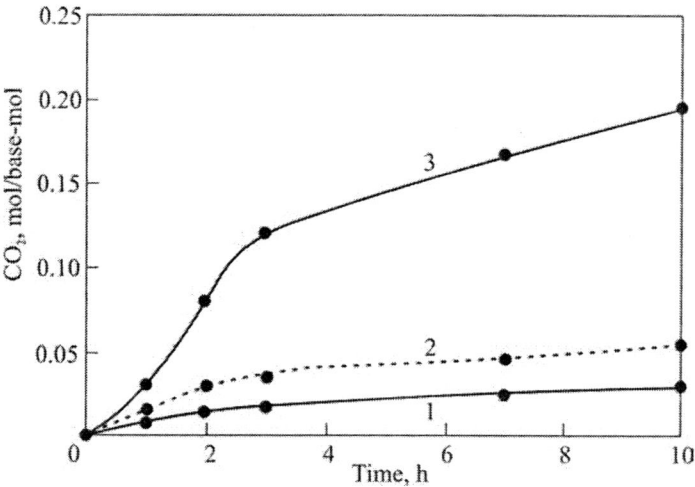

Figure 6. Carbon dioxide release kinetics at KI-75 thermal oxidation at 300 (1), 320 (2) and 350°C (3) in air

This process proceeds intensively at 350°C. IR-spectroscopy data [21 - 23] show that initial changes happen in the absorption range of ester aromatic fragments: absorption band intensity at $v_{C=O}$ = 1740 cm^{-1}, v_{C-O} = 1270 и 1160 cm^{-1}, $v_{C=C}$ = 1600 and 1500 cm^{-1}, $\delta_{C=C}$ = 720 cm^{-1} is reduced. At maximal exposure (thermal oxidation at 350°C during 10 h) only ether absorption bands at v_{C-O-C} = 1080 cm^{-1} and aromatic structure bands are preserved. The spectral background significantly decreases, which is caused by formation of intermolecular crosslinks. At 350°C a great amount of oligomers is formed. They precipitate in the ampoule near the reaction zone at temperature ≈ 150°C. The structure and ratio of oligomers with appropriate end groups, identified by ^{13}C NMR technique in the oligomer degradation products, are shown below:

The amount of oligomers was estimated by the ratio of reflex squares with appropriate chemical shifts: 118.5 and 115.82 ppm (*product a*); 115.43 ppm (*product b*), and 120.72 ppm (*product c*). According to these data *product a* gave 15 mol%, *product b* – 31 mol%, and *product c* – 10 – 12 mol%.

The observation of *p*-oxybenzoic acid in thermal oxidation products of neighboring units allows a suggestion about simultaneous proceeding of copolycondensation and homopolycondensation of *p*-acetoxybenzoic acid. Free *p*-oxybenzoic acid output is 3 – 4 times higher than in the bound state in the form of end groups of oligomers. Apparently, free *p*-oxybenzoic acid is formed in thermal reactions at degradation of labile bonds in structural *p*-oxybenzoic blocks.

Analysis of kinetics and LCP degradation products in the processing temperature range allowed detection of some general features, observed in degradation behavior of heat resistant polyheteroarylenes [24]: structure graphitization, H_2 release, thermal oxidation stability increase at transition metal injection, etc. The idea of their stabilization is based on the following suggestions about degradation mechanisms:

- classical radical-chain thermal oxidation mechanism;
- formation of a molecular complex with oxygen;
- molecule transition to electronically excited state.

Injection of additives is the common method for investigating the mechanism of chemical reactions. It was found that of high effectiveness is the mixture stabilization by the triple system of copper compound, phenol antioxidant and phosphite in polyalkanimide, polyphthalamide and other heat-resistant polymers. The idea of such mixture is based on the action mechanism of such additives:

- phenol antioxidant inhibits thermal oxidation by interacting with peroxy radicals $ROO^\bullet$;
- phosphite, secondary antioxidant, destroys hydroperoxides;
- Cu^{2+}-containing compound acts differently, for example, forms complexes during inhibition of peroxy radical or macrochain defects.

Table 3 shows comparative data on thermal oxidative stability for non-stabilized and stabilized KI-75 LCP.

Table 3. Thermal oxidation of KI-75 LCP in air at 350°C during 30 min

Stabilization compounding	Oxygen absorption, mol/monomeric unit	CO_2 release, mol/monomeric unit
Nonstabilized	0.22	0.033
0.005% $CuSO_4$ + 0.3% *Irgafos 126* + 0.1% *Irganox 1010*	0.08	0.019

The results obtained show that stabilizer injection causes a significant (over two times) deceleration of thermal oxidation in LCP. Of interest is the effect of additives on polymer morphology, determined during studying stabilized and non-stabilized samples (before and after thermal oxidation in air) by the X-ray diffraction analysis. It is found that crystalline reflex is preserved in stabilized polymers, whereas it disappears in non-stabilized samples. The stabilization effect on the physical structure of polymers was not studied well with respect to chemistry of degradation processes. Only complex consideration of the problem (chemistry + change of physical permolecular structure) may cause the increase of thermal stability of prepared product and extension of the material lifetime in articles".

REFERENCES

[1] *US Patent No. 2,944,993*, Glass-filled thermoplastic composites derived from linear polypyromellitimides, Publ. Jun. 14, 1955 (cl. 260-37).

[2] Semenov N.N., *On Some Problems of Chemical Kinetics and Reactivity (Free Radicals and Chain Reactions)*, 2nd Ed., Moscow, AN SSSR, 1958, 686 p. (Rus)

[3] Kandratiev V.N. and Nikitin E.E., *Chemical Processes in Gases*, Moscow, Nauka, 1981, 262 p. (Rus)

[4] Voevodsky V.V., *Physics and Chemistry of Elementary Chemical Processes*, Moscow, Nauka, 1969, 414 p. (Rus)

[5] Neiman M.B., *Aging and Stabilization of Polymers*, Moscow, Nauka, 1964, 332 p. (Rus)

[6] Kuz'minsky A.S., *Oxidation of Caoutchoucs and Rubbers*, Moscow, Goschimizdat, 1957, 319 p. (Rus)

[7] Grassi N., *Chemistry of Polymer Degradation Processes*, Moscow, Inostrannaya Literatura, 1959, 252 p. (Rus)

[8] Madorsky S., *Thermal Degradation of Organic Polymers*, Moscow, Mir, 1967, 320 p. (Rus)

[9] Emanuel N.M., Denisov E.T., and Maizus Z.K., *Chain Reactions of Hydrocarbons Oxidation in the Liquid Phase*, Moscow, Nauka, 1965, 375 p. (Rus)

[10] Emanuel N.M. and Buchachenko A.L., *Chemical Physics of Polymer Aging and Stabilization*, Moscow, Nauka, 1982, 359 p. (Rus)

[11] Shlyapnikov Yu.A., Kiryushkin S.G., and Mar'in A.P., *Antioxidant Stabilization of Polymers*, Moscow, Khimia, 1986, 252 p. (Rus)

[12] Popov V.A., Rapoport N.Ya., and Zaikov G.E., *Oxidation of Oriented and Stressed Polymers*, Moscow, Khimia, 1987, 232 p. (Rus)

[13] Zaikov G.E. and Moiseev V.V., *Chemical Resistance of Polymers in Aggressive Media*, Moscow, Khimia, 1979, 216 p. (Rus)

[14] Minsker K.E. and Fedoseeva G.T., *Degradation and Stabilization of Polyvinylchloride*, 2nd Ed., Moscow, Khimia, 1979, 272 p. (Rus)

[15] Mark H., Atlas S.M., and Ogata N., *J. Polym. Sci.*, 1962, vol. 61, p. 849.

[16] *Liquid-Crystal Polymers*, Ed. N.A. Plate, Moscow, Khimia, 1988. (Rus)

[17] Slonim I.Ya. and Urman Ya.G., *NMR-Spectroscopy of Heterochain Polymers*, Khimia, 1982, 232 p. (Rus)

[18] Johnson L.F., *Carbon-13 NMR Spectra*, N.Y., Wiley, 1972.

[19] Breitmaier E., Haas G., and Voelter W., *Atlas of Carbon-13 NMR Data*, Heyden, London, 1979.

[20] Formacek V., Desnoyer L., Kellerhals H.P., Keller T., Clerc J.T., ^{13}C Data Bank, *Bruker Physik*, Karlsruhe, 1976.
[21] Nakanisi K., *IR-Spectra and Structure of Organic Compounds*, Moscow, Mir, 1965, 215 p. (Rus)
[22] Hummel/Scholl, *Atlas of Polymer and Plastics Analysis*.
[23] Dekhant I., Danu V., Cimmer V., and Schmolke R., *Infrared Spectroscopy of Polymers*, Moscow, Khimia, 1976. (Rus)
[24] Kalugina E.V., NovoTortseva T.N., and Andreeva M.B., 'Thermal Oxidation Features of Heat-Resistant Heterochain Polymers', *Obzor Polim. Mater.*, 2001, No. 6, pp. 29 - 37. (Rus)

Chapter 4

LINEAR FREE ENERGY RELATIONSHIPS IN CHEMISTRY OF SOLUTIONS

R. G. Makitra, A. A. Turovsky and G. E. Zaikov[1]

L.V.Pisarevsky Institute of Physical Chemistry Ukranian National Academy of Sciences, 3A Nauka str., L'viv 79053, Ukraine
[1] N.M.Emanuel Institute of Biochemical Physics, 4 Kosygin str., Moscow 119991, Russia

1. LINEAR FREE ENERGY (LFE) PRINCIPLE

The way out of the above situation was found due to application of the linear free energy principle (LFER – linear free energy relationships), which had been primarily suggested by Broensted in acid-basic catalysis investigations and developed by Hammett [1, 2] in generalization of substituting agent influence on equilibrium states and rates of chemical reactions. Generally, LFER is reduced to a suggestion that any deviations from the standard free energy (thermodynamic potential) of a compound are induced by various independent perturbing factors (medium, substituting agent, temperature, etc. effects) and total value ΔG is obtained by summing up these independent energy contributions:

$$G = G_0 + \sum_{i=1}^{n} \Delta G_n \tag{1}$$

LFE relationships are a manifestation of the so-called extra thermodynamic relationships, suggested interaction models combined with the notions of thermodynamics. Though the LFE principle is not strongly valid from positions of thermodynamics, nevertheless, it may give useful information about actuality of the suggested interaction model and the origin of connections in it [3]. In chemistry of solutions, LFER is reduced to an assumption that a solute may interact with the medium by several mechanisms (solvation types), and final solute behavior (equilibrium constants, reaction rates, distribution, and enthalpy) is defined by the linear sum of energy effects of all mentioned interactions. The type of interactions in the medium – solute system and their significance depend on the interaction model suggested. Anyway, combined consideration of both specific and nonspecific interactions and summing up their contributions are necessitated. This summation is implemented using multiparameter linear equations of the following type:

$$y(\Delta G, \ln k, \ldots) = a_0 + \sum a_i x_i, \qquad (2)$$

where a_0 is the studied value without interaction (often equal it in the gas phase; however, in the presence of this "free term" all calculation errors are also summed up); xa are effects of separate types of substrate interaction with the environment, where a is the contribution (intensity) of effects to the current process; a_i and x are solvent parameters.

According to R. Dundell, the solvation free energy change at substance dissolution is given by summation of four components defined by properties of particular solvent, specifically, cavity formation energy for solute particle disposition, solvent orientation energy around solvated particle, isotropic (nonspecific) interaction energy which is the sum of electrostatic, polarization and dispersive interactions in the dissolution structure and energy of anisotropic, specific interaction of definite solvent areas and the solute (charge transfer complex – CTC, and hydrogen bond formation, etc.) [4].

To a certain extent, this approach is based on formulae suggested by Buckingham in 1957 [5] and Muirhead-Gould and Laidler in 1967 [6] for solvation energy of ions, E:

$$E = E_1 + E_2 + E_3 + E_4 + E_5 + H_{solv}, \qquad (3)$$

where E_1 is total energy of ion-dipole and ion-quadrupole interactions; E_2 is the energy of ion-induced dipole interaction; E_3 is the dispersion interaction energy; E_4 is the repulsion energy; E_5 is the energy of particle interaction in the solvation complex; H_{solv} is the ion solvation enthalpy by the Born equation.

For kinetics of organic reactions, the medium effects were first separated to components in 1957 by Fainberg and Winstein [7] in consideration of *tert*-butyl chloride solvolysis. Taking into account the presence of hydrogen bonds in ionizing media, they suggested representing ionizing capacity (Y) parameters as a sum of two terms: one induced by the medium polarity and another based on electrophilic solvation of the substrate via hydrogen bond formation. This results in the following expression:

$$\lg k = \lg k_0 + m_1 Y_1 + m_2 Y_2, \qquad (4)$$

where m are contributions of separate types of solvation to the total process or, to put it differently, sensitivity of the current compound to solvolysis caused by some solvation effects.

In 1971, Fowler and Katritzky [8] studied the medium influence on spectral parameters of solutes and the rates of some processes. They suggested separation of the medium effects on the sum of two terms induced by polarity and chargeability of the medium, i.e. a specific modification of Mac-Ray, Bakshiev or other formulae with variable constants with respect to the process under consideration. Such approach was found satisfactory for generalization of data on low-polar media. To obtain satisfactory solvent properties and process indices relationships for polar media, they suggested a linear combination of three parameters: n, ε and E by Dimroth-Reichardt.

To a considerable extent, further developments were inspired by the Mallican quantum-chemical model of the donor-acceptor bond represented by a linear combination of two wave functions, which reflect covalent and electrostatic interactions:

$$\psi = a_1 \psi_{coval} + a_2 \psi_{el}. \qquad (5)$$

In 1971, basing on broad experimental material on complexes with hydrogen bonds and considering the solvent influence on spectral characteristics of solutes, A.V. Iohansen [9] suggested the so-called multiplicity rule:

$$H(\Delta v, \sqrt{\Delta I}) = \Delta H_{stand}(\Delta v_{at}, \sqrt{\Delta I}_{at}) P_i E_j. \qquad (6)$$

According to this rule, hydrogen bon formation enthalpy (H), absorption band shifts (X-H) or changes of their intensities (I) are defined by acidity and basicity of interacting components. The authors have presented tables showing P and E factors for the sequence of X-H acids and n- and π-donors of electrons. The scales were conditionally divided from 0 to 1, therefore, acidity of C-H group in CH_3 or $=CH_2$ is taken for 0 and acidity of –OH in phenol is equaled 1. However, sometimes Arrhenius acids possess $P > 1$: for example, $P = 1.25$ for $CH_2ClCOOH$. In this connection, CCl_4 with $E = 0$ and diethyl ether with $E = 1$ are taken for the control for bases.

Clearly, the multiplicity rule is based on the LFE principle. However, disadvantage of this approach is its limitation only for X-H acids and spectral events, as well as moderate accuracy of forecasting due to abandoning consideration of nonspecific solvation.

V.A. Terentiev has also based his work on this multiplicity rule: spectrum-determined thermodynamic characteristics (H^{AD}, lgK_{eq}) of donor-acceptor systems equal the product of corresponded donor and acceptor characteristics. This statement, primarily proposed for a system with the hydrogen bond [10], was then extended to complexes with the charge transfer [11, 12]. For a general case, he suggested the following equation:

$$E = (1 - p)a \cdot d + pC_A C_D + p(I_A I_D), \qquad (7)$$

where C_A, C_D, I_A, I_D are covalent ionic components of the interaction; a and d are dipole components; p is the contribution of the ionic structure. However, such equation is too complicated for practical use. Therefore, it was reduced to a simpler form as follows:

$$E = pC_A \cdot C_D + pI_A I_D$$

or $J = ad$,

where J is the interaction intensity. Hence it follows that

$$\Delta H = \Delta H_A \cdot \Delta H_D;$$
$$\Delta v = \Delta v_A \cdot \Delta v_D;$$

$$K = K_A K_D. \qquad (8)$$

These works show average characteristics of some donors and acceptors (D and A, respectively), obtained by statistical processing of a broad experimental material. Similar to the previous concept, the current one is of high predictive force; however, good conformity to the experiment and high-quality linear correlations are obtained only for systems with strong interactions, in which contribution of nonspecific interaction is negligibly low. For low energy effects with low A and D, deviation of the results is much greater.

Terentiev's equation approaches R. Drago concept [13, 14], according to which the donor-acceptor interaction enthalpy may be represented by the sum of two contributions:

$$H = C_A C_D + E_A E_D \qquad (9)$$

Parameters C and E computed on the basis of broad experimental results and normalized in the range from 0 to 1 are also shown in the works by this author. They are based on enthalpy of substance interaction with iodine and phenol. Technically, products $C_A C_D$ and $E_A E_D$ might be considered to be electrostatic and covalent contributions to the complex formation enthalpy. However, the absence of direct relationship between C and E and physical parameters of donors (acceptors) and the absence of a simple logical interpretation of these parameters are emphasized: "These parameters must not be considered as related to properties of donor or acceptor in the ground state" [14, p. 110]; "Parameters E and C are complex values" [14, p. 114].

Drago's two-parameter equation, applied to H only, gives better results rather than various single-parameter dependencies using Gutmann donor numbers (DN) e.g. enthalpies of substance mixing with $SlCl_5$, for example. However, there are cases when Drago's equation gives results significantly deviating from the experiment, specifically for systems containing boron compounds. Some examples of nonconformity of calculation and experimental data are shown [49].

Such deviations may be explained by both shortcomings of formally developed mathematical model and the probability of appreciable errors of C and E determination. By way of example, note that for JCl data on selection of eight enthalpies gave $C_A = 0.830$ with maximal error equal 0.334 kcal/mol; for parachlorophenol $C_A = 0.478$ with the error equal 0.17, etc. Nevertheless, the predictive strength of Drago's equation allowed its use for generalization of data on several hundred systems; therefore, the equation received wide recognition. Critical analysis of this equation and other, most popular models and their comparison were performed in [15].

Table 1. Comparison experimental and calculation (equation (9)) complex formation enthalpies [14]

Acceptor + donor	H_{exp}, kcal/mol	H_{calc}, kcal/mol
$B(CH_3)_3 + N(CH_3)_3$	17.6	25.4
$Al(CH_3)_3 + N(C_2H_5)_3$	26.5	32.5
$BF_3 + (CH_2)_4 S$	5.2	16.1

Various solvation types at ion transfer from standard to experimental medium are discussed by Mayer [16, 17]. As deduced from the LFE principle, the general change in thermodynamic potential at ion solvation equals the sum of separate contributions:

$$\Delta G_{solv} = \Delta G_{cav} + \Delta G_{con} + \Delta G_{sp} + \Delta G_{dp}, \tag{12}$$

where ΔG_{cav} is the energy expenditure for formation of ion disposition cavity (das Hohlraum); ΔG_{con} is the concentration-dependent nonspecific energy contribution; ΔG_{sp} is the effect of its short-range specific solvation; ΔG_{dp} id the effect of dielectric polarization of the medium.

In the systems under author's consideration, ΔG is changed with ion transfer from standard solvent (acetonitrile) to another organic solvent:

$$\Delta\Delta G_{solv} = \Delta G_i + \Delta G_{MeCN} = \Delta\Delta G_{cav} + \Delta\Delta G_{con} + \Delta\Delta G_{sp} + \Delta\Delta G_{dp}. \tag{11}$$

In the case of polar solvents, $\Delta\Delta G_{con}$ and $\Delta\Delta G_{dp}$ are approximately equal zero, and $\Delta G_{cav} \approx \Delta H_{vap}$. Therefore, specific solvation is associated with the acid-base interaction determined by DN and AN by Gutmann. Finally:

$$\Delta\Delta G_{solv} = a\Delta DN + b\Delta AN + c\Delta\Delta G_{vap}. \tag{12}$$

Thus the Mayer equation differs by consideration of nonspecific solvation influence in it; in most instances the term $\Delta\Delta G_{vap}$ is found statistically valueless. The equation allows proper generalization of data by both solubility of salts (KCl, KBr, etc.) in organic solvents and complex formation ΔG. It also generalizes rate constants (lg$k$) for definite kinetic series in various solvents, for example, p-nitrofluorobenzene interaction with NaN$_3$. The disadvantage of this model is that, on the one hand, it is not extended to low-polar media; on the other hand, for proton solvents (H$_2$O, HCONH$_2$, alcohols) some calculation data noticeably (by 1.5 – 2 times) deviate from experimental values.

The Fawcett-Krygowski model [18 – 20] also takes into account specific interactions and characterizes solvent acidity via E_T parameter:

$$Q = Q_0 + \alpha E_m + \beta DN. \tag{13}$$

The difference from the previous studies concludes in the use of factor analysis and normalized regression indices for validation of statistical significance of the model that allows determination of the percentage contributions of separate solvation types. The model was found suitable for description of some scales which characterize the solvent (G, Z, B, etc.), some kinetic, spectral or other data. However, correlation of data is not always satisfactory and obtained values of the coefficient of multiple (total) correlation, R, are often unacceptably low: 0.90 – 0.95. For example, for $tert$-butyl chloride solvolysis R = 0.909, for enthalpy of hydrogen bond formation by p-fluorophenol R = 0.936 and so on. The authors explain this fact by apparent effect of unconsidered contribution of nonspecific solvation and the process entropy alteration. The model gives acceptable results ($R \geq 0.95$) in 55% of studied cases of the medium influence on ion-ion and ion-dipole interactions. For 35% of cases, R falls within the range of 0.9 – 0.95.

The advantage of the Fawcett-Krygowski (and Mayer) model is its simplicity: the presence of two factors only that makes interpretation of the equation easier. However, for dipole-dipole interaction, acceptable correlation ($R > 0.9$) is observed in 75% of studied cases. Thus the forecasting ability of these two models is lower. For 90% of cases, examination using vector analysis shows that variable Q may be described by two vectors. However, sometimes the third vector should be introduced which is caused by the presence of experimental errors. Usually, one of the vectors is connected to E_t or DN ($R \geq 0.85$) and correlation of the second vector with specific solvation factors is poorer [18]. Direct connection of calculated vectors and nonspecific solvation factors is not observed, too. However, it should be noted that all these arguments are deduced from analysis of influence of a small number of solvents (6 – 10), and data for structured, protogenic solvents deviate from them.

Swain *et al.* [21] suggested solvating ability scales of the solvents basing on their attitude to anion (acidity) and cation (basicity). Basically, their equation is similar to these suggested in Fawcett-Krygowski or Mayer models:

$P_{ij} = aA + bB + c$.

However, these parameters are not directly connected to any particular physical and chemical properties of solvents but are obtained from statistical generalization of experimental results for 77 reaction series:–by rate constant $\lg k$, spectral shifts, equilibrium $\lg K$, etc.

2. KOPPEL-PALM AND KAMLET-TAFT-ABRAHAM SOLVATION MODELS

By now, two linear multiparameter equations by Koppel-Palm and Kamlet-Taft-Abraham are the most commonly encountered. Owing to combined consideration of factors of both specific and nonspecific solvation, they appear practically universal for generalization of data on solvent influence on kinetics of heterolytic reactions and spectral characteristics of solutes (frequency shifts in UV-, NMR and IR-spectra), as well as equilibrium constants of complex formation and some other processes. Further introduction of additional terms to these equations extends their application field to any processes in the liquid phase. V.A. Palm has started his works in the early 1960's.

The fundamental work by V.A. Palm and A.A. Koppel [22] was published in 1972. They suggested characterizing the influence of solvents of solute behavior by a four-parameter LFE equation. According to the Koppel-Palm model, all types of interaction must be taken into account. It is suitable to divide them to nonspecific, stipulated by the solvent influence as a continuum on the solute, and specific ones, stipulated by formation of complexes from solvent and solute molecules. Nonspecific solvation is associated with the medium polarity and ability to polarize. V.A. Palm has suggested expression of the medium polarity by the Kirkwood function:

$$\frac{\varepsilon-1}{2\varepsilon+1},$$

and chargeability by the following expression:

$$\frac{n^2-1}{2n^2+2}.$$

However, other functions may also be used, for example:

$$\frac{1}{\varepsilon} \text{ or } \frac{n^2-1}{2n^2+2},$$

because they are almost mutually proportional.

It is suggested to determine specific interaction as the sum of acid and base contributions, because with respect to co-agent the majority of solvents may represent an acid and a Lewis base, simultaneously.

Primarily, in accordance with Gordy [24], the measure of solvent basicity was represented by OD band shift in the spectrum of methanol. Further on, this parameter was replaced by OH band shift in phenol IR-spectrum in the presence of appropriate donor in CCl_4 solution compared with phenol dissolved in pure CCl_4 because of high sensitivity of this value [25]. The solvent ability to electrophilic solvation is characterized by the Dimroth-Reichardt parameter, E_t, [3] improved by subtraction of nonspecific interaction influence in accordance with the following equation:

$$E = E_T - 25.10 - 14.84\frac{\varepsilon-1}{\varepsilon+2} - 9.52\frac{n^2-1}{n^2+2}. \tag{14}$$

In the Koppel-Palm model, total effect of the medium on physical and chemical properties of the solute is determined by the four-parameter equation as follows:

$$X(\lg k, \Delta v) = a_0 + af(n) + a_2 f(\varepsilon) + a_3 B + a_4 E, \tag{15}$$

where a_0 formally conforms to $\lg k$ or Δv in the gas phase which is sometimes experimentally proved. Applicability of the Koppel-Palm equation is validated [22, 25] for 70 different reaction series, mostly reaction rate and spectral shifts in various solvents. Among them 60 series were found dependent on 2 or 3 parameters, mostly a combination of nonspecific solvation and one of specific solvation types. In several cases only all four parameter were required for complete description. Further on, the effectiveness of Koppel-Palm equation was confirmed by many authors on various chemical objects. For instance, studies of the solvent effect on carbon acid interaction with diazodiphenylmethane carried out by Chapman, Shorter et al. [26].

Examination of statistical value of coefficients at separate terms in the regression equation (and examination of total correlation R decrease at sequential elimination of separate terms [26]) allows exclusion of low-valued parameters and obtaining of reduced equations. The latter gives an opportunity to conclude adequately about solvation effects on chemical reaction and the reaction chemistry.

Later on, it has been shown [27] that E and E_T values (barring some cases) are proportional; therefore, it has been suggested to use in calculations E_t as the primary value, directly determined in experiments.

Sometimes the Koppel-Palm model is criticized because of many parameters in it that, supposedly, decreases statistical value of calculation results, specifically in the case of small number of points (solvents). However, it should be noted that for the majority of cases the four-parameter equation is reduced to a shape with two or three significant parameters. Moreover, the generalization ability, predictive strength and wide abilities to interpret results of this equation are already justified on many examples.

The validity of the four-parameter model suggested by Koppel and Palm is proves by results of the factor analysis [18] with the only exclusion for electron demand (E) replacement by the acceptor number (AN) and replacement of direct solvation parameters by their normalized orthogonal values correlating with the initial parameters, possessing $R > 0.9$.

Since 1976, two series of reviews by Kamlet and Taft et al. were published: "The solvatochromic comparison method" и "Linear solvation energy relationships", in which systematic studies of the medium influence on physical and chemical properties of solutes were performed, primarily, spectra and kinetics of reactions. At the beginning, it has been suggested that, depending on the type of solute and solvent, the solvent effect may be described by one of three scales as follows: basicity (β) for solvents—hydrogen bond acceptors [28], acidity (α) for solvents – proton donors [29], and bipolarity (π^x) for consideration of combined effect of polarity and medium chargeability [30]. However, it has been found that, generally, three parameters shall be applied, simultaneously:

$$XYZ = XYZ_0 + a\alpha + b\beta + \pi^x. \tag{16}$$

Thus the Kamlet-Taft model is distinguished from the Koppel-Palm one only by incorporation of polarity ($f(\varepsilon)$) and chargeability ($f(n)$) to a single parameter. Later on, for some solvents it seemed desirable to introduce a correction for "excessive chargeability" ($d\delta$), where $\delta = 0$ for aliphatic compounds, $\delta = 0.5$ for polyhaloalkanes, and $\delta = 1$ for aromatic solvents. Coefficient d is determined from experimental data by the step-by-step approach technique.

Further on [31], description of the medium influence on various (including thermodynamic) physical and chemical properties of solute using the following generalized equation was suggested:

$$XYZ = XYZ_0 + S(\pi^x + d\delta) + a\alpha + b\beta + h\delta_H + e\zeta, \tag{17}$$

where δ_H is the Hildebrand solubility parameter; ζ is the empirical correction for coordination covalence of P=O, C=O, S=O, R$_2$O, and R$_3$N groups. At processing of data on solvent influence on spectroscopic parameters, it was also suggested [32] to add the Kamlet-

Taft formula by a parameter considering viscosity of solvents. However, this suggestion was not widely used.

Note also that parameter π^x, at least for aprotic solvents, appeared associated with n and ε, namely, linear dependent on the expression

$$\frac{\varepsilon-1}{2\varepsilon+1} \cdot \frac{n^2-1}{2n^2+1} \text{ [79],}$$

MacRay function or some other expressions [33 – 37].

Perhaps, it seems desirable to finish this brief review of various solvent influences on physical and chemical behavior of solutes and its description by multiparameter equation by H. Reichardt citation [38]: "In many instances the use of multiparameter equations in favor of one-parameter ones leads to a sharp improvement of correlation between solvent-dependent process (reaction rate, light adsorption) and independent solvent properties, i.e. they take into account the diversity of substrate-solvent interactions instead of the medium polarity only. Although the way and the model grounds for better separation of polarity of solvation ability of the solvent into several, wherever possible, independent complementary interaction parameters are not clear yet. In this connection, multiparameter equations have not been finally approved yet".

3. THE RELATIONSHIPS OF SOLVENT SOLVATION SCALES

As mentioned above, linear dependencies between separate "polarity" (solvation capacity) scales are observed. Comparison of some dozens of various scales [39] indicated more or less satisfactory correlation only between those characterizing the same property (acidity or basicity of the medium), for example Z and E_T with $R = 0.998$, Z and γ_B with $R = 0.989$, OD and DN with $R = 0.984$; therefore, for AN and E_T $R = 0.918$ only. Scales characterizing different parameters possess no correlation. The absence of satisfactorily linear correlation between empirical scales and physical parameters of the medium, such as dielectric constant, dipole moment, etc. should be specially emphasized. Usually, obtaining of even symbate dependencies requires separate consideration of proton and aprotic solvents, though with numerous omissions.

It has been suggested [52 – 55] that there is a linear correlation between E_T and a combination of polarity and chargeability functions as aliphatic and aromatic solvents are considered separately. However, such correlation is observed for low-polar solvents [66]: solvents of higher polarity ($CHCl_3$, nitriles, etc.) and, the more so, proton containing ones, deviate from such dependencies. It is not surprising taking into account that polarity is the decisive factor in this case. It also has been found [39] that E_T and Kirkwood function display not linear but symbate curvilinear correlation.

The effectiveness of application of multiparameter equations indicates comparison advisability of solvation scales, most commonly used in such equations, for the purpose of determination of their apparent, harmony and substitutability.

As indicated [25], OH band shifts in IR-spectra of H-acids, caused by solvents, are proportional. However, for extremely weak acids (for example, $CHCl_3$ [86]) or some alcohols (cyclohexanes, n-butanol) [87], obtaining an adequate correlation with the medium basicity ($B \cong \Delta v_{PhOH}$) requires application of poly-parameter equations. It is found also that the donor number scale is proportional to B_{PhOH} [42], the correlation level is low:

$$B_{PhOH} = 15.82 + 11.20 DN; R = 0.852. \tag{18}$$

Therefore, satisfactory correlation may be reached at the application of multiparameter equation with respect to the donor chargeability [43]:

$$DN = 8.2 + 0.0876B - 35.734 \frac{n^2 - 1}{n^2 + 2}; R = 0.973; S = 3.32. \tag{19}$$

Several "basicity scales" were comprehensively studied [44]: B, DN, β [28], Drago C_B and E_B [14], Iohansen E-factor [9]. All these values may be satisfactorily transformed to one another only by multiparameter dependencies (based on the Koppel-Palm equation. Obviously, this results from many methods of their obtaining: IR-spectroscopy for B, UV-spectroscopy for β, calorimetry for DN, data averaging on complex formation constants and enthalpies for C_B, E_B and E-factor. However, of importance is that they all are symbate and characterize the ability of substances to act as Lewis bases. For example, Iohansen E-factors and basicity B is weakly connected [61]:

$$E\text{-factor} = (0.320 \pm 0.076) + (0.0237 \pm 0.0087)B;$$

$$n = 27; R = 0.868; S = 0.203 \tag{20}$$

However, at consideration of other solvation factors the connection becomes stronger:

$$E\text{-factor} = -0.097 + 1.030 \frac{\varepsilon - 1}{2\varepsilon + 1} + 0.00255B;$$

$$n = 18; R = 0.970; S = 0.115. \tag{21}$$

Analysis of corresponded expressions characterizes some scales under consideration. For example, the Drago C_B [14] is generally associated with the substance basicity:

$$C_B = -0.867 + 9.277 \frac{n^2 - 1}{n^2 + 2} - 13.393 \delta^2 + 0.124B;$$

$$n = 21; R = 0.964; S = 0.645 \tag{22}$$

at $r_{OB} = 0.894$. At the same time, the electrostatic scale E_B, suggested in the same work, is mostly associated with the solvent polarity.

Of interest is that the Kamlet-Taft scale β [28] is bound to the substance basicity B only for aprotic solvents, even under consideration of nonspecific solvation:

$$\beta = 0.257 - 1.243 \frac{n^2-1}{n^2+2} + 0.593 \frac{\varepsilon-1}{2\varepsilon+1} + 1.192\delta^2 + 0.00081B; \tag{23}$$

$n = 18; R = 0.965; S = 0.069$.

For alcohols, β is defined by their polarity and electron demand:

$$\beta = -2.058 + 7.740 \frac{\varepsilon-1}{2\varepsilon+1} - 0.521\delta^2 - 0.061E;$$
$$n = 7; R = 0.981; S = 0.07 \tag{24}$$

that most likely approaches to the Reichardt "polarity" scale E_T. However, it was reported [44] about a necessity to "divide" the scale β into two groups: alcoholic and unassociated solvents.

The conclusion about mutual nonlinearity of scales B and β is confirmed [45], and an equation binding the value PhOH = B for aprotic substances to Drago parameters C_B and E_B:

$$\Delta v_{PhOH} = 333.1E_B + 49,71C_B - 228.5. \tag{25}$$

It is found [27, 45] that correlation between scales β, B and DN is low. Parameter β is associated with the proton affinity (PA) of solvents; however, β - PA curves display two branches: linear dependence for proton solvents and curvilinear for aprotic ones.

Based on band C-J shift in IR-spectrum of acetylene derivatives J-C≡C-X (X = iPr, Ph, CN, etc.) [47], a "soft" basicity scale correlating well with corresponded enthalpies of iodine complex formation was suggested. Unfortunately, any information about probable application range of this scale is absent.

Similar to basicity scales, there is a series of data on correlation between separate electron demand ("polarity") scales of solvents. Some correlations, for example, between scales E_T and Z, Y and others, are discussed in the review [39]. However, the works on comparison of newer electron demand scales AN, α and the most widespread E_T are scanty. The solvent effect on spectral shifts of pyridine N-oxide (determined by Z scale) was compared with the Kamlet-Taft equation parameters [48]. No linear dependence between these values was observed, but as chargeability π^X (the polarity parameter) was taken into account, an excellent correlation was obtained (except for acetic acid):

$$v_{max} = 35.42 + 0.61\pi^X + 2.49\alpha; n = 30; R = 0.991; S = 0.15 \tag{26}$$

with five-fold dominance of π^X in the influence of electron demand α.

Though different models of the medium influence amenably necessitate consideration of the medium acidity (electron demand), the problem of scale comparison should be modified. While α, Z and E_T scales were obtained on the basis of UV-spectroscopy data and NMR signal shifts gave AN scale, the fundamental possibility of obtaining an additional calorimetric scale of electron demand similar to DN scale was confirmed [49]. The enthalpy of the strongest base (ethylene diamine) mixing with solvents is associated with their electron demand by the following equation:

$$\Delta H = 24.9 - 90.9 \frac{n^2-1}{n^2+2} + 3.90\delta^2 - 0.38 E_T;$$
$$n = 15; R = 0.957; S = 2.58. \tag{27}$$

Here the determining role is played by electron demand of substances. A particular correlation index by this parameter equals $r_{OE} = 0.924$.

Heretofore, the problem of nonspecific solvation parameters is ambiguous. In some works, investigators consciously abandon the consideration of this factor, suggesting it to be low-valuable, in order to obtain simpler mathematical pattern: Fawcett and Krygowski [18 – 20], Mayer [16, 17]. In other models, the nonspecific solvation effects is somewhat distributed by mathematically formal parameters characterizing different solvents: Drago [14], Iohansen [9], Terentiev [10, 11], Swain [21]. Unfortunately, this generally results in decrease of accuracy and predictive strength of suggested equations. That is why Drago [45] added the two-parameter equation by the third term which takes into account the possibility of nonspecific interaction:

$$\Delta X = E_A E_B + C_A C_B + sD. \tag{28}$$

The same change was made to the initial Kamlet-Taft concept: the authors added the bipolarity parameter π^X by the term δ which reflects increased chargeability of some solvents – aromatic compounds, polyhaloalkanes, etc. For the matter of fact, this approach is identical to the Koppel-Palm one [23]. However, parameter d in the expression ($\pi^X + d\delta$) must be determined empirically on the basis of generalization of experimental data.

The works by V. Bekarek et al. [50 – 53] are devoted to nonspecific solvation contribution into physical and chemical events. Based on application of the Onsager reaction field, these works represent better quantitative description of the medium effect upon physicochemical processes under consideration of molecule deformation, induced by the medium polarization degree. In this connection, the authors have suggested to replace α, β and π^X parameters in the Kamlet-Taft equation by new, "modified" scales, deduced by dividing corresponded parameters by $\frac{n^2-1}{2n^2+1}$ factor. In particular,

$$\pi^X_{\text{mod}} = \frac{\pi^X}{f(n^2)} + 3.15.$$

Over 50 systems (chemical kinetics, equilibrium constants, spectral parameters) were studied, in which a significant increase of the correlation index at application of "modified" parameters compared to primary ones by Kamlet and Taft were observed. However, the question about possible importance of the "additional chargeability factor" δ in the Kamlet-Taft equation is still unclear in this approach. For aliphatic solvents, Drago [45] equates the parameter of nonspecific interaction D to π^X:

$$D = 0.161 + 2.86\pi^X; R = 0.997.$$

The authors of the present monograph have brought up the problem of quantitative description of the liquid phase effect upon thermodynamic processes of gas dissolution, substance distribution between two liquid phases, etc. They were based on Mayer [16, 17] and Pierotti [54] concepts of combined consideration of the medium interaction with a molecule in it and energy expenditure for cavity formation in the medium for molecule disposition in it:

$$\Delta C = \Delta C_{int} + \Delta C_{cavity}. \tag{29}$$

If the second among these values is theoretically proportional to the cohesive energy density, thermodynamic potential of interaction between components will be most likely, similar to kinetics, be divided into separate, specific (acid-base) and nonspecific components. This should be made in accordance with the principle of linear free energies (LFE) relationships. To the authors' point of view, such consideration is most clearly observed in the Koppel-Palm model, validated on over a hundred examples. As compared with Mayer, Fawcett-Krygowski *et al.* two-parameter equations, it favorably considers nonspecific solvation that increases its accuracy and many substances, characterized by basicity B and electron demand E scales. As compared to three-parameter (or even more at present) Kamlet-Taft equation and its Bekarek modification, separate consideration of medium polarity and chargeability contributions seems more logical to the authors rather than their summing up in a single parameter characterizing bipolarity with further incorporation of corrections for excessive chargeability. With respect to the process and substrate type, a broad material considered on application of the Koppel-Palm equation clearly indicates different contributions of these two factors, up to full negligibility of one of them, their separate interpretation is more desirable.

The desirability of medium polarity consideration was criticized [27], because the function

$$\frac{\varepsilon - 1}{2\varepsilon + 1} \text{ at } \varepsilon > 10$$

changes insignificantly to cause a sufficient effect on the studied value. However, this expression, observed in some cases, opposes the linear dependence between $\lg K$ or $h\nu$ and $\frac{1}{\varepsilon}$ (or other ε functions). It may be possibly explained by a decrease of macroscopic constant ε

in close vicinity to ions or dipoles and decreases. Other deviations from the Borne theory also are of definite importance. In this connection, the effect of high ε values is probably leveled: a solute particle is more sensitive to relatively low ε values. This phenomenon must be studied in more details.

The selection of electron demand scale is also ambiguous. Primarily, Koppel and Palm have used the Reichardt electron demand scale E_T. However, later on [22], they suggested to "let this parameter free" from the influence of nonspecific solvation by subtracting corresponded corrections, obtained from data on E_T solvents incapable of specific interactions:

$$E = E_T - 25.57 - 14.39 \frac{\varepsilon - 1}{\varepsilon + 2} - 9.08 \frac{n^2 - 1}{n^2 + 2}. \tag{30}$$

However, it is shown [27] that for solvents with $E > 42$ (i.e. with the exception of hydrocarbons, ethers and some other substances), good linear dependence with $R = 0.984$ between E and E_T is observed. Therefore, for electrophilic substances these scales are adequate. That is why application of E_T scale is more desirable, because it is deduced in the experiment and possesses no calculation errors. The disadvantage of E scale is that for some substances values in it are negative, which has not been considered by the authors. This, for example, resulted in characterization of different substances by the same electron demand $E = 0$ (diethyl ether, benzene and nitrobenzene).

An attempt was made [55, 56] to determine a correlation between E_T and nonspecific solvation factors. A definite correlation was found between these factors for aprotic compounds, only if aliphatic and aromatic compounds were considered separately.

For aliphatic compounds:

$$E_T = 21.4 + 50.3 \frac{\varepsilon - 1}{2\varepsilon + 2} - 22.2 \frac{\varepsilon - 1}{2\varepsilon + 2} \cdot \frac{n^2 - 1}{2n^2 + 2}; \tag{31}$$

$n = 32; R = 0.956; S = 1.40;$

for aromatic compounds:

$$E_T = 26.0 + 38.0 \frac{\varepsilon - 1}{2\varepsilon + 2} - 22.5 \frac{\varepsilon - 1}{2\varepsilon + 2} \cdot \frac{n^2 - 1}{2n^2 + 2}; \tag{32}$$

$n = 10; R = 0.986; S = 0.60.$

The absence of convincing theoretical justification of deduced formulae, the necessity to divide substances to separate groups and noticeable (up to 2-3 units) deviations in calculated and experimental E_T values for some substances (i-Pr$_2$O, Et$_2$O, THF, MeCN, VeNO$_2$) shows low probability of E_T parameter correlations with ε and n. Anyway, a definite contribution of ε and n into E_T is possible [22], which is confirmed in [65].

To a high degree of reliability, a linear correlation between parameters π^X, E_T and S with function $\sqrt{0.5 - \dfrac{\varepsilon - 1}{2\varepsilon + 1}}$ ($R > 0.97$) was determined. However, it was observed only for so-called "favorite" solvents, i.e. for solvents unable of specific solvation.

Hence, one more question is brought up, if the cohesive energy parameter δ^2 correlates with solvation parameters in the Koppel-Palm equation, i.e. if it is statistically independent. Seemingly, physical properties of this model give not grounds for such suppositions; however, there is a calculation [22, p. 279] showing satisfactory correlation between δ and parameters in the Koppel-Palm equation for about 30 compounds:

$$\delta = (0.68 \pm 1.04) + (21.91 \pm 4.40)\dfrac{n^2 - 1}{n^2 + 2} +$$

$$+ (5.92 \pm 2.25)\dfrac{\varepsilon - 1}{2\varepsilon + 1} + (5.07 \pm 0.05)E; \qquad (33)$$

$R = 0.95$; $s\% = 5.3$,

where $s\% = 5.3$ is the relation of standard error to maximal change of variable δ (in %).

This calculation has two disadvantages. Firstly, theories of regular solutions and NTRL deal not with the solubility parameter δ but with δ^2 (or $(\delta_1 - \delta_2)^2$ of two components) which is quite logical: δ^2 is of the energy dimensionality, and δ is the square root from energy. Another disadvantage is a limited selection of solvents, mainly hydrocarbons, halohydrocarbons, alcohols and ethers, at full absence of bipolar, etc. solvents. Actually, it was found that as the list under consideration is added by such solvents (Table 2), the correlation between δ or δ^2 and parameters in the Koppel-Palm equation is abruptly reduced [66] below acceptable Jaffe correlation [58]. Since the electron demand E is used in [22], reliable comparison it was also used in calculations below for more preferable E_T parameter. For initial 25 solvents listed in the Table, corresponded to [22], the equations are deduced:

$$\delta = -0.966 + 29.75\dfrac{n^2 - 1}{n^2 + 2} + 4.15\dfrac{\varepsilon - 1}{2\varepsilon + 1} + 0.00021B + 0.694E; \qquad (35)$$

$R = 0.949$; $S = 1.14$,

e.g. nearly identical to the equation for previous 30 substances. However, the situation changes abruptly with consideration of all 38 solvents from Table 3:

Table 3. Comparison densities of cohesive energy for various solvents

No.	Solvent	δ, (cal/ml)$^{0.5}$	δ^2, cal/ml
1	Heptane	7.38	54.46
2	Cyclohexane	8.20	67.24
3	Benzene	9.14	83.54
4	Toluene	8.92	79.57
5	CCl_4	9.05	81.90
6	$CHCl_3$	9.10	82.81
7	CH_2Cl_2	9.75	95.00
8	Dichloroethane	10.05	101.00
9	Chlorobenzene	9.70	94.09
10	Diethyl ether	7.44	55.35
11	Isopropyl ether	7.25	52.56
12	Tetrahydrofuran	8.72	76.00
13	Dioxane	10.49	110.00
14	Ethyl acetate	8.85	78.32
15	Acetone	9.68	93.70
16	Anisole	10.53	110.90
17	Acetonitrile	11.98	143.40
18	Water	24.33	592.00
19	Methanol	15.00	225.00
20	Ethanol	12.95	167.6
21	Isopropanol	11.60	134.6
22	Ethylene glycol	21.93	318
23	Pyridine	10.20	104
24	Nitromethane	12.92	167
25	Nitrobenzene	10.63	113
26	Acetic acid	9.27	86
27	Aniline	10.42	108.6
28	Dimethylaniline	8.88	78.85
29	Quinoline	10.63	113
30	Formamide	20.00	400
31	Dimethylformamide (DMFA)	14.07	198
32	Dimethyl acetamide	14.11	199
33	Dimethyl sulfoxide	15.00	225
34	Tributyl phosphate	6.33	40
35	Benzonitrile	10.84	117.4
36	Triethylamine	7.53	56.70
37	Cyclohexanone	10.15	103.5
38	Hexamethylphosphorotriamide	9.80	96

$$\delta = 1.613 + 15.97 \frac{n^2 - 1}{n^2 + 2} + 12.14 \frac{\varepsilon - 1}{2\varepsilon + 1} + 0.00039B + 0.593E; \qquad (34)$$

$R = 0.826 \;(!); \; S = 2.32.$

Similar situation is also observed for δ^2:

$$\delta^2 = 173.88 + 833.85 + 21.90 + 0.0195B + 21.617E; \qquad (36)$$

$R = 0.915; S = 52.45,$

and for all 38 solvents:

$$\delta = -126.97 + 401.46\frac{n^2-1}{n^2+2} + 240.26\frac{\varepsilon-1}{2\varepsilon+1} + 0.0008B + 15.799E; \qquad (37)$$

$R = 0.812; S = 71.89.$

As the solvent selection is extended by substances of other classes, so abrupt decrease of the correlation degree between δ (δ^2) and solvation parameters indicates independence of cohesive energy density and competence of tits application to calculations. The validity of this parameter is also confirmed indirectly [2] by generalization of data on gas dissolution by four-parameter equations, also characterized by low correlation degree: Bunsen absorption coefficient (K_H) logarithm for SO_2 (25 solvents) gives $R = 0.884$; $\lg K_H$ for NMe_3 (14 points [59]) gives $R = 0.815$ and for p-NH_2PhCh_2Cl (8 points [59]) $R = 0.860$.

Thus the solvent influence on thermodynamic phenomena is generally considered on the basis of the Mayer-Koppel-Palm generalized model using equation (2.38), which considers both energetic effects of specific and nonspecific solvation and energy expenditure for cavity formation:

$$X = \Delta G_{int} + \Delta G_{cavity} = a_0 + a_1\frac{n^2-1}{n^2+2} + a_2\frac{\varepsilon-1}{2\varepsilon+1} + a_3B + a_4E + a_5\delta^2. \qquad (38)$$

This equation also possesses high effectiveness in consideration of spectral and kinetic processes.

Such approach is performed in studies by Kamlet, Taft and Abraham *et al.* A generalized equation, taking into account the solvent influence on spectral shifts (UV, IR, NMR, ESR), rate and equilibrium constants of chemical reactions, distribution in GLC processes, and fluorescence time, is suggested [31]:

$$XYZ = XYZ_0 + s(\pi^X + \delta d) + a\alpha + b\beta + h\delta_H + e\xi. \qquad (39)$$

Compared to previously suggested equation, the last one possesses new regression terms: Hildebrandt solubility (δ_H) and "coordinate covalence" (ξ) parameters, the latter accepted for a series of groups and compounds, similar to δ, basing on consecutive concordances of theoretical and experimental data. Besides randomness and empirical type of δ and ξ corrections, the following notes may be imposed on this equation:

(1) the authors report on unreliability of β parameters for amphitropous (bipolar) solvents;
(2) direct application of cohesive energy density parameter δ_H to energy calculations is ineligible, because it is of energy square root dimensionality: $(kcal/l)^{0.5}$ or $(kJ/l)^{0.5}$, whereas $\lg k$, $\lg K$, $\Delta h \nu$ and other parameters are proportional to ΔG of kcal/mol or kJ/mol dimensionality. Therefore, application of the cohesive energy density itself (δ^2) is more correct. This disadvantage was eliminated in later investigations.

Equations binding some broadly applied solvent parameters to Kamlet-Taft equation parameters have been suggested [50]. For example, for 32 aprotic solvents:

$$\Delta \nu(E_T) = 10.60 + 5.12(\pi^X - 0.23\delta); R = 0.972; \tag{40}$$

$$\Delta \nu(E_T) = 10.60 + 5.12(\pi^X - 0.23\delta) + 5.78\alpha; R \text{ is absent;}$$

$$\Delta \nu(E_T) = 10.60 + 5.12(\pi^X - 0.23\delta); R = 0.972; \tag{41}$$

$$AN = 1.04 + 15.4(\pi^X - 0.08\delta) + 32.6\alpha; n = 22; R = 0.994; s = 0.17; \tag{42}$$

$$\Delta \nu_{PhOH} \equiv B = -34.5 + 512\beta + 313\xi; n = 43; R = 0.989.$$

Unfortunately, classes of compounds and exclusions for application of these equations are not mentioned in the cited work.

Wide selections of physical and chemical properties and empirical parameters are shown in different monographs [3, 23, 61, 62] and reviews [2, 29, 37, 63, 64], in which over 340 aprotic solvents are studied.

REFERENCES

[1] Hammett L.P., 'The effect of structure upon the reactions of organic compounds. Benzene derivatives', *J. Am. Chem. Soc.*, 1937, vol. 59, pp. 96 - 108.
[2] Johnson K., *Hammett Equations*, Moscow, Mir, 1977, 240 p. (Rus)
[3] Reichardt C., *Solvents And Solvent Effects In Organic Chemistry*, 3rd Edition, Weinheim 2003, Wiley – VCH, pp. 5 - 57, 381 - 402.
[4] Reichardt C., 'Die Lusungsmitteleinfluss auf chemische Reaktionen', In: *Chemie in unserer Zeit*, 1981, Bd. 15(5), S. 139 - 148.
[5] Buckingham A.D, *Disc. Faraday Soc.*, 1957, vol. 24, p. 151 - 157.
[6] Muirhead-Gould J.S. and Laidler K.J., *Trans. Faraday Soc.*, 1967, vol. 63(4), pp. 944 - 952.
[7] Fainberg A.H. and Winstein S., *J. Am. Chem. Soc.*, 1957, vol. 79, pp. 1608 - 1612.
[8] Fowler F.W. and Katritzky A.R., *J. Chem. Soc.*, Ser. B, 1971, No. 3, pp. 460 - 469.
[9] Iohansen A.V., *Teor. Eksper. Khimia*, 1971, vol. 7(3), pp. 302 - 311. (Rus)
[10] Terentiev V.A., *Zh. Fiz. Khim.*, 1972, vol. 46(8), pp. 1918 - 1920. (Rus)

[11] Terentiev V.A., *Donor-Acceptor Bond Thermodynamics*, Izd. Saratovskogo Universiteta, 1981, 276 p. (Rus)
[12] Terentiev V.A., 'Charge-transfer complexes and donor-acceptor properties of molecules', In Coll.: *Structure And Properties Of Molecules*, No. 3, Izd. Kuibyshevskogo Universiteta, 1978, pp. 3 – 44. (Rus)
[13] Drago R.S. and Wayland B.B., *J. Am. Chem. Soc.*, 1965, vol. 87(16), pp. 3571 - 3572.
[14] Drago R.S., *Structure And Bonding*, 1973, No. 15, pp. 73 - 139.
[15] Burger K., *Solvation, Ionic Reactions And Complex formation In Solutions*, Moscow, Mir, 1984, 256 p. (Rus)
[16] Mayer U., *Monatsh. Chemie*, 1978, Bd. 109(2), S. 421 - 433.
[17] Mayer U., *Pure Appl. Chem.*, 1979, vol. 51(8), pp. 1697 - 1712.
[18] Fawcett W.R. and Krygowski T.M., *Canad. J. Chem.*, 1976, vol. 54(20), pp. 3283 - 3292.
[19] Krygowski T.M. and Fawcett W.R., *J. Amer. Chem. Soc.*, 1975, vol. 97(8), pp. 2143 - 2148.
[20] Fawcett W.R. and Krygowski T.M., *Austral. J. Chem.*, 1975, vol. 28(10), pp. 2115 - 2124.
[21] Swain C.G., Swain M.S., Powell A.L., and Alunni S., 'Solvent effects on chemical reactivity. Evaluation of anion and cation solvation components', *J. Am. Chem. Soc.*, 1983, vol. 105(3), pp. 502 - 513.
[22] Koppel I.A. and Palm V.A., 'The influence of the solvent on organic reactivity', In: *Advances In Linear Free Energy Relationships*, Ed. N.B. Chapman and J. Shorter, London, Plenum Press, 1972, pp. 203 - 280.
[23] Palm V.A., *The Fundamentals Of Quantitative Theory Of Organic Reactions*, Leningrad, Khimia, 1972, 360 p. (Rus)
[24] Gordy W., *J. Chem. Phys.*, 1941, vol. 9(3), pp. 215 - 223.
[25] Koppel I.A. and Payu A.I., *Reakts. Spos. Org. Soed.*, 1974, vol. 11(1), pp. 121 - 138. (Rus)
[26] Chapman N.B., Dack M.R.J., Newman D.J., Shorter J., and Wilkinson R., 'The influence of the solvent on organic reactivity', *J. Chem. Soc.*, 1974, Perkin Trans. 2, No. 5, pp. 962 - 971.
[27] Krygowski T.M., Milczarek E., and Wrona P.K., 'An extension of the Kamlet-Taft basicity scale of solvents', *J. Chem. Soc.*, 1980, Perkin Trans.2, No.11, pp. 1563 - 1568.
[28] Kamlet M.J. and Taft R.W., *J. Amer. Chem. Soc.*, 1976, vol. 98(2), pp. 377 - 383.
[29] Taft R.W. and Kamlet M.J., *J. Amer. Chem. Soc.*, 1976, vol. 98(10), pp. 2886 - 2894.
[30] Kamlet M.J., Abboud J.L., and Taft R.W., *J. Amer. Chem. Soc.*, 1977, vol. 99(18), pp. 6027 - 6038.
[31] Kamlet M.J., Abboud J.L.M., Abraham M.H., and Taft R.W., *J. Org. Chem.*, 1983, vol. 48(17), pp. 2877 - 2787.
[32] Kupfer M. and Abraham W., *J. Prakt. Chemie*, 1983, Bd. 325(1), S. 95 - 103.
[33] Bekarek V., *J. Phys. Chem.*, 1981, vol. 85(6), pp. 722 - 723.
[34] Kolling O.W., *Trans. Kansas Acad. Sci.*, 1981, vol. 84, pp. 32 - 38.
[35] Abboud J.L. and Taft R.W., *J. Phys. Chem.*, 1979, vol. 83(3), pp. 412 - 419.
[36] Brady E.B. and Carr P.W., *J. Phys. Chem.*, 1982, vol. 86(16), pp. 3053 - 3057.
[37] Svoboda P., Pytela O., and Vecera M., *Collect. Czech. Chem. Commun.*, 1983, vol. 48(11), pp. 3287 - 3306.

[38] Reichardt C. and Dimroth K., 'Fortschr.Chem.', *Forschung*, 1966, Bd. 11(1), S. 1 - 73.
[39] Griffiths T.R. and Pugh D.C., *Coord. Chem. Rev.*, 1979, vol. 29(23), pp. 129 - 211.
[40] Makitra R.G., Pyrig Ya.N., Tsikanchuk Ya.M., and Zhukovsky V.Ya., *Ukr. Khim. Zh.*, 1980, vol. 40(7), pp. 729 - 734. (Rus)
[41] Makitra R.G. and Pyrig Ya.N., *Ukr. Khim. Zh.*, 1984, vol. 50(7), pp. 763 - 766. (Rus)
[42] Makitra R.G., Pyrig Ya.N., Sendega R.V., and Turkevich O.E., DAN URSR, 1976, Iss. 11, pp. 998 – 1001. (Ukr.)
[43] Makitra R.G. and Pyrig Ya.N., *Reakts. Spos. Org. Soed.*, 1979, vol. 16(1), pp. 103 - 107. (Rus)
[44] Minesinger R.R., Jones M.E., Taft R.W., and Kamlet M.J., *J. Org. Chem.*, vol. 42(11), pp. 1929 - 1934.
[45] Doan P.E. and Drago R.S., *J. Am. Chem. Soc.*, 1982, vol. 104(17), pp. 4524 - 4529.
[46] Kolling O.W., *Anal. Chem.*, 1982, vol. 54(2), pp. 260 - 264.
[47] Laurence C., Quignec-Cabanetos M., and Dziembowska T., *J. Am. Chem. Soc.*, 1981, vol. 103(10), pp. 2567 - 2573.
[48] Vorkunova E.I. and Levin Ya.A., *Zh. Obshch. Khim.*, 1984, vol. 54(6), pp. 1349 - 1352. (Rus)
[49] Makitra R.G., Pyrig Ya.N., and Tsvetkov V.G., *Zh. Obshch. Khim.*, 1984, vol. 54(4), pp. 833 - 836. (Rus)
[50] Bekarek V., *J. Chem. Soc.*, 1983, Perkin Trans. 2, No. 9, pp. 1293 - 1296.
[51] Bekarek V., Buchtova M., and Stolarova M., *Acta Univ. Palack., Olomouc, Fac. Rer. Natur. Chem.*, 1983, v. 76(22), pp. 121 - 129.
[52] Stolarova M. and Bekarek V., *Acta Univ. Palack., Olomouc, Fac. Rer. Natur. Chem.*, 1983, vol. 76(22), pp. 131 - 135.
[53] Bekarek V. and Stolarova M., *Collect. Czech. Chem. Commun.*, 1983, vol. 48(5), pp. 1237 - 1240.
[54] Pierotti R.A., *J. Phys. Chem.*, 1963, vol. 67, pp. 1840 - 1845.
[55] Bekarek V., *Acta Univ. Palack., Olomouc, Fac. Rer. Natur. Chem.*, 1982, vol. 73(21), pp. 45 - 50.
[56] Bekarek V. and Jurcina J., *Collect. Czech. Chem. Comm.*, 1982, vol. 47(4), pp. 1060 - 1068.
[57] Samoshin V.V. and Zefirov N.S., *Doklady AN SSSR, Ser. Khim.*, 1982, vol. 264(4), pp. 873 - 875. (Rus)
[58] Jaffe H.H., *Chem. Rev.*, 1953, vol. 53(2), pp. 191 - 261.
[59] Abraham M.H., 'Substitution of saturated carbon. P. 8. Solvent effects on the free energy of trimethylamine, nitrobenzylchloride and trimethylaminonitrobenzylchloride transition states', *J. Chem. Soc.*, Ser. B, 1971, No. 2, pp. 299 - 308.
[60] Entelis S.G. and Tiger R.G., *Reaction Kinetics In The Liquid Phase*, Moscow, Khimia, 1973, 412 p. (Rus)
[61] Shmid R. and Sapunov V.M., *Nonformal Kinetics*, Moscow, Mir, 1985, 264 p. (Rus)
[62] Kamlet M.J., Abboud J.L.M., and Taft R.W., In: *Progr. Phys. Org. Chem.*, 1981, vol. 13, pp. 485 - 630.
[63] Makitra R.G., Pyrig Ya.N., and Kivelyuk R.B., *The Most Important Characteristics Of Solvents Used In LFE Equations*, Dep. VINITI 1986, No. 628-V86, 33 p. (Rus)
[64] Abboud I.L.M. and Notario R., *Pure Appl. Chem.*, 1999, vol. 71(4), pp. 645 - 718.

[65] Makitra R.G., Pyrig Ya.N., and Kivelyuk R.B., *Zh. Obshch. Khim.*, 1990, vol. 60(10), pp. 2209 – 2215. (Rus)

[66] Makitra R.G. and Pyrig Ya.N., *Zh. Obshch. Khim.*, 1986, vol. 56(3), pp. 657 – 665. (Rus)

In: Chemistry as Art
Editors: L. Liu and G. E. Zaikov, pp. 81-137

ISBN 1-59454-585-5
© 2006 Nova Science Publishers, Inc.

Chapter 5

COPOLYMERS WITH CYCLIC FRAGMENTS IN DIMETHYSILOXANE BACKBONE

O.V. Mukbaniani and G. E. Zaikov[*]

Sukhumi State University, 12 L.Djikia str., Tbilisi-380087, Georgia
[*]N.M.Emanuel Institute of Biochemical Physics, 4 Kosygin str., Moscow 119991, Russia

It is known [1, 2] that introduction of different components or groups of different chemical origin or structure into the backbone is one of efficient methods for modifying linear organosiloxane polymers. Introduction of alien fragments is resulted in the variation of a spiral-shaped structure of dimethylsiloxane polymer, which causes variation of their physical and chemical properties [3].

In particular, introduction of aromatic fragments to linear polyorganosiloxanes [4, 5] hampers the chain transfer reaction accompanied by ring separation which, in its turn, increases stability of the mentioned polymers.

There are some information in the literature about synthesis of polymers with alternating diorganosiloxane and organosilsesquioxane units [6 – 10] in cyclolinear two-chain macromolecule of the polymer [11]. The first series of works has used co-hydrolysis products of dimethyldichlorosilane with phenyltrichlorosilane [6, 7], dimethyl dichlorosilane with methyltrichlorosilane [8] or methylphenyldichlorosilane with phenyltrichlorosilane [9] for synthesis of the polymers.

As a result of anionic polymerization of co-hydrolysis product at equimolar ratio of diorgaosiloxane and organosilsesquioxane units, 3D-polymers were synthesized. Polymerization of bicyclodimethylsiloxanes with various lengths of dimethylsiloxane chain between two cyclotetracilocane rings has given spatially cross-linked polymers [10]; copolymerization of octamethylcyc;lotetrasiloxane with polyphenylsilsesquioxane leads to formation of soluble low-molecular polymers [11].

However, all above-mentioned co-hydrolysis reactions lead to formation of low-molecular compounds with statistic disposition of linear and cyclic fragments, in which statistic disposition of siloxane (*D*) and silsesquioxane (*T*) units is preserved.

Discussed in this Chapter is synthesis of cyclolinear copolymers with regular disposition of monocyclic fragments in dimethylsiloxane backbone using HFC and hydride polyaddition reaction as the methods for synthesis of polymers.

1. SILOXANE CYCLOLINEAR COPOLYMERS WITH ORGANOCYCLOTETRA (PENTA-, HEXA-) SILOXANE FRAGMENTS IN DIMETHYLSILOXANE BACKBONE

HFC reaction of bifunctional organocyclosiloxanes with α,ω-dichloro(dihydroxy, dimethylamino)dimethylsiloxanes of different length (n) proceeding with formation of cyclolinear copolymers possessing regular disposition of aromatic fragments in dimethylsiloxane chain is discussed in the current Section.

Studying HFC reaction, the authors of works [12 – 17] have used organocyclotetrasiloxanes with different disposition (1,3- and 1,5-positions by edge and diagonal) of functional groups in cyclotetrasiloxane, namely:

where Cl, OH; R = Me, Ph.

The effect of substituents at atom of silicon in bifunctional organocyclotetrasiloxanes on reactivity of haloid and hydroxyl groups interacting with α,ω-dichloro(dihydroxy)dimethylsiloxanes was studied. Polycondensation of dichlorohexaorganocyclotetrasiloxanes were performed at room temperature in 70% solution of dry toluene or benzene both with acceptor and without it.

When HFC reaction was performed without acceptor, the run of hydrogen chloride extraction was searched for. Figure 1 shows dependence of hydrogen chloride extraction on time at 20°C, whence it follows that as the length of dimethylsiloxane unit increases (n = 2 – 6), conversion by hydrogen chloride decreases from 45% (n = 2, R = Me) to 33% (n = 6, R = Me) (Figure 1, curves 1 – 3).

Substitution of methyl side group by phenyl one at silsesquioxane atom of silicon induces abrupt decrease of hydrogen chloride extraction rate from 20 – 25% (n = 2, R = Ph) to 5% (n = 4, R = Ph), which is displayed by curves 4 and 5 in Figure 1.

Gas-liquid analysis of the product obtained by polycondensation of 1,5-dichlorohexaphenylcyclotetrasiloxane with 1,3-dihydroxytetramethyldisiloxane has shown that initial compounds are absent in it and that octamethylcyclotetrasiloxane, which would be formed by homocondensation of disiloxanediol in acidic medium is also absent, but products with higher boiling point, i.e. products of partial intramolecular condensation are present. In the presence of pyridine, HFC reaction between 1,5-dichlorohexaphenylcyclotetrasiloxane and α,ω-dihydroxydimethylsiloxanes proceeds by analogy (at low values of n).

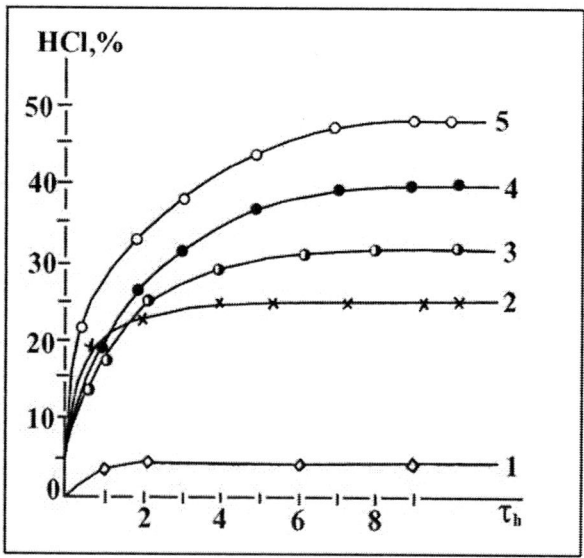

Figure 1. Dependence of polycondensation degree between 1,5-dichloro-1,5-diorganotetraphenylcyclotetrasilocane and α,ω–dihydroxydimethylsiloxanes: curves 1, 2, 3 – at R – Me, n = 2, 4, 6, respectively; curves 4, 5 – at R = Ph, n = 2, 4, respectively

Thus basing on data obtained in the study of HFC reaction between dichloro (dihydroxy)organocyclotetrasiloxanes with α,ω–dihydroxy(dichloro)dimethylsiloxanes it has been shown [14] that the reaction is both intermolecular forming cyclolinear copolymer and intramolecular giving dicyclic structures (at low values of n). As a consequence, the current reaction proceeds in accordance with the following scheme:

where X = Cl, OH; Y = OH, Cl; R = Me, Ph; n = 2, 4, 5, 6, 12, 25, 51, 70, 101.

$$xCl\text{[cycle with Ph,Ph,Ph,Ph]}Cl + xY(SiMe_2O)_{n-1}SiMe_2Y \xrightarrow[-2xPy\cdot HCl]{2xPy} HO\left[\text{[cycle with Ph,Ph,Ph,Ph]}O(SiMe_2O)_n\right]_x H$$

III   (2)

where $n = 2, 4, 6, 12, 25, 70$.

Reactivity of functional groups at their different disposition in the ring (1,5- or 1,3-) was estimated in the reaction with α,ω-dihydroxydimethylsiloxane with the polymerization level $n \approx 25, 70$ at 180°C in block. From experimental data a conclusion was made [18] that 1,3- and 1,5-positions of chlorine atoms at silicon ones in dichlorohexaphenylcyclotetrasiloxane interacting with α,ω-dihydroxydimethylsiloxanes cause an insignificant effect on their reactivity during polycondensation (Figure 2), which runs counter to the conclusion in the ref. [19].

At room temperature, polycondensation of 1,5-dihydroxyhexaphenylcyclotetrasiloxane with α,ω–dichlorodimethylsiloxanes in 70% solution of anhydrous toluene proceeds at deeper level, and conversion by hydrogen chloride reaches 60%, whereas in HFC reaction of 1,5-dichlorohexaphenylcyclotetrasiloxane with α,ω–dichlorodimethylsiloxanes it is below 30% (Figure 3).

As the screening effect only of substituents at atoms of silicon is considered, they display equal values of it. Thus, for estimating reactivity of functional groups in current reactions [18], both the screening effect of substituents at atoms of silicon linked to reactive groups and the inductive action of this radical on the reactive group are considered. Because maximal conversion by hydrogen chloride did not exceed 50 – 60%, the acceptor (pyridine) was introduced for the purpose of increasing the depth of HFC reaction proceeding.

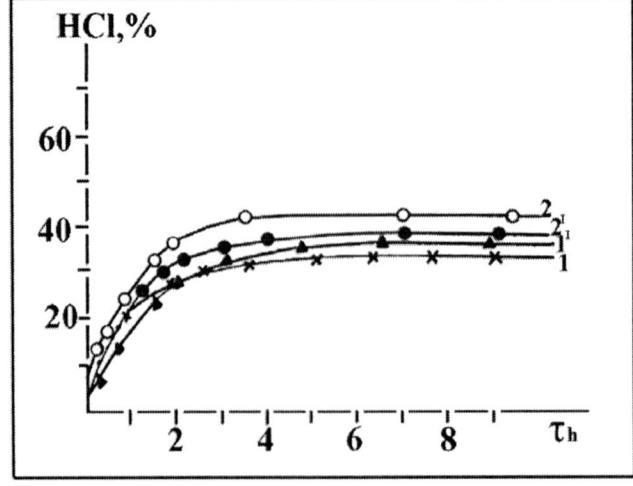

Figure 2. Dependence of hydrogen chloride extraction rate in polycondensation reaction between dichlorohexaphenylcyclotetrasiloxane and α,ω–dihydroxydimethylsiloxanes: curves 1 and 2 – with 1,5-disposition of chlorine atoms and at $n \approx 25, 70$, respectively; curves 1' and 2' - with 1,3-disposition of chlorine atoms and at $n \approx 25, 70$, respectively

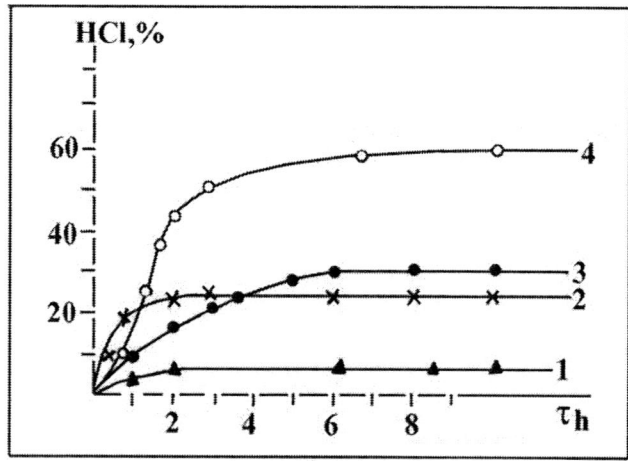

Figure 3. Hydrogen chloride conversion in polycondensation of 1,5-dichloro- and 1,5-dihydroxyhexaphenylcyclotetrasiloxanes with α,ω-dihydroxy(dichloro)dimethylsiloxanes ($n = 2, 4$): curves 1 and 2 - for 1,5-dichlorohexaphenylcyclotetrasiloxane with $n = 4, 2$, respectively; curves 3 and 4 – for 1,5-dihydroxyhexaphenycyclotetrasiloxane with $n = 4, 2$, respectively

It is shown that under conditions of the experiment at 23 – 25°C condensation of α,ω-dihydroxydimethylsiloxanes by hydroxyl groups does not proceed, which correlates with previous conclusions [20]. In this work it is indicated that homofunctional condensation of 1,3-dihydroxytetramethyldisiloxane does not proceed even in the presence of more basic amines.

As HFC proceeds in 60 – 70% solution of anhydrous toluene in the presence of pyridine, the yield of copolymers is increased to 72 – 93%, and copolymers synthesized after redeposition represent colorless or light-yellow transparent solids of viscous substances, well soluble in usual organic solvents with $\eta_{spec} \approx 0.1 - 0.5$.

Study of the reaction (3.1) has indicated that at short length of linear dimethylsiloxane unit the yield of copolymers after redeposition is much lower. After removing solvent from the mother solution, a viscous product with molecular mass of ~830 was obtained. Such molecular mass can be displayed by the product of intramolecular condensation with structure II only. To prove the possibility of compound with structure II formation, the authors of the work [21] have performed direct synthesis of this compound. Results of the current synthesis were also used in estimation of cis- and trans-isomers' contents in the initial 1,5-dichlorohexaphenylcyclotetrasiloxane. Namely, HFC reaction of a mixture of cis- and trans-isomers of 1,5-dichlorohexaphenylcyclotetrasiloxane with 1,3-dihydroxytetramethyldisiloxane was studied (equimolar ratio of initial reagents, 5% anhydrous toluene solution, 0°C, in the presence of acceptor – pyridine). In the case of cis-isomer, formation of polycyclic structure compound is the most probable, whereas trans-isomers give polymeric products. The reaction proceeds in accordance with the scheme as follows:

$$\begin{array}{c} \text{Ph} \quad \text{Ph} \\ \text{R} \diagdown \quad \diagup \text{R} \\ \text{Cl}^{\prime\prime\prime} \diagup \quad \diagdown {}^{\prime\prime\prime}\text{Cl} \\ \text{Ph} \quad \text{Ph} \end{array} + \text{HO(SiMe}_2\text{O)}_2\text{H} \xrightarrow[-2\text{Py·HCl}]{2\text{Py}} \text{II} + \text{IV} \quad (3)$$

Composition and structure of individual polycyclic compounds, synthesized by fractionation in high vacuum, were determined on the basis of elementary analysis by determination of molecular mass, IR and ^{1}H NMR spectral data. Based on this reaction, the authors have concluded that 1,5-dichlorohexaphenylcyclotetrasiloxane used in HFC reactions represents a mixture of cis- and trans-isomers. The cis-form promotes for structures II and IV formation, total yield of which equals ~50%; the rest 50% are given by oligomeric structures, formed from trans-forms of 1,5-dichlorohexaphenylcyclotetrasiloxane. Spectral data of ^{1}H NMR analysis have indicated the regular structure of cyclolinear copolymers, which correlates well with data from the literature [22]. Some physical and chemical parameters of copolymers with cyclotetrasiloxane fragments in the backbone are shown in Tables 1 – 3.

For copolymers, thermomechanical studies were carried out in accordance with the technique, described in ref. [23]. Expectedly, basing on data from the literature [24, 25] the studies have shown that the presence of inclusions of any chemical origin in the polydimethylsiloxane backbone (framing groups different from methyl ones, rings of different structure, branching points) causes the loss of the ability of copolymers to crystallize, if the distance between neighboring inclusions equals about 30 siloxane units.

Copolymers with $n = 2 - 25$ studied do not crystallize, and the ones with $n \approx 70$ behave themselves as crystallizing polydimethylsiloxanes (PDMS). The temperature range corresponded to glass transition in copolymers regularly expands towards higher temperatures with decrease of n. Of special attention is the fact that disposition of cyclotetrasiloxane fragment in polydimethylsiloxane backbone (1,5- and 1,3-positions) causes an insignificant effect on T_g of copolymers (Figure 4).

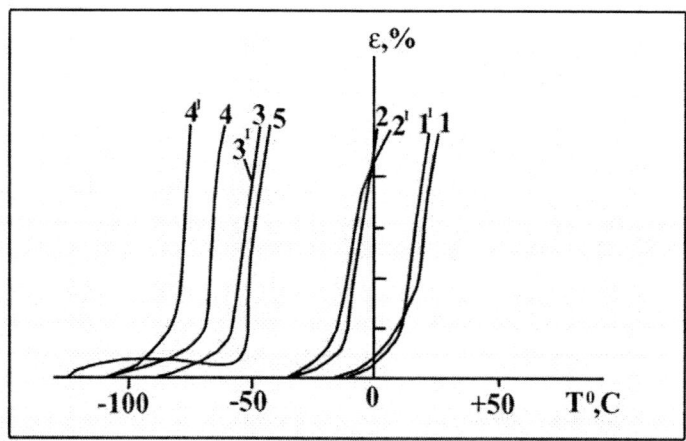

Figure 4. Thermomechanical curves for copolymers with 1,3- and 1,5-disposition of cyclotetrasiloxane fragments in PDMS backbone: 1, 2, 3, 4, 5 – for copolymers with 1,3-disposition of hexaphenylcyclotetrasiloxane fragment in the backbone at $n \approx$ 2, 4, 12, 25, 70, respectively; 1', 2', 3', 4' for copolymers with 1,5-disposition of hexaphenylcyclotetrasiloxane fragment in the backbone at $n \approx$ 2, 4, 12, 25, respectively (100 g load, 5 deg/min heating rate)

Table 1. Some physical and chemical parameters and yield of cyclolinear copolymers of structure I with 1,5-disposition of cyclotetrasiloxane fragment in dimethylsiloxane backbone

No	Copolymer	R	n_{SiO}	Yield, %	T_g, °C	η_{spec}*
1		Me	2	72	-54	0.15 - 0.16
2		Me	4	83	-57	0.09 - 0.16
3**		Me	5	86	-	0.20
4		Me	6	86	-71	0.10 - 0.22
5		Mc	12	90	-94	0.24 - 0.27
6**		Me	25	90	-117	- 0.33
7		Me	51	91	-123	- 0.40
8		Me	70	93	-123	0.42 -
9		Me	101	96	-123	- 0.49
10		Ph	2	72	-10÷-20	0.06 - 0.11
11		Ph	4	80	-34	0.10 - 0.14
12		Ph	6	93	-55	0.10 - 0.21
13		Ph	12	93	-83	0.19 - 0.26
14		Ph	25	95	-110	0.19 - 0.34
15		Ph	51	96	-123	- 0.40
16		Ph	70	92	-123	0.15 -
17		Ph	101	97	-123	- 0.42

Note: * Hereinafter, second values of viscosity are obtained by high-temperature polycondensation of 1,5-dihydroxyorganocyclohexasiloxanes with α,ω-bis(dimethylamino)dimethylsiloxanes
** For these samples, viscosities in benzene are determined

Table 2. Some physical and chemical parameters and yield of cyclolinear copolymers of structure II with 1,3-disposition of cyclotetrasiloxane fragment in dimethylsiloxane backbone

No	Copolymer	n_{SiO}	Yield, %	T_g, °C	η_{spec}
1	HO−[−(Ph)(Ph)Si—Si(Ph)(Ph)−O−(SiMe₂O)ₙ−]ₓ−H (cyclotetrasiloxane with Ph, Ph / Ph, Ph framing)	2	83	-23	0.06
2		4	90	-35	0.10
3		6	91	-	0.11
4		12	92	-83	0.31
5		25	93	-107	0.37
6		70	92	-123	0.49

Of special importance is the fact that introduction of rings containing phenyl framing groups into the linear chain is accompanied by T_g increase compared with cyclolinear carbosiloxane polymers containing methyl framing groups [26]. In the case of cyclotetrasiloxanes with phenyl groups, T_g increase is higher than in the action of network points, which do not change T_g of PDMS networks at $n \approx 20$ [27].

The latter is determined by a significant effect of bulky phenyl groups, the presence of which in any place of PDMS backbone significantly increases T_g [28]. Substitution of phenyl side group by methyl one in organosilsesquioxane fragment of organocyclotetrasiloxanes decreases the glass transition temperature by ~15 - 20°C. The effect of cyclotetrasiloxane fragment is detected at dimethylsiloxane unit $n \approx 25$ long and lies above T_g of PDMS. Thermogravimetric studies of cyclolinear copolymers indicate that 1,3- and 1,5-dispositions of hexaphenylcyclotetrasiloxane fragment cause not difference in mass losses of copolymers (refer to Figure 5).

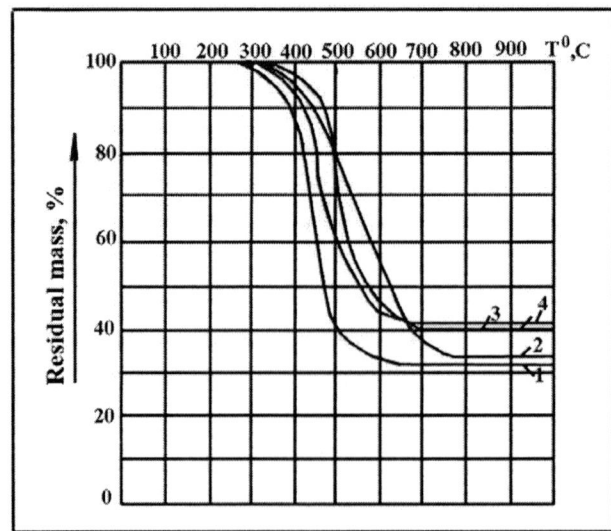

Figure 5. Thermogravimetric curves of copolymers with 1,3- and 1,5-disposition of hexaphenylcyclotetrasiloxane fragments in PDMS backbone: curves 1 and 3 – for copolymer with 1,5-disposition at $n \approx 70$, 6, respectively; curves 2 and 4 – for copolymer with 1,3-disposition at $n \approx 70$, 6, respectively (in air, 5 deg/min heating rate)

As a result, by HFC reaction between 1,5-dihydroxy-1,5-bistrimethylsiloxytetraphenylcyclotetrasiloxane and α,ω-dichloromethylsiloxanes in the presence of pyridine in 60% anhydrous toluene solution at 20 - 25°C cyclolinear copolymers with regular disposition of 1,5-bistrimethylsiloxytetraphenylcyclotetrasiloxane in the dimethylsiloxane backbone, completely dissoluble in organic solvents, were synthesized [29]. The reaction proceeds in accordance with the general scheme as follows:

$$x \begin{array}{c} \text{HO} \quad \text{Ph}_2 \quad \text{OH} \\ \diamond \\ \text{Me}_3\text{SiO} \quad \text{Ph}_2 \quad \text{OSiMe}_3 \end{array} + x\ \text{Cl}(\text{SiMe}_2\text{O})_{n-1}\text{SiMe}_2\text{Cl} \xrightarrow[2x\text{PyHCl}]{2x\text{Py}}$$

$$\longrightarrow \text{HO} \left[\begin{array}{c} \text{Ph}_2 \\ \diamond \\ \text{Me}_3\text{SiO} \quad \text{Ph}_2 \quad \text{OSiMe}_3 \end{array} \text{O}-(\text{SiMe}_2\text{O})_n \right]_x \text{H} \qquad (4)$$

where $m = 1, 2, 3, 10$.

Depending on the length of dimethylsiloxane unit, copolymers represent crystalline ($T_g \approx$ 100°C, $n = 1$), rubbery ($T_g \approx$ 30°C, $E_{elong} \approx 400\%$) and quite strong ($\delta_{break} = 40$ kg/cm^2 at 24°C) products. Of interest is that dielectric constant of this polymer decreases linearly with temperature increase.

A suitable method has been suggested [12] for obtaining cyclolinear copolymers, which concludes in high-temperature HFC reaction between dihydroxyorganocyclotetra(penta, hexa)siloxanes and α,ω–disdiamino(dimethylamino)dimethylsiloxanes. The method suggested is simpler than the above-discussed reactions 1 and 2, because synthesized copolymers possess low molecular mass. Moreover, obtaining of pure copolymers in accordance with the above-shown schemes 1 and 2 requires application of additional stage of polymer purification from secondary product which is amine hydrochloride.

For the purpose of synthesizing cyclolinear copolymers with regular disposition of organocyclotetra-, -penta- and –hexasiloxane fragments in linear dimethylsiloxane backbone, HFC reaction of 1,5-dihydroxy-1,5-dimethyl(diphenyl)phenylcyclotetra(penta, hexa)siloxanes and 1,7-dihydroxy-1,7-dimethyl(diphenyl)octaphenylcyclohexasiloxane with α,ω–bis(dimethylamino)(dichloro)dimethylsiloxanes was studied [15 – 17].

In this case, the reaction proceeds in accordance with the general scheme as follows:

$$x \begin{array}{c} R \quad \text{O(SiPh}_2\text{O})_m \quad R \\ \text{Si} \quad \quad \text{Si} \\ Z \quad \text{O(SiPh}_2\text{O}) \quad Z \end{array} + x\ Y(\text{SiMe}_2\text{O})_{n-1}\text{SiMe}_2 Y \longrightarrow \left[\begin{array}{c} R \quad \text{O(SiPh}_2\text{O})_m \quad R \\ \text{Si} \quad \quad \text{Si}-(\text{SiMe}_2\text{O})_n \\ \text{O(SiPh}_2\text{O})_1 \end{array} \right]_x$$

I, V-VII

(5)

where Z = OH, Cl; Y = Me$_2$N, OH*; $m = l = 1$ - structure I; $m = 1$, $l = 2$ - structure V; $m = 1$, $l = 3$ - structure VI; $m = l = 2$ - structure VII; R = Me, Ph.

HFC was studied at 20 - 150°C in dry nitrogen or argon flow both in block and in solution. As the reaction proceeds in block, the reaction mixture is heated up to 50 - 60°C until a homogeneous mixture is formed; thereafter, the reaction is continued in vacuum at P = 3–5 mm Hg and in the temperature range of 120 - 150°C up to constant viscosity. Application of reactive α,ω–bis(dimethylamino)dimethylsiloxanes allowed performance of the initial stage of the reaction in solution at 20 - 50°C up to formation of a homogeneous mixture. For more full removal of amine, extracted in the reaction, the mixture is aerated by dry inert gas. Polymers synthesized in accordance with this technique require no purification, because amines obtained as secondary reaction products possess very low boiling point and are completely removed from the reaction mixture. In as much as a mixture of dichloro(dihydroxy)organocyclosiloxanes' isomers was used in HFC, atactic copolymers were synthesized.

In the case of HFC between 1,5-dihydroxyhexaphenylcyclotetrasiloxane and α,ω–bis(diethylamino)dimethylsiloxanes, gel formation proceeds after diethylamine conversion reaching ~5%. Previously, it was informed about a possibility of siloxane bond break in trimethyltriphenylcyclotrisiloxane by diethylamine and formation of a complex with charge transfer [30]. In accordance with the authors' opinion [15 – 17], immediate removal of diethylamine produced in the reaction with α,ω–bis(diethylamino)dimethylsiloxanes could be hardly guaranteed. In its turn, dimethylamine interacts with the siloxane bond in aromatic fragment and forms bipolar zwitter-ion [31]. The HFC reaction proceeding with α,ω–bis(diethylamino)(diamino)dimethylsiloxanes [12] forms copolymers completely dissoluble in organic solvents, because dimethylamine and ammonium extracted during synthesis cause no breakage of siloxane bonds in aromatic fragments of the polymeric backbone.

It is proved that at short lengths of dimethylsiloxane unit, HFC reaction proceeds in two directions: intramolecular ring formation giving polyaromatic products and intermolecular with formation of cyclolinear copolymers. Formation of polyaromatic products is proved by the direct synthesis.

To prove the fact of formation of copolymers with regular disposition of aromatic fragments in the macromolecular backbone, some copolymers were fractioned into several fractions. Results of the elementary analysis have indicated that values detected for fractions coincide with calculated ones, which represents direct proof of the regular structure of copolymers.

Many investigators have studied dissolved solutions of polyorganosiloxanes with different side groups at atoms of silicon in the macromolecule backbones [32 - 35]. These works show results of the studies of the effect of the side groups' origin, their disposition and the influence of hydrodynamic and conformations parameters of macromolecules.

The effect of introduction of regularly disposed aromatic fragments into the macromolecular backbone on conformational and hydrodynamic parameters is also studied [30]. For this purpose, copolymers 3 and 4 (Table 1) were fractioned into twelve fractions from the benzene (solvent) – methanol (precipitator) system. The influence of aromatic groups, introduced into the polymer backbone, on rigidity parameters was determined by direct computer modeling of macromolecular coil with the help of the Monte-Carlo method.

The process was modeled for the copolymer of structure I with $n = 1$, 5 and 10 in the structural unit of dimethylsiloxane chain, and the Cune segment A and $<h^2f>/nM_0$ were calculated, where M_0 is the molecular mass of the structural unit, $<h^2f>$ is the mean-square distance between chain ends at free rotation around virtual bonds[1], n is the number of structural units.

Structural unit of the copolymer was modeled on the basis of literary data on the structure the simplest molecules, close by the composition and structure to monomeric units. Shown below is the geometrical structure of repeated structural unit in polyorganosiloxane at $n = 1$. Table 3 shows values of bond lengths and angles between them.

Table 3. Bond lengths and angles between them in polyorganocyclosiloxanes of structure I (n = 1)

Bond	Bond length, Å	θ_1^*, deg	Rotation conditions
l_1	4.25	54.5	Prohibited
l_2	1.63	37.0	Free
l_3	1.63	70.0	Free
l_4	1.63	37.0	Free
l_5	1.63	54.5	Free

Note: * For polymeric backbone, this coefficient is exclusively associated with the structure of macromolecules and potentials of internal rotation around internal bonds of the backbone. This dependence gives information about the energy of chain conformation and mutual transitions

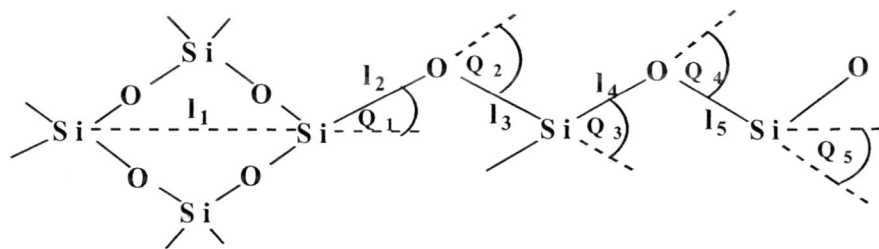

Rotation around virtual Si-------Si (l_1) bond is prohibited to avoid the ring twisting. Table 4 shows calculation results, which represent theoretical values of the Cune segment (A) and $<h^2f>/nM_0$ for polymers of structure I ($n = 1$; $n = 5$; $n = 10$), And for comparison, analogous values for PDMS [37].

These theoretical data show that the influence of the eight-membered siloxane ring is displayed at $n = 1$ only. Saturation is observed already at transition from $n = 5$ to $n = 10$, and increase of the quantity of ≡Si-O- groups causes no effect on the coil size.

For the fraction of copolymer 3 of structure I and $n = 5$, experimental values of the Cune segment and $<h^2f>/nM_0$ are shown in Table 1.

---

[1] By the term of virtual bond the area of the chain is meant, rotation around which is possible. In particular case, it may be a simple valence bond –(l2-l5); in the general case, it may contain not only valence bonds, but also rings, or may determine the distance between atoms in the ring as (l1) does.

Table 4. Calculation values of the Cune segment (A) and $<h^2f>/nM_0$

Polymer of structure I	n_{SiO}	$A = <h^2f>/nl_0A$	$<h^2f>/nM_0$
HO-[Me-Ph₂-Me / Ph₂]-O-(SiMe₂O)ₙ-H]ₓ	1	15	0.28
	5	10	0.24
	10	10	0.22
PDMS	10	10	0.24

One of the methods for estimation of experimental thermodynamic flexibility of the polymer backbone is determination of parameters, associated with the size of isolated macromolecule coil. At the present time, methods of macromolecular coil parameter estimation under ideal and non-ideal conditions. In the first case, characteristic viscosity, $[\eta]$, of the fraction with known molecular mass in ideal θ-solvent was measured. This method is based on the known Flory-Fox relation [38]. The second method represents measurement of fraction $[\eta]$ in good thermodynamic solvents and extrapolation of experimental data in accordance with the known techniques [39 - 41]. Because in the current work [36] all measurements were performed in good thermodynamic solvent, unperturbed dimensions of macromolecules, $<h_\theta^2>$, were determined by graphic extrapolation in accordance with the Stockmayer-Fixman, suggested by the authors for flexible macromolecules, in $[\eta]M^{1/2} - M^{1/2}$ coordinates (Figure 6).

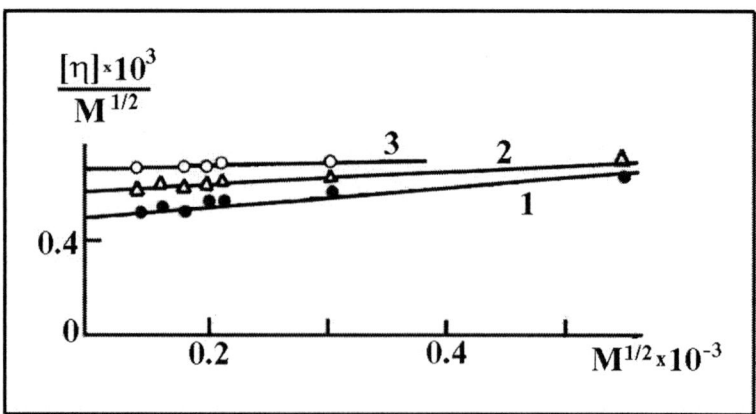

Figure 6. Determination of unperturbed dimensions by extrapolation in accordance with the Stockmayer-Fixman method for copolymers 3 (Table 1): 1 – at 25°C; 2 – at 40°C; 3 – at 50°C

Table 5 shows mean-weight molecular masses, $\bar{M}_\omega$, and $[\eta]$ for the fraction of copolymer 3 ($n = 5$) (Table 3.1) with structure I at 25, 40 and 50°C in toluene.

Table 5. Molecular masses and characteristic viscosities of copolymer 3 fractions (Table 1, structure I) at different temperatures

$$HO-\left[\begin{array}{c} Me \quad Ph_2 \quad Me \\ \diamond \\ Ph_2 \end{array} - O-(SiMe_2O)_5 \right]_x - H$$

Fraction No.	$\bar{M}_\omega \times 10^{-3}$	[η] in toluene (дл/g) at temperature, °C		
		25	40	50
1	9.0	0.04	-	-
2	17.0	0.06	0.07	0.13
3	22.0	0.07	0.09	0.11
4	26.0	0.08	0.10	0.15
5	33.0	0.09	0.11	0.12
6	42.0	0.12	0.13	0.14
7	44.0	0.11	0.13	0.15
8	72.0	0.15	0.15	0.12
9	85.0	0.16	0.17	0.18
10	92.0	0.19	0.20	0.21
11	141.0	0.26	0.23	0.38
12	239.0	0.33	0.38	-

Table 6 shows experimental values of conformational parameters for the Cune segment, $A = <h_0^2>/nl_0$ (where l_0 is the structural unit length), hindrance factor of rotation around the single bond, $\sigma = (<h_0^2>/<h^2_f>)^{1/2}$, and constants of the Mark-Cune-Hauvink equation, $K_\theta \approx [\eta]/M^{1/2}$, at different temperatures.

Table 6. Experimental conformational parameters of copolymer 3 (structure I, Table 1)

No.	T, °C	$K_\theta \times 10^3$	$<h_0^2>/nM_0$	A, Å	$\sigma = (<h_0^2>/<h^2_f>)^{1/2}$
1	2	3	4	5	6
1	25	0.46	0.303	16.0	1.12
2	40	0.58	0.340	18.0	1.18
3	50	0.67	0.380	20.0	1.25

Experimental data on conformational parameters of copolymer 3 (Table 1), shown in Table 6, exceed those calculated in the supposition of free rotation around virtual and valence bonds.

Comparison of experimental and calculated values of the above-mentioned parameters indicates existence of hindrance of rotation around valence bonds ≡Si-O- and ≡Si-Ph due to their short lengths, because distances between neighboring atoms Si----Si and O----O are shorter or close to the sum of Van-der-Waals radii of these atoms. Data from Table 6 indicates also that experimental values of hindrance factors are low.

As shown by Flory, these low values of σ, i.e. an insignificant increase of PDMS molecule dimension compared with dimensions, calculated in the supposition of free rotation, are associated with dominant containing of the plane trans-chain type conformation rather than the absence of rotation hindrance, compared with coiled conformations. Hence, low σ values for these chains may be realized at both low and high differences of energies of rotational isomers under the condition of energy gain of the trans-shape [42, 43]. On the other hand, low σ must be combined with positive temperature coefficient of unperturbed dimensions of macromolecules[2], because relative content of the trans-shape must decrease with temperature growth [42, 43]. For linear PDMS, studies of the temperature coefficient of unperturbed dimensions of macromolecules were carried out in the work [44], and the value obtained equaled $(0.78 \pm 0.06) \times 10^{-3}$ deg^{-1}.

For the purpose of determination of the temperature coefficient for unperturbed dimensions of copolymer 3 (Table 1), $[\eta]$ values were measured in the same solvent (toluene) at different temperatures. In accordance with the Stockmayer-Fixman method, $K_\theta = (<h_0^2>/M)^{1/2} \times F_0$ values was determined using the least-square technique. The temperature coefficient of unperturbed dimensions was calculated from the values obtained at different temperatures (Table 6) using the relation, suggested in the work [45]: $\mathrm{dln}<h_0^2>/dT = 2/3 \times \mathrm{ln}K_\theta/dT$.

The coefficient of unperturbed coil dimension ($\mathrm{dln}<h_0^2>/dT$), determined for copolymer 3 (Table 1), equals 0.85×10^{-3} deg^{-1}. Figure 7 shows that this value is positive and quite close to the temperature coefficient of unperturbed PDMS dimensions [44].

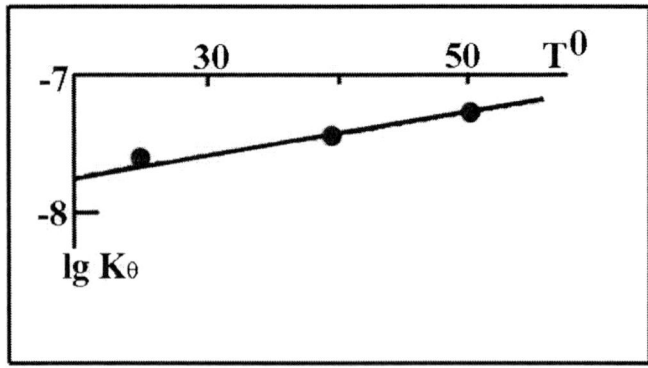

Figure 7. Temperature dependence of $\mathrm{lg}K_\theta$ for copolymer 3 (Table 1)

Thus, data shown in Table 6 indicate that thermodynamic rigidity, mean-square dimensions and hindrance of copolymer 3 chains (Table 1) are increased with temperature. These facts as well as positive sign of the temperature coefficient of unperturbed dimensions testify about predominance of energetically preferable for copolymer 3 chains, similar to PDMS, compared with coiled trans-shaped isomers. As temperature increases, the fracture of plane trans-chains decreases, which, as a result of transition from trans- to gauche-state, proceeds with an increase of unperturbed dimensions [41].

---

[2] For polymeric chain, this coefficient is associated with the structure of macromolecules and potentials of internal rotation around internal bonds in the backbone. Such dependence gives information about energetic conformation of the backbone at mutual transitions.

Summing up all above-discussed, one may conclude that, compared with PDMS, introduction of eight-membered aromatic fragments with symmetrically disposed phenyl side groups /substituents/ causes no change of internal rotation conditions, as it takes place for polymethylphenylsiloxane with asymmetric side phenyl groups [43]. However, already for the ratio 1 : 5 (1 aromatic group per 5 siloxane units), i.e. at the decrease of the distance between aromatic fragments in the backbone, their introduction into dimethylsiloxane backbone induces a significant increase of thermodynamic rigidity of macromolecules. The rigidity increase of copolymer 3 is also displayed in thermomechanical properties of copolymers. For example, T_g of copolymer 3 ($T_g \approx -50°C$) is much higher than T_g of PDMS.

The study of hydrodynamic behavior of copolymer 6 (Table 1) indicates that, apparently, macromolecules of this copolymer represent branched backbones.

In the Mark-Cune-Hauvink equation, parameter α for copolymer 6 solution in toluene at 25°C equals 0.30, which is typical of branched macromolecules. For cyclolinear polymer 3 under the same conditions, this parameter equals 0.62 (Figure 8). Molecular masses vary within the range from 4×10^3 to 565×10^3.

Mean-weight molecular masses, $\overline{M}_\omega$, were measured on Sofica device. Moreover, molecular masses below 10,000 were also determined by ebullioscopy method ($\overline{M}_n$). Good coincidence between $\overline{M}_\omega$ and $\overline{M}_n$ was obtained. The reason for branching may be the fact that initial oligomers (α,ω-dichlorodimethylsiloxanes) for synthesizing cyclolinear copolymers (polyorganocyclotetrasiloxanes) with $n = 5$ and 10 were produced by partial hydrolysis reaction of dimethyldichlorosilane, and oligomers with $n = 25$ were telomerized in autoclave [46], which may cause formation of copolymers containing units, capable of branching. Additional proof of branching is the loss of copolymer 6 solubility with time. However, behavior of copolymer 6 under θ-conditions different from the behavior of branched macromolecules was observed. Parameter α equaled 0.53, which exceeded $\alpha \approx 0.25$ typical of branched copolymers under θ-conditions. For copolymer 6, θ-conditions were set on a device of polymer temperature precipitation. Isopropyl alcohol at 22°C was used as θ-solvent (Figure 9).

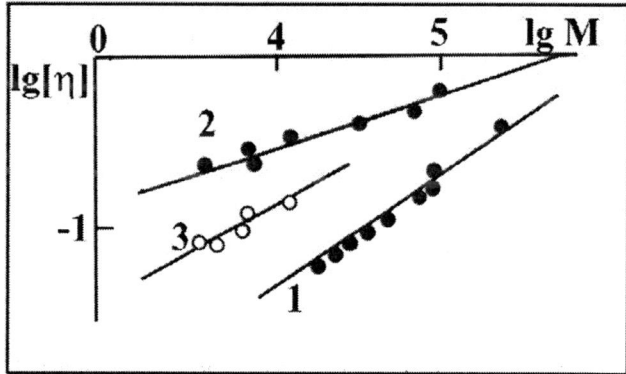

Figure 8. Dependence of lg[η] on lg$M$: 1 – for copolymer 3 in toluene at 25°C, [η] $\approx 1.394 \times 10^{-4} M^{0.62}$; 2 – for copolymer 6 at 25°C, [η] $\approx 1.79 \times 10^{-2} M^{0.30}$; 3 – for copolymer 6 in isopropyl alcohol at 22°C, [η] $\approx 9.86 \times 10^{-4} M^{0.53}$

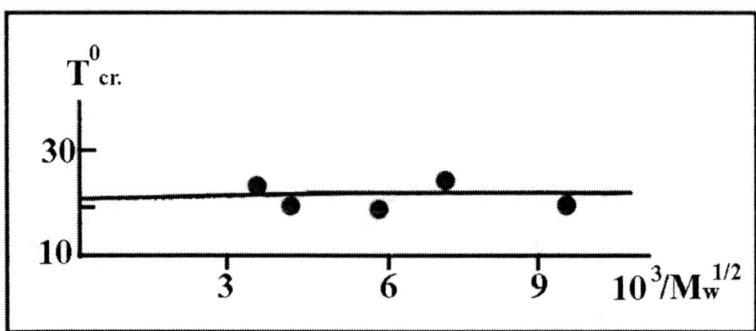

Figure 9. Dependence of critical precipitation temperature, T^0_{cr}, of copolymer 6 fraction (Table 1) on molecular mass

The attempt to prove correctness of θ-conditions selection for copolymer 6 by checking equality to zero of the virial coefficient, A_2, has given interesting results. It has been found that purification of polymer solution from dust by filtration via dense Shot filter (No. 5) causes a sharp increase of light scattering intensity, which does not allow measurement of molecular mass and A_2. Repeated filtering induces much more sharp increase of light scattering intensity. The precipitation temperature of filtered solution is increased. This fact contradicts to the supposition that copolymer 6 represents a branched structure. Possible reason for this effect may be solution structuring during long-term filtration. To check this suggesting [36], copolymer solutions (filtered and non-filtered) were studied on optical microscope. Films obtained from 1% filtered solution displayed oriented chains and aggregates from them, which was not observed in films from non-filtered solution. Studies carried out with the help of electron microscope have not allowed definite conclusions, because the technique requires dilution to 0.01 g/ml concentration of the solution, i.e. an order of magnitude lower than that, at which molecular mass is determined.

Thermogravimetric studies of copolymer 2 and 4 (Table 1) have determined the influence of regularly disposed organocyclotetrasiloxane fragments in linear dimethylsiloxane chain on thermal stability of the studied copolymers. Pyrolytic spectrum (Figure 10) indicates that thermal degradation of polymers 2 and 4 below 300 and 350°C, respectively, displays full absence of organosilicon and organic compounds in the pyrolysis products.

At 350 - 400°C, thermal degradation rate is considerably increased, and the reaction proceeds with benzene isolation. In 380 - 440°C temperature range, degradation products display cyclosiloxane components of D_n composition, $D_4 < 1.5\%$. Note also that pyrolysis temperature increase induces sharp increase of CH_4, D_3 and D_4 compounds extracted. Extraction of benzene and methane is due to $\equiv$Si-C and C-H bond break, which induces cross-linking at the expense of methyl and phenyl groups [47].

It has been shown before [47 – 51] that degradation proceeds by the radical mechanism, mainly forming oligomeric products instead of organocyclosiloxanes. Such phenomenon was observed in studies of thermal degradation of organocyclosiloxane cyclolinear copolymers [49]. This gives an opportunity to suggest that thermal pyrolysis of copolymers 2 and 4 is guided by the backbone break in compliance with the law of randomness with formation of oligomeric products. Simultaneously with backbone degradation, radical cross-linking reactions by methyl and phenyl groups proceed, accompanied by extraction of hydrogen, methane and benzene. Introduction of organosiloxane cyclic fragments into linear dimethylsiloxane backbone induces a considerable decrease of the degradation process of such copolymers by depolymerization mechanism at 320 - 340°C, whereas unblocked PDMS is polymerized by almost 95% at these temperatures.

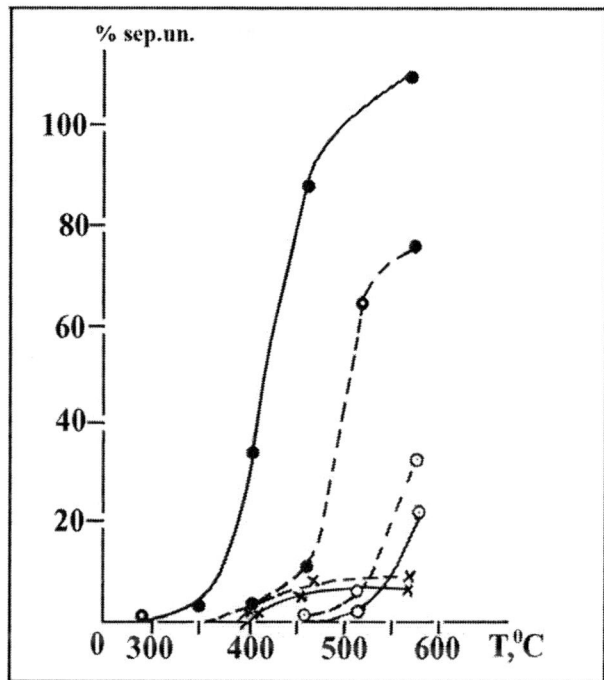

Figure 10. CH_4, C_6H_6, D_3 and D_4 extraction curves for cyclolinear copolymers 2 and 4 (structure I, Table 1): ● - C_6H_6, ○ - CH_4, × - D_3 and D_4 (<1.5%). Here — is corresponded to copolymer 2 and ---- to copolymer 4

The general feature of all copolymers is decrease of their thermooxidative stability and increase of final mass losses with dimethylsiloxane unit length increase. Figure 11 gives comparative estimation of thermooxidative stability of copolymer, which proves negligible effect of 1,5- or 1,7-disposition of cyclohexasiloxane fragments (similar to organocyclotetrasiloxane fragments) on thermooxidative stability of copolymers.

Furthermore, an increase aromatic fragment volume, i.e. at transition from organocyclotetrasiloxane fragment to organocyclopenta- and organocyclohexasiloxane ones, at the same length of dimethylsiloxane unit induces a decrease of thermooxidative stability of the copolymers, which, in turn, is explained by increase of the mass part of phenyl groups.

Comparative thermomechanical studies of copolymer with organocyclotetra-, organocyclopenta- and organocyclohexa siloxane (1,5- and 1,7-positions) fragments in the chain have been implemented. Data on T_g of copolymer, shown in Table 7, indicate that increase of dimethylsiloxane unit length leads to obligate decrease of T_g of copolymers. It has been found that both cases of hexaphenylcyclotetrasiloxanes (1,3- and 1,5-positions) and organocyclohexasiloxanes 1,5- or 1,7-disposition of aromatic fragments with the same framing groups at atom of silicon in silsesquioxane and equal length of the linear dimethylsiloxane unit causes negligible effect on T_g of copolymers.

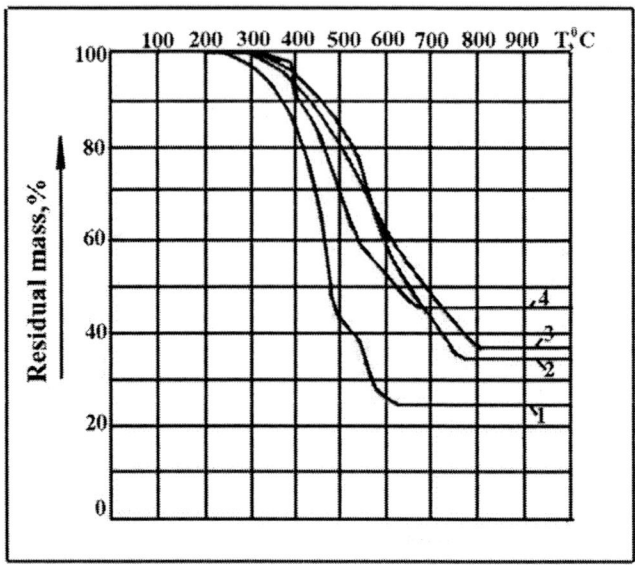

Figure 11. Thermogravimetric curves of cyclolinear copolymers in air: 1 – copolymer with 1,5-disposition of octaphenylcyclopentasiloxane fragment, $n = 8$; 2 - copolymer with 1,5-disposition of decaphenylcyclohexasiloxane fragment, $n = 4$; 3 - copolymer with 1,7-disposition of decaphenylcyclohexasiloxane fragment in dimethylsiloxane backbone, $n = 4$; 4 - copolymer with 1,5-disposition of octaphenylcyclopentasiloxane fragment in dimethylsiloxane backbone, $n = 4$

Table 7. Some physical and chemical parameters of copolymers with organocyclopentasiloxane (structure V) fragments in the backbone

No	Copolymer	R	n_{SiO}	Yield, %	η_{spec}	T_g, °C
1		Me	2	77	0.08	+15
2		Me	4	81	0.13	-14
3		Me	6	93	0.14	-38
4		Me	8	96	0.18	-
5		Me	12	96	0.20	-65
6		Me	40	94	0.25	-123
7		Me	51	97	0.32	-123
8		Ph	2	79	0.06	+15
9		Ph	4	83	0.08	0
10		Ph	6	93	0.11	-25
11		Ph	8	93	0.14	-
12		Ph	12	93	0.18	-50
13		Ph	40	96	0.23	-123
14		Ph	51	98	0.29	-123

Table 8. Some physical and chemical parameters of copolymers with 1,5-disposition of organocyclohexasiloxane (structure VI) fragments in the backbone

No	Copolymer	R	n_{SiO}	Yield, %	η_{spec}	T_g, °C
1	2	3	4	5	6	7
1		Me	2	75	0.09	+10
2		Me	4	80	0.14	-1
3		Me	8	83	0.17	-47
4		Me	34	88	0.26	-123
5		Me	51	90	0.30	-123
6		Ph	2	76	0.10	+21
7		Ph	4	81	0.13	+8
8		Ph	8	82	0.18	-38
9		Ph	34	87	0.24	-123
10		Ph	51	91	0.29	-123

Table 9. Some physical and chemical parameters of copolymers with 1,7-disposition of organocyclohexasiloxane (structure VII) fragments in the backbone

No	Copolymer	R	n_{SiO}	Yield, %	η_{spec}	T_g, °C
1	2	3	4	5	6	7
1		Me	2	85	0.06	+11
2		Me	4	92	0.08	0
3		Me	6	93	0.17	-
4		Me	12	94	0.21	-50
5		Me	51	94	0.26	-123
6		Me	101	96	0.37	-123
7		Ph	2	82	0.05	+24
8		Ph	4	90	0.07	+14
9		Ph	6	93	0.13	-35
10		Ph	12	93	0.19	-54
11		Ph	51	94	0.22	-123
12		Ph	101	96	0.35	-123

Figure 12 shows T_g dependence on dimethylsiloxane unit length, which indicates that increase of T_g of copolymers with the number of phenyl groups and volume of aromatic fragment. Partial substitution of methyl groups at silicon atom in silsesquioxane fragment by phenyl ones will increase T_g by ~10°C. Similar dependence was detected for all copolymers with aromatic fragments in dimethylsiloxane backbone.

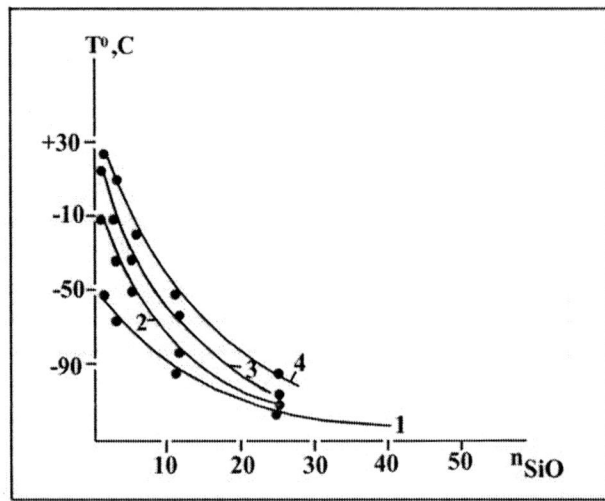

Figure 12. Dependence of T_g on the number of dimethylsiloxane units for cyclolinear copolymers: 1 – copolymers with 1,5-disposition of dimethyltetraphenylcyclotetrasiloxane fragments; 2 – copolymers with 1,5-disposition of hexaphenylcyclotetrasiloxane fragments; 3 - copolymers with 1,5-disposition of dimethylhexaphenylcyclopentasiloxane fragments; 4 - copolymers with 1,7-disposition of decaphenylcyclohexasiloxane fragments

X-ray investigations were carried out for cyclolinear copolymers. Figure 13 displays copolymers to be amorphous systems. Independently of length of single-unit dimethylsiloxane chain, diffraction patterns display two amorphous halos. It is known that the first amorphous halo, d_1, characterizes the average interchain distance in amorphous polymer [52], whereas d_2 is more complicated and corresponds to both intrachain and interchain or interatomic distances.

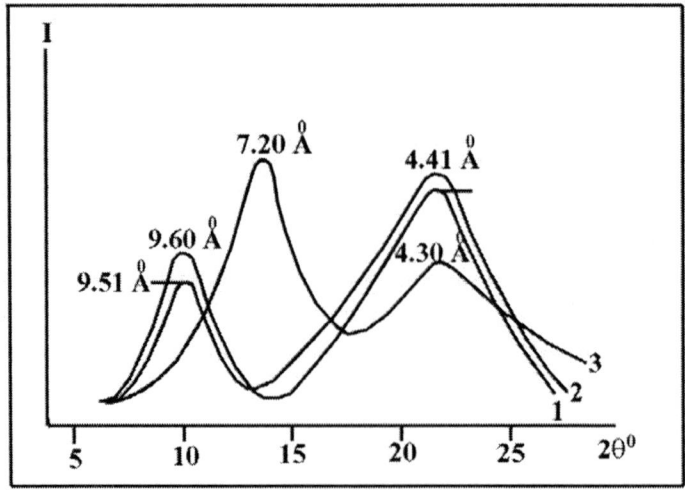

Figure 13. Diffraction patterns for cyclolinear copolymers: 1 – copolymer with 1,5-octaphenylcyclopentasiloxane fragment in dimethylsiloxane backbone (n = 2); 2 – copolymer with 1,5-decaphenylcyclohexasiloxane fragment in dimethylsiloxane backbone (n = 2); 3 – copolymer with 1,5-dimethyloctaphenylcyclohexasiloxane fragment in dimethylsiloxane backbone (n ≈ 51)

For copolymers with cyclopentasiloxane and cyclohexasiloxane fragments in dimethylsiloxane backbone, difference in d_1 values is negligible. Maximal interchain distance ($d_1 \approx 9.60 - 9.54$ Å) is observed at short lengths of dimethylsiloxane chain ($n = 2, 4$), and d_1 is decreased to 7.20 Å, i.e. to PDMS interchain distance, with increase of the length, n.

It is known from the literature that occurrence of the mesomorphous state in cyclolinear siloxane polymers (CLSP) is mainly determined by the size of cyclosiloxane fragment, i.e. by the number of diorganosiloxane groups in the structure of cyclosiloxane fragment, and its symmetrical disposition in the polymeric backbone [53], molecular mass [54] and tacticity of the polymeric backbone [55]. Mesomorphous state of CLSP occurs at sufficient length of symmetrical alkyl substituents (R = C_nH_{2n+1}, $n \geq 2$) irrespective of the backbone structure. Cune segment increase [56] and considerable increase of thermal stability of the mesophase for cyclolinear atactic hexaalkylcyclotetrasiloxanes (ACTS) [57, 58] happen in consequence of minimal increase of chain rigidity at transition from linear polyorganosiloxanes to atactic polyhexaorganocyclotetrasiloxanes. Atactic methyl-substituted ACTS, as well as its linear analogue (PDMS) does not possess mesomorphous properties [57 – 59]. Further increase of polyorganosiloxane chain rigidity by increasing the fracture of trans-units in ACTS macromolecules induces qualitative changes: occurrence of mesomorphous properties for methyl-substituted ACTS [59 – 61].

For the purpose of broadening conceptualization of mesomorphous properties display in polymers with tetrasiloxane rings in the backbone [62] copolymers derived from ACTS with alkyl substituents in the ring and different length of dialkylsiloxane junction were synthesized by HFC reaction between 1,5-dihydroxyhexaalkylcyclotetrasiloxanes and α,ω–dichloropolydialkylsiloxanes in the presence of hydrogen chloride acceptors in various solvents in accordance with the scheme as follows:

(6)

where R = Me, Et, Pr; $m = 1, 2, 3$. α,ω–Dichloropolydialkylsiloxanes were synthesized by partial hydrolysis of diethyldichlorosilanes and di-n-propyldichlorosilanes. Physical and chemical parameters of initial polydialkylsiloxanes and abbreviations and characteristics of synthesized copolymer are shown in Tables 10 and 11, respectively.

Table 10. Physical and chemical parameters of α,ω–dichloropolydialkylsiloxanes, structural formula $Cl(SiR_2O)_{m-1}SiR_2Cl$

No	R	m	T_{boil}, °C P (Pa)	n_D^{20}	^{29}Si chemical shifts, m.p.
1	Et	2	215-217 (10^5)	1.4370	8.40
2	Et	3	139-141 (700-900)	1.4375	5.57 -17.20
3	н-Pr	2	100-102 (133)	1.4420	5.87
4	н-Pr	3	164-166 (133)	1.4412	2.92 19.32

Table 11. Physical and chemical characteristics of PACS

No	Copolymer	R	m	$[\eta]$*, дл/g	Yield %	Cis : trans units relation in copolymer	^{29}Si chemical shifts, $\delta_{m.p.}$ (I_{rel})** D_c	D_l	T
1	MCTS***	Me	0	0.17	80	88:12	-19.27 -19.31	-	-65.74 (0.11) -65.53 (0.78) -65.77 (0.10)
2	MCTSS-1	Me	1	0.15	81	93:7	-19.51 (0.83) -19.58 (0.17)	-21.15 (0.03) -21.23 (0.26) -21.33 (0.70)	-65.12 (0.85) -65.30 (0.15)
3	MCTSS-2	Me	2	0.16	75	95:5	-19.64	-21.56	-65.30
4	MCTSS-3	Me	3	0.13	70	95:5	-19.67	-21.64	-65.39
5	ECTS***	Et	0	0.26	-	70:30	-19.66 -19.65	-	-66.61 -66.65 -66.97 -66.99
7	ECTS-1	Et	1	0.08	75	70:30	-20.20 (0.65) -20.23 (0.35)	-21.92 (0.10) -21.95 (0.40) -21.95 (0.50)	-66.63 (0.70) -66.89 (0.30)
8	ECTS-2	Et	2	0.09	80	70:30	-20.16	-22.07	-66.71 (0.70) -67.01 (0.30)
9	ECTS-3	Et	3	0.09	47	70:30	-20.28	-22.23 (0.33) -22.37 (0.67)	-66.91 (0.70) -67.17 (0.30)
10	PCTS***	n-Pr	0	0/23	60	70:30	-21.86 (0.85) -21.98 (0.15)	-	-68.14 (0.15) -68.33 (0.85)

Notes: * In toluene at 25°C.
** Dc and Dl are chemical shifts for R2Si groups in the structure of rings (cyclic fragments) and linear fragments, respectively; T is chemical shift for RSiO1.5 groups
*** Corresponded polyhexaalkylcyclotetrasiloxane copolymers are shown for a comparison

It is found that cyclolinear hexapropylcyclotetrasiloxane copolymers with di-*n*-propylsiloxane units are not formed under current reaction conditions. In accordance with the results of gas-liquid chromatography studies, primary compounds enter the reaction completely, but reaction conditions (temperature and diluter concentration) do not provide for sufficient chain propagation.

All synthesized ACTS copolymers are completely soluble in usual organic solvents. The structure of copolymers was proved by data of ultimate analysis, ESR and ^{29}Si NMR spectra, chemical shifts of which are shown in Table 11.

ACTS copolymers were studied by differential scanning calorimetry (DSC) method. Figure 14 shows DSC curves for copolymers, which indicate that MCTSS-1 and MCTSS-3 are incapable of regulation. MCTSS-2 crystallizes and transits into an isotropic melt above $T_{melt} = +5°C$. As concerns ECTSS-1 and ECTSS-2, they are in the mesomorphous state below the isotropization temperature, T_{iso}, transiting into mesomorphous glass below glass transition temperature, T_g.

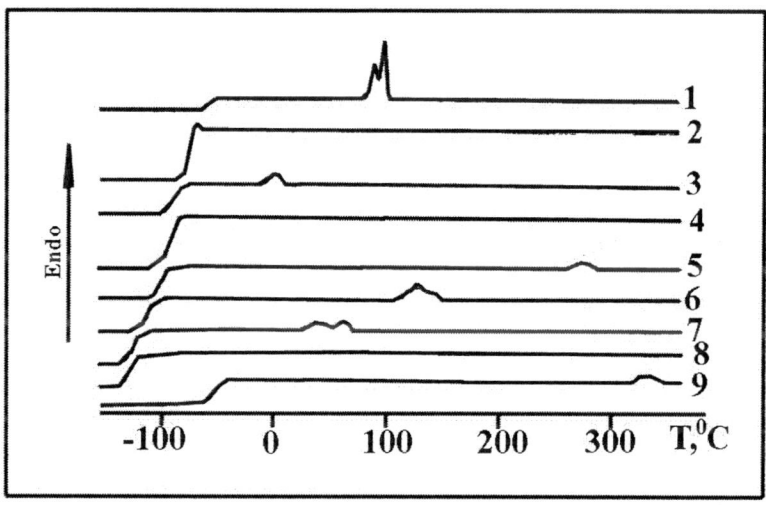

Figure 14. DSC curves for MCTS (1), MCTSS-1 (2), MCTSS-2 (3), MCTSS-3 (4), ECTS (5), ECTSS-1 (6), ECTSS-2 (7), ECTSS-3 (8), and PCTS (9)

Flexible junction length increase, *m*, decreases thermal stability of the mesophase, and in the case of *m* = 3 (ECTSS-3), to suppression of copolymer ability to transit into the mesophase, i.e. ability of copolymer to long-range intermolecular regularity is completely suppressed due to the loss of ability to intramolecular regulation. However, sufficient length of flexible junction and considerable intermolecular overlapping make possible occurrence of close regularity between aromatic fragments both at intramolecular and intermolecular levels. This conclusion is based on occurrence of amorphous dispersion in the angular area, close to $2\theta_{ll}$ value for polyethylcyclotetrasiloxane, on diffraction patterns of ECTSS-3 at low temperatures. To put it differently, a tendency to close regulation by type of nematic diskophase, N_D, is observed for ECTSS-3 [63].

MCTSS-1 and MCTSS-3 diffraction patterns, obtained in the temperature range of -130 - +40°C, and MCTSS-2 one, obtained above +25°C, are qualitatively identical to the diffraction pattern of homopolymeric atactic MCTS. They contain two amorphous halos: the first in intensive one in the angular area of $2\theta = 9 - 12$ ($1/2\Delta = 1.2$) with the maximum at $2\theta_1$ and the second weak diffuse one in the area of $2\theta = 16 - 25$. It is common knowledge [64] that for polyorganosiloxanes $2\theta_1$ value characterizes the average intermolecular distance in amorphous component or isotropic melt. It is shown that interplane distances, d_m, in copolymers decreases with increase of flexible junction length, m, and increases with the length of side framing, n. As is introduced and the length of flexible junction increases, angular location of the maximum of the first amorphous halo ($2\theta_1$) is shifted to the area of larger angles, and the glass transition temperature (T_g) of MCTSS decreases, approaching T_g of its linear analogue (PDMS) at $m = 3$. This means that decrease of polymeric chain rigidity with m increase proceeds simultaneously with decrease of average intermolecular distances.

As suggested in ref. [62], tendency of ACTS copolymers to regulation is displayed on two interconnected levels: intramolecular and intermolecular. On the one hand, cylindrical fragments tend to realize a conformation with the shortest intramolecular distances. On the other hand, this conformation provides for a disposition of side radicals, which may induce rather strong interaction between them to affect intramolecular regularity and stabilize a definite revolute configuration. Moreover, if isomers of organic polymers are stabilized by van-der-Waals detachment of valence unbonded atoms or radicals only, in the case of considered compounds, this mechanism is added by a factor, stipulated by the following feature ACTS copolymer macromolecules: different nature of the backbone and side framing. Similar to linear analogues [65 – 70], side radicals in ACTS copolymers will tend to form organic cover for coiled siloxane chain.

Thus, lack of anisodiametry of the aromatic fragment is compensated by two interconnected factors – configuration features of the backbone and dephility of the backbone and side framing. These factors promote for formation and stabilization of configurative trans-tactic isomers, large sequences of which may be considered as peculiar conformational mesogens.

For previously synthesized polymethylcyclosiloxanes with aromatic pentasiloxane and heptasiloxane fragments, no mesophase was observed. However, it should be noted that cyclolinear polyorganocyclotetrasiloxanes with aliphatic substituents form a mesomorphous state, substituent length increase fro methyl to propyl fragment inducing mesophase isotropization temperature increase, and at sufficient molecular mass polyhexa-*n*-propylcyclotetrasiloxane is present in the mesophase in the range from glass transition to decomposition temperature transiting into the mesophase glass below the glass transition temperature [57].

These results gave grounds [72] to suggest that introduction of aliphatic substituents (ethyl, propyl, etc.) into polyorganosiloxanes with aromatic pentasiloxane fragments in the linear backbone promotes mesophase formation and increases thermal stability. For this purpose, HFC reaction of 1,5-dichlorooctaethylcyclopenta siloxane with 1,3-dihydroxytetraethyldisiloxane in the presence of pyridine, proceeding in argon flow in the temperature range from 25 to 50° in accordance with the scheme shown below, was studied:

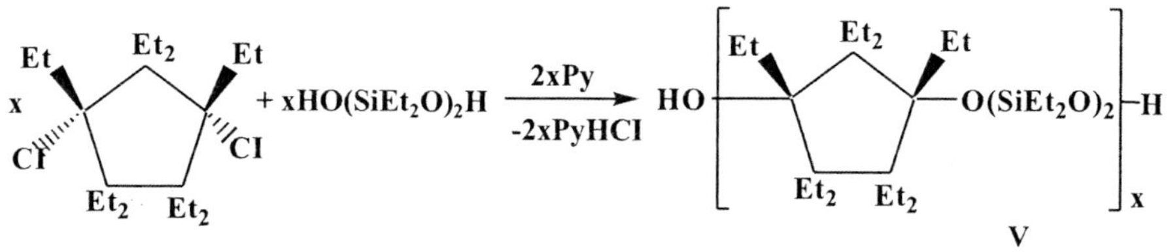

(7)

After precipitation, polymer with $[\eta] \approx 0.07$ with 36% yield was obtained. The structure of synthesized polymer was detected based on spectral data.

^{29}Si NMR spectra of polymer solution display D^{Et} (diethylsiloxane group) and T^{Et} (ethylsesquioxane group) signals: signals in the D-range at -22.39, -22.47, -22.52, and -22.56 m.p. and 3 : 4 : 2 : 1 ratio and in the T-range at -68.50 and -68.56 m.p. and 1 : 5 ratio. Complexity of these signals is stipulated by spatial nonequivalence and for D^{Et}, in addition, by positional nonequivalence of silicon atoms.

X-ray and calorimetric studies of the polymer shows T_g = -127°C, i.e. introduction of a flexible junction decreases T_g compared with polyhexa-n-propylcyclotetrasiloxane bead-shaped polymer [57] and approaches it to T_g of linear polydiethylsiloxane (PDES) [73]. Above T_g, diffraction patterns of the polymer display a single sharp peak at 50°C: 2θ= 10.13° ($\Delta \sim 30'$). Dispersion remains unchanged up to isotropization at 35°C. The data obtained correlate with the results of DSC studies and thermal analysis. Taking into consideration temperature dependencies of angular dispositions of observed reflexes and atactic structure of the backbone, one may suggest that for synthesized polymer long-range positional order in the packing of gravity centers of macromolecules.

The effect of methylcyclosiloxane fragments in dimethylsiloxane backbone on properties of copolymers and formation of the mesophase state was studied [74] by HFC reaction between 1,7-dichloro(dihydroxy)-1,7-diorganooctamethyl cyclohexasiloxane and α,ω–dihydroxy(dichlor)dimethylsiloxanes in the equimolar ratio of initial components in the presence of amines. The reaction proceeds as follows:

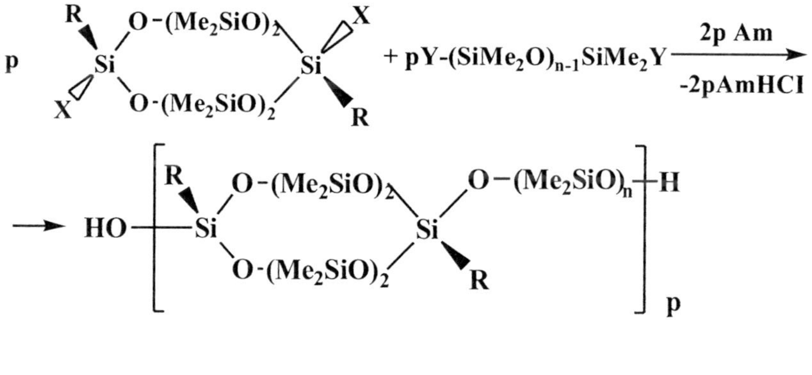

(8)

where R = Me, Ph; n = 0, 1, 2, 3.

In synthesized cyclolinear siloxane polymers (CLSP) the dimethylsiloxane unit was taken as a flexible spacer [74]. Table 10 shows temperatures and heat transitions in relation to the flexible spacers' lengths. These polymers were found to exist in the crystalline and thermotropic states, like the trans-tactic polydecaorganocyclohexasiloxanes, but temperature range of the mesomorphous state is considerably narrower; n increase to 3 gives completely amorphous polymers. This sharp decrease of isotropization temperature in CLSP with flexible spacers is not unexpected, since addition of dimethylsiloxane spacer instead of the hydrocarbon one has similarly led to a sharp decrease of isotropization temperature in liquid-crystalline polymers with classic mesogenic groups [75]. The authors studied the structure of decaorganocyclohexasiloxane polymer until with the help of Stuart-Briegleb models and compared reflex variation with temperature at $2\theta = 9 - 11°$. Figure 15 shows that CLSP unit resembles a disc with $d_1 = 11.5$ Å and height $h = 7.8$ Å assembled into a polymeric chain from disc-shaped thin plates comparable by thickness to interplanar distance $d = 8.4$ Å, calculated from the experimental maximum position in Figure 16.

Existence of the long-range order in the packing with the lower limit of coherence area from $0.5h \geq 400$ Å indicates that the packing shown in Figure 3.15 only is possible with these intermolecular distances.

Table 10. Temperatures and heat transitions of cyclolinear polymethylphenylsiloxanes with flexible spacers

No	Backbone	R	n_{SiO}	$[\eta]$ in toluene dl/g	T_g, °C	T_m, J/g	H_m, K?	T_{iso}, °C	H_i, J/g
1		Me	0	0.15	-91	-53	5.5	49	10.0
2			1	0.13	-88	-7--12	17.6	67-97	3.6
3			2	0.14	-91	-	-	7-57	1.3
4			3	0.15	-90	-	-	-	-
5		Ph	0	0.20	-41	187	13.8	427	3.0
6			1	0.19	-59	82-107	16.8	137-167	1.3
7			2	0.15	-67	52	14.7	-	-
8			3	0.14	-80	-	-	-	-

It should be noted that distances of close values are realized in the ring itself; in this case, however, existence of so high long-range order is impossible, since intra- and intermolecular distances would be alternating: first, they are different and secondly, differ. As a consequence, it is impossible to retain an unchanged long-range order in the wide temperature range of the mesophase. The existence of a single peak only follows from the suggested model. Thus, the order in the basic plane is realized in one direction only – perpendicular to the plane of the ring that is polymer chain unit, which is simultaneously perpendicular to the main axis of the polymer macromolecule.

Compared data shown in Figure 15 with d values from the diffraction pattern in Figure 16 indicate (Figure 16a) that the mesomorphic component (a narrow reflection) d is independent of the polydecamethylcyclohexasiloxane chain tacticity. Hence, irrespective of the trans-tactic polymer ability for crystallization, the mesophase temperature range changes insignificantly.

At lower temperature of –73°C only, there appears a bend on the temperature dependence curve, $d(T)$, of the mesophase peak.

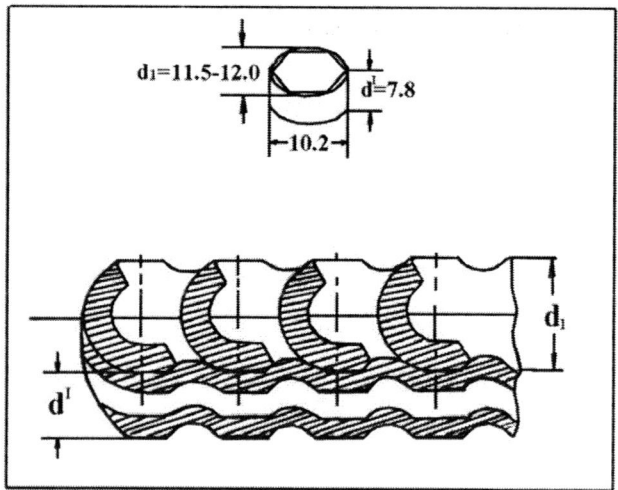

Figure 15. Hypothetical model of decaorganocyclohexasiloxane polymer chain

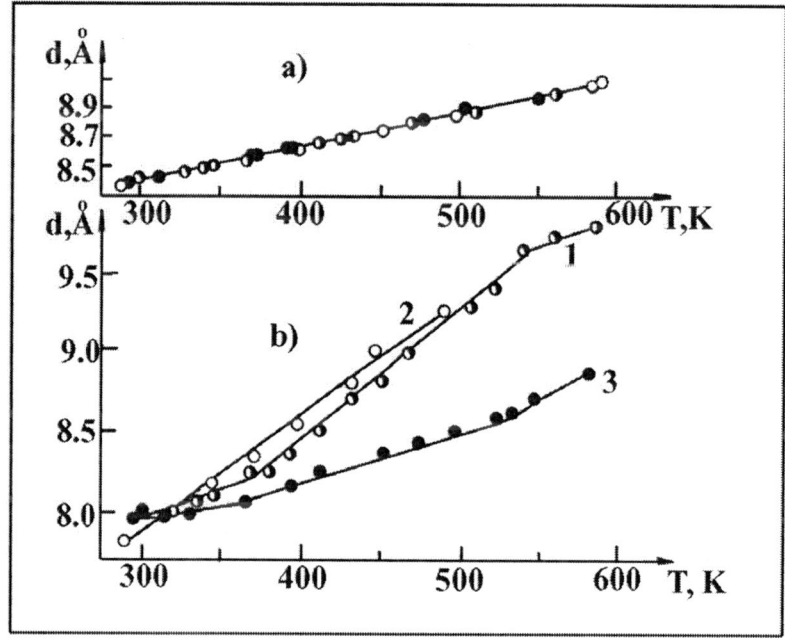

Figure 16. Interplanar distance d as a function of temperature for decamethylcyclohexasiloxane polymer: a – for mesomorphic component; b – for amorphous component, depending on the polymer chain tacticity: 1 – atactic, 2 – microtactic, 3 – trans-microtactic

Varyation of $d(T)$ dependence at the bend is accompanied with a considerable increase of the narrow reflection integral half-width and decrease of the integral intensity. In this temperature range, the interplanar distance, d, is comparable with the disc –

decamethylcyclohexasiloxane unit – height (Figure 15) as a consequence of close packing, leading to decreased long-range order areas.

Using the methods of amorphous halo separation from the narrow reflection, the dependence of $d(T)$ position are presented in Figure 16b for different tacticity polydecamethylcyclohexasiloxanes. It follows from Figure 16b that tacticity variation affects the values and shape of $d(T)$ dependence of amorphous halo. At 20°C, the difference of the studied samples in d positions is insignificant, but at 197 - 247°C the difference of amorphous halo approaches 1 Å, approximately. The bend observed at the beginning of active melting of the mesophase is indicated by abrupt intensity drop of the mesophase peak.

Thus, as follows from the analysis of $d(T)$ dependence of amorphous dispersion, the polymer chain tacticity breakdown causes changes of packing in two directions only, where the long-range order is not realized, but the interplanar spacing is unaffected.

As follows from $d(T)$ dependence (Figure 16) of the mesophase peak and amorphous halo, introduction of phenyl substituents to organosilsequioxane fragments in the organocyclohexasiloxane unit leads to the differences depending on the polymer chain tacticity, not only in the amorphous dispersion, but also in the mesophase peak position.

Thus, the above data show the main factors responsible for mesomorphous state formation, which are: first, the increased conformational parameters and Kuhn segment (A) of CLSP as a consequence of certain conformation of organocyclosiloxane unit, most likely, the folded ring conformation or any other ensuring, approximately, co-planar conformation of these cyclosiloxanes; secondly, inclusion of organocyclosiloxane unit with symmetrical position into the polymer chain by para-position type; thirdly, the use of organic substituents ensuring amplification of both intra- and intermolecular interactions simultaneously capable of increasing the rigidity of the backbone.

The factors responsible for the mesomorphous state temperature range are: first, the polymer chain tacticity, perfect for the mesophase emerging in CLSP is the trans-tactic structure enabling to achieve the highest isotropization temperatures at low degrees of polymerization; secondly, the growth of CLSP molecular weight leading, irrespective of the CLSP unit structure, to extended mesomorphous state temperature range in CLSP.

Thermal decomposition of branched chain methylsiloxane polymers has been studied before [5, 76-79]. Methylsiloxane cycles and polycycles were detected as volatile products of the methylsiloxane resin pyrolysis [5, 76-79].

The results indicate that decomposition of methylsiloxane resins by pyrolysis is similar to that observed for methylsiloxane chain polymers [4]. Relation between the polymer structure and thermal degradation products can be interpreted in terms of this decomposition mechanism, but a resin of irregular structure does not provide proper system for studying this relation.

Thermal decomposition products [80] of two methylsiloxane polymers of strictly regular and known structure were examined to clarify the relations mentioned. The polymers investigated include di- and tri-functional structural units (i.e. $(CH_3)_2SiO_{2/2}$ denoted as -D- and $CH_3SiO_{3/2}$ denoted by -T<) and involve cyclolinear chain structures, as follows:

$(T_2D_6)_n$:

$$-\left(T\begin{smallmatrix}D\\ \diamond\\ D\end{smallmatrix}T-D-D-D-D\right)_n-$$

and

$(T_2D_9)_n$:

$$-\left(T\begin{smallmatrix}D\\ \diamond\\ D\end{smallmatrix}T-D-D-D-D-D-D-D\right)_n-$$

These structures are ensured by the method of synthesis used.

Thermal decomposition of investigated cyclolinear methylsiloxane polymers at 350°C gives cyclic methylsiloxane oligomers. Generally, the polymer $(T_2D_6)_n$ decomposes to D_3, and polymer $(T_2D_9)_n$ to D_3, D_4, D_5 etc., similar to poly(dimethylsiloxane), but the degree of conversion is very low at this temperature. At higher temperatures several polycyclic compounds are also produced from both polymers. Figure 17 shows relative amounts of products, produced by pyrolysis at 550°C due to GLC results. The gas chromatographic peak areas are related to the peak area of the most abundant product. The product distribution in the temperature range of 450 - 600°C does not vary significantly with pyrolysis duration or pyrolysis temperature. The author identified ten of the fifteen main products by comparing gas chromatographic retention times with these of pure standard substances on three different GC stationary phases. Polycyclic siloxane standards were obtained from methylsiloxane resin by pyrolysis followed by preparative gas chromatography.

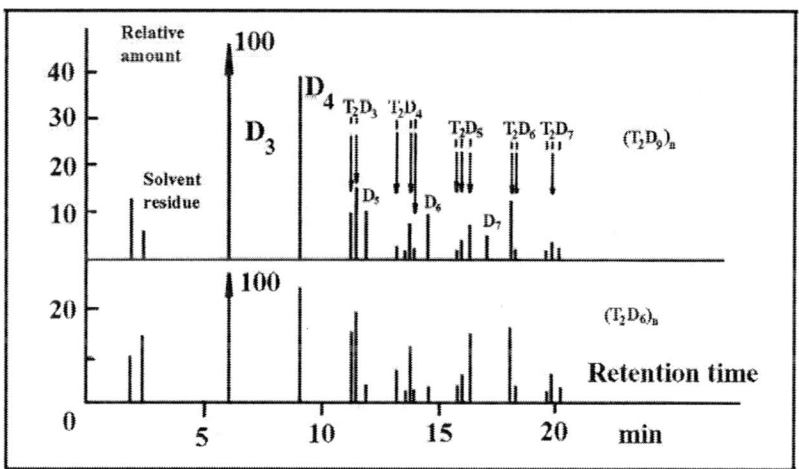

Figure 17. Cyclolinear methylsiloxane polymers' pyrolysis products separated in OV-1 gas chromatographic phase

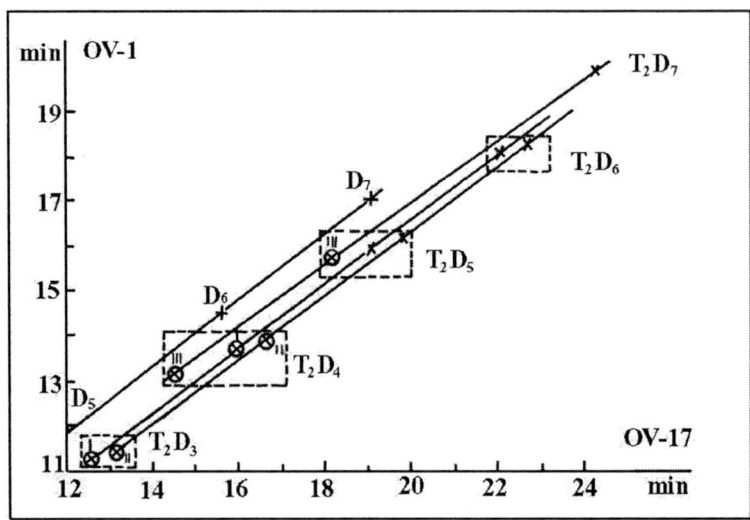

Figure 18. A plot of retention time against retention time for pyrolysis products on temperature programmed columns coated with OV-1 and OV-17 phase, respectively

Their identities were proved by mass spectrometry, ^{29}Si NMR spectroscopy and X-ray diffraction patterns [76-79]. Standards are not available for some important products, but these seem to be homologous with the products identified. This is indicated by the straight lines which join the points given by the retention time values of the products measured on two different stationary phases with a linear temperature program (Figures 18 and 19). The identified products are represented by circled points in the Figures.

On examining the structures and amounts of the pyrolysis products we can see some correlation between the polymer structure and the pyrolysis products.

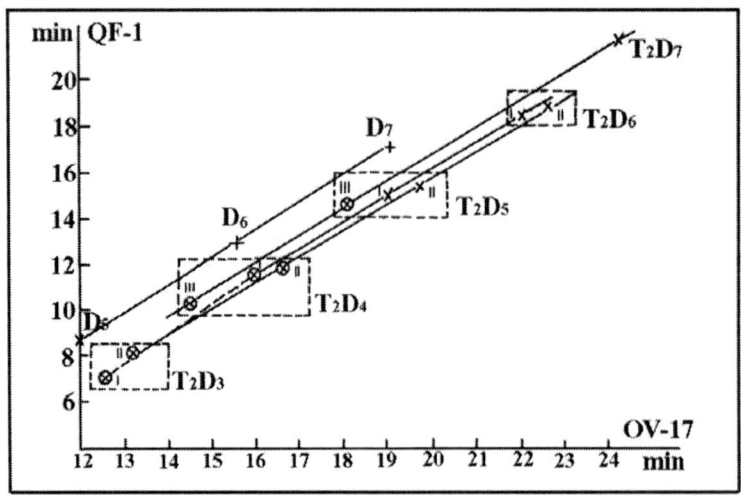

Figure 19. A plot of retention time against retention time for the pyrolysis products on temperature programmed columns coated with QF-1 and OV-17 phase, respectively

Table 11. Homologous series of pyrolysis of cyclolinear methylsiloxane copolymers

Composition	Series I	Series II	Series III
T_2D_3	(structure)	(structure)	–
T_2D_4	(structure)	(structure)	(structure)
T_2D_5	(structure)	(structure)	(structure)
T_2D_6	(structure)	(structure)	(structure)
T_2D_7	–	–	(structure)

Pyrolysis products are either cyclic or bicyclic methylsiloxanes ranging from 3 to 9 siloxane units. Pyrolizate of the polymer $(T_2D_9)_n$ displays a high quantity of cyclic products than in that of $(T_2D_6)_n$. The relative quantities of cyclic products D_3, D_4 and D_5 in the pyrogram of $(T_2D_9)_n$ are similar to these found in the pyrogram of poly(dimethylsiloxane), but a smaller quantity of D_4 and D_5 from $(T_2D_6)_n$ was observed.

All the important pyrolysis products belong to a homologous series of siloxane units (see Fig. 18 and 19). Three such series have been found for bicycles shown in Table 11. None of the products include more than two T units.

Quantitative evaluation of the total amounts of T and D units, found in the pyrolyzate, gives a T/D ratio corresponded to that in the original polymer. For the conversion of peak area values into weights authors used reflex factors for the Flame Ionization Detector [81].

Structures of the volatile degradation products show that all of these molecules were detached from the cyclolinear polymer by the mechanism first suggested by Thomas [4] for poly(dimethylsiloxane). In this mechanism, degradation starts with the formation of an intermediate four-centered structure, and then two siloxane bonds involved are rearrange. The resulting volatile product is an oligomer cycle in the case of the poly(dimethylsiloxane) chain. Similarly, D_3, D_4, D_5 cyclic oligomers are detached from the linear part of investigated cyclolinear methylsiloxanes. When the siloxane bond rearrangement takes place between two linear segments, bonded to the same cycle of the polymer, a bicycle of the Series 1 can be detached as follows:

In T and D symbols:

$$\text{(diagram)} \quad (9)$$

may result from such inter- or intramolecular bicycles of Series II may form similarly, but the macromolecular segment, in which the siloxane bond rearrangement is taking place has to include a three-membered siloxane cycle:

$$\left(\cdots T\underset{D}{-}T-\cdots\right)$$

This modified molecular segment may result from such an inter- or intramolecular rearrangement, as follows:

$$\text{(diagram)} \quad (10)$$

As a consequence, Series II products are assumed to be the products of three consecutive rearrangement steps.

Distinct bicycles were also found among the pyrolysis products. Their formation can be described in terms of bond rearrangement between the linear part and the adjacent cycle of the polymer, as follows:

$$\text{(diagram)} \quad (11)$$

In reaction 3 two cleavages occur, simultaneously. The absence of products with more than two T units indicates that none secondary reactions of the primary products take place.

Under the pyrolysis conditions, the authors used stoichiometric composition of total pyrolyzate similar to that of the polymer. In addition, no change in functionality of siloxane units takes place confirming with the authors' earlier result [77].

Accordingly, the structure of degradation products is closely connected with the macromolecular one. The original linear segments of the macromolecule appear in the pyrolyzate as cyclic oligomers, and the branching points are detached with adjacent fragments as polycycles. The ratio of monocyclic and bicyclic products' quantities characterizes the average frequency of branching points in the polymer network. The exact description of structures of bicyclic products may help in distinguishing distribution of the branching points, and thus the microstructure of methylsiloxanes.

The above discussion indicates that synthesized CLSP with phenyl framing groups are characterized by high thermal oxidative stability and easy processing to articles. That is why these substances were suggested for production of specific rubbers [12].

For the purpose of increasing selectivity of immovable phase for complete and selective separation of organic compounds and complex mixtures of organosilicon compounds by gas-liquid chromatography method, CLSP of the following general formula was suggested [82]:

$$\text{HO} \left[\begin{array}{c} \text{Me} \quad \text{Ph} \quad \text{Ph} \\ \diagdown \diagup \\ \diagup \diagdown \\ \text{Ph} \quad \text{Ph} \quad \text{Me} \end{array} \text{O-(SiMe}_2\text{O})_n \right]_x \text{H}$$

I

where $n = 3 - 20$; $x = 10 - 100$.

As the immovable phase, CLSP was applied on N-AW chromaton carrier in 20% quantity of its mass. The column used was 2 m long and 2 mm of internal diameter. Figure 20 shows chromatogram of difficultly separable compound separation: cyclohexanol (1), cyclohexylamine (2), phenol (3), and aniline (4) at 100°C. Moreover, the possibility for suggested phase to separate compounds from the naphthene sequence from compounds of standard structure is illustrated on the example of hexane (5) separation from cyclohexane (6) (Figure 20).

The phase peculiarity to alcohols of the fatty sequence is shown on the example of synthetic alcohols $C_{10} - C_{21}$ fraction separation in the temperature range from 100 to 270°C (at 4 deg/min rate) without preliminary transfer to volatile derivatives.

In spite of the known phases, the advantage of suggested selective immovable phase is the possibility of selective analysis of various classes of organic and organosilicon compounds. Of special attention should be peculiarity of the phase to separation of aromatic and naphthene hydrocarbons, amines and alcohols with preservation of high thermal stability. Table 12 shows comparative data on Mac-Reynolds constants for a series of universal non-polar phases and high parameters of suggested phase.

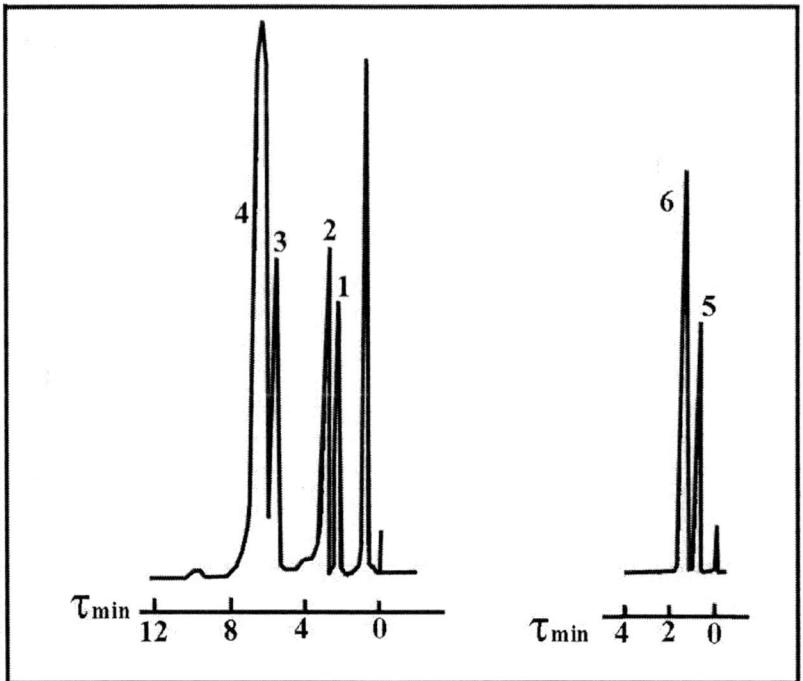

Figure 20. Chromatographic curves for compounds

Table 12. Comparative data on Mac-Reynolds constants

Immovable phase	Mac-Reynolds constants				
	X'	Y'	Z'	I'	S'
SE-30	15	53	44	64	41
SKT	17	57	46	67	45
SE-54	33	72	66	99	67
ДФОК	86	130	134	181	154
Suggested CLSP naphthene phase ($n = 3$)	95	135	120	185	145

Note also that in spite of suggested phase, considered immovable phases give no opportunity to separate many classes of chemical compounds. In particular, amines and alcohols can be separated on these immovable phases, if transferred to volatile derivatives only; aromatic and naphthene amines and alcohols cannot be separated, and complete separation of complex mixtures of organosilicon compounds cannot be performed.

2. CYCLOLINEAR COPOLYMERS WITH CYCLIC CARBOSILOXANE FRAGM,ENTS IN DIMETHYLSILOXANE BACKBONE

Cyclolinear copolymers containing cyclic carbosiloxane fragments in dimethylsiloxane backbone were synthesized by HFC reaction of dichlorine-containing organocyclocarbosiloxanes (described in Chapter 1) with α,ω–dihydroxymethylsiloxanes.

Copolymers were synthesized in 60 – 70% solution of anhydrous toluene both in the presence and in the absence of pyridine (hydrogen chloride acceptor) at 20 - 25°C. It is found that at HFC proceeding without acceptor, increase of dimethylsiloxane unit length and volume of the cyclic fragment induce conversion decrease by hydrogen chloride. In both cases, the reaction proceeds by the scheme as follows [14, 83]:

$$(12)$$

where $m = 1$ (VIII), 2 (IX); $n = 2, 4, 10, 20, 37$.

When precipitated from toluene solution with methanol, synthesized copolymers are solid or viscous substances with $\eta_{spec} \approx 0.07 - 0.30$, well soluble in usual organic diluters, with regard to dimethylsiloxane unit value.

Comparatively reduced yield at low lengths of dimethylsiloxane unit ($n = 2, 4$) in reaction (12) is explained by partial proceeding of HFC reaction by intramolecular cyclization mechanism with formation of polycyclic structures. Similar to HFC reaction (1) a product of intramolecular condensation was extracted from the mother solution and additionally derived by direct synthesis. Table 14 shows inherent viscosity, molecular mass and some physical and chemical properties for copolymer with structure VIII ($m = 1, n = 2$), determined in benzene solution. Table 13 shows yield, T_g and viscosity parameters of cyclolinear carbosiloxane copolymers.

Table 13. Physical and chemical properties for copolymer with structure VIII ($m = 1, n = 2$)

$[\eta]$, dl/g	Diffusion coefficient (D), cm^2/s	Sedimentation parameter, S	Specific volume (γ), cm^3/g	Refraction index increment (dn/dc), cm^3/g	M_{SD}
0.07	28×10^{-7}	2.6	0.87	0.28	16,500

Table 14. Physical and chemical properties of cyclolinear carbosiloxane copolymers with structures VIII and IX

No	Copolymer	m	n_{SiO}	Yield, %	η_{spec} toluene at 25°C	T_g, °C	d_1, Å
1		1	2	78	0.08	-25	8.82
2		1	4	79	0.11	-60	-
3		1	10	86	0.18	-90	-
4		1	20	87	0.25	-105	7.24
5	HO—Si(R)(C₂H₄)—Si(R)—O—(SiMe₂O)ₙ—H ; O—Si(Ph₂)—O]ₘ)ₓ	1	37	90	0.29	-123	7.21
6		2	2	77	0.09	-10	9.80
7		2	4	80	0.12	-40	-
8		2	10	85	0.18	-75	-
9		2	20	87	0.26	-90	7.25
10		2	37	89	0.30	-123	7/24

Thermogravimetric studies of cyclolinear copolymers have displayed their sufficiently high thermal oxidative stability. For short lengths of dimethylsiloxane unit, initial mass loss is observed at 260 – 280°C and 5% loss at 300°C. Above 750°C no mass loss variation is observed, which is apparently associated with formation of secondary structures. Thermal oxidative stability of copolymers with carbocyclosiloxane fragments in the backbone is similar to that of their pure siloxane analogues. Moreover, thermal oxidative stability of copolymers is decreased with increase of the volume of cyclic carbosiloxane fragment.

Figure 21 shows dependence of the glass transition temperature of copolymers on cyclic carbosiloxane fragment content in dimethylsiloxane backbone. Similar to copolymers from poly(dimethylphenylsiloxane) sequence [53], the effect of non-dimethylsiloxane units on T_g of copolymers with cyclic carbosiloxane fragments in the backbone is primarily observed at ~3% concentration of the latter. The Figure shows also that the effect of cyclic carbosiloxane fragments with $m = 2$ on T_g of studied copolymers is higher rather than for copolymers with $m = 1$; T_g of copolymers is increased with mol% concentration of cyclic carbosiloxane fragments.

Diffraction patterns of copolymers with cyclic carbosiloxane fragments in the backbone are characterized by the presence of two diffraction maximums typical of amorphous polymers. Data in Table 15 show that interchain distances, d_1, decrease with the increase of dimethylsiloxane backbone length and at $n \approx 37$ reach the value typical of PDMS. Simultaneously, the increase of cyclic fragment volume in copolymers induces a considerable raise of d_1.

One more interesting feature of these copolymers is that initial bifunctional cyclic organocarbosiloxanes can be derived easily, and copolymers themselves contain reactive cyclic carbosiloxane fragments capable of cross-linking without extracting volatile compounds [84].

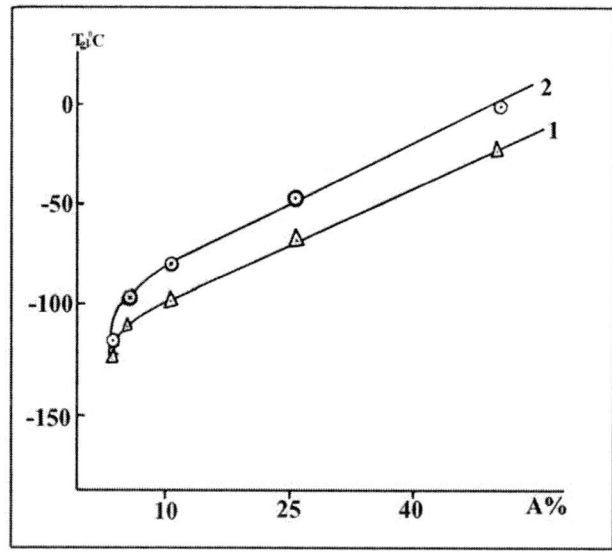

Figure 21. Dependence of glass transition temperature of copolymers on mol% concentration of cyclic carbosiloxane (A) fragments: 1 – copolymers with $m = 1$; 2 – copolymers $m = 2$

3. ORGANOSILOXANE COPOLYMERS WITH HETEROCYCLIC FRAGMENTS IN DIMETHYLSILOXANE BACKBONE

It has been shown in the literature [85] that introduction of arylene fragments into the dimethylsiloxane backbone increases thermal oxidative stability of copolymers. For the purpose of synthesizing and studying properties of copolymers with 1,3-diorgano-1,3-disila-2-oxaindane fragments in the backbone [85, 86], HFC reaction between 1,3-dichloro-1,3-diphenyl(dimethyl)-1,3-disila-2-oxaindane with α,ω–dihydroxydimethylsiloxanes, dihydroxydiphenylsilane and 1,4-bis(hydroxydimethylsilyl)benzene both in the presence and in the absence of pyridine (hydrogen chloride acceptor) was investigated [87].

Initial 1,3-dichloro-1,3-diphenyl(dimethyl)-1,3-disila-2-oxaindanes were synthesized by high-temperature pyrolysis of 1,3-dichloro-1,3-dimethyl(diphenyl)-1,3-diphenyldisiloxane [88] in the inert medium in accordance with the scheme (1). HFC reaction between 1,3-dichloro-1,3-diphenyl-1,3-disila-2-oxaindane and dihydroxydiphenylsilane or 1,4-bis(hydroxydimethylsilyl)benzene proceeded with different ratios of reagents, which gave dihydroxy-containing oligomers with different transformation level ($m \approx 6 - 18$). Several parameters of these oligomers are shown in Table 14. Further on, they were used for deriving block-copolymers of $(AB)_m C_n$ type.

HFC reaction between 1,3-dichloro-1,3-diorgano-1,3-disila-2-oxaindane and α,ω–dihydroxydimethylsiloxanes proceeded at equimolar ratio of initial components. In both cases, it proceeds in accordance with the scheme as follows:

$$yCl-\underset{R}{\underset{|}{Si}}\overset{O}{\diagup}\underset{R}{\underset{|}{Si}}-Cl + xHO-A-OH \xrightarrow{-HCl} HO\left[-\underset{R}{\underset{|}{Si}}\overset{O}{\diagup}\underset{R}{\underset{|}{Si}}-O-A-O-\right]_m H$$

XI

(13)

where A = -Me$_2$Si-C$_6$H$_4$-SiMe$_2$-: R = Ph, $x = y$, $m \approx 18$ (XI^1); A = (R'$_2$SiO): R = R' = Ph; $x \neq y$; $m \approx 6, 9$ (XI^2); A = (R'$_2$SiO): R = Me, Ph; R' = Me, ($x = y$), $n = 2, 4, 8, 20, 37$ (XI^3).

If the reaction proceeds in 60 – 79% toluene solution at 20 – 25°C in the absence of acceptor, final conversion by hydrogen chloride decreases from 70% ($n = 2$) to 54% ($n = 4$) with increase of the length of dihydroxydimethylsiloxane unit.

Methyl radical substitution by phenyl one at the atom of silicon in silaindane ring also decrease conversion by hydrogen chloride from 62% ($n = 2$) to 43% ($n = 4$). High conversion by hydrogen chloride at low n of dimethylsiloxane backbone can be explained by increased reactivity of ≡Si-Cl bond, which is apparently associated [89] with a decrease of valence angle of ≡Si-O-Si≡ bond in 1,3-disila-2-oxaindane. Lower reactivity of silaindane fragment with phenyl framing at atoms of silicon is explained by both steric effect and inductive influence of framing groups. That is why for synthesizing high-molecular copolymers, HFC reactions were studied in the presence of pyridine. Re=precipitated copolymers are transparent or slightly opalescent products, well soluble in different organic solvents.

Several physical and chemical parameters and yields of synthesized copolymers are shown in Table 15.

Data shown in Table 16 indicate that the yields of copolymers – the products of HFC reaction without acceptor, are rather low. Moreover, experimental values, detected by the ultimate analysis, are somewhat different from theoretical ones. This fact proves that besides basic HFC process, side homofunctional condensation of initial α,ω–dihydroxydimethylsiloxanes proceeds under the reaction conditions, which distorts the ratio and decreases molecular mass. Slight underestimation of yields of copolymers with disilaoxaindane fragments in the backbone at short lengths of dimethylsiloxane unit ($n = 2, 4$) is caused by simultaneous proceeding of intermolecular and intramolecular cyclization during HFC, the latter synthesizing bicyclic compounds. To prove regular structure of obtained copolymers, copolymer 11 (Table 16) was fractioned into four fractions. The ultimate analysis of all four fractions gave values similar to calculated ones. ^{1}H NMR spectrum of the copolymer displays a singlet signal for methyl protons with chemical shift at $\delta \approx 0.15$ m.p., and the ratio of methyl and phenyl protons found coincides well with the calculated one. In turn, this proves the reaction proceeding with formation of copolymers of regular structure. These data coincide with experimental results [90].

Thermogravimetric analysis of copolymers has displayed 5% mass loss at 330 - 360°C. Basic degradation process proceeds in the temperature range of 420 - 550°C, and above 620°C mass change curves become supersaturated. Thermal oxidative stability of copolymers with disilaindane fragments in dimethylsiloxane backbone is higher than for copolymers with cyclic carbosiloxane fragments in the backbone.

Table 15. Physical and chemical parameters and yields of (oligomers) copolymers with 1,3-diorgano-1,3-disila-2-oxaindane fragments in the backbone with structure XI

No	Copolymer (oligomer) unit	R	n (m)	Yield, %	η_{spec} at 25°C	T_g, °C	d_1, Å	$M_\omega \times 10^{-3}$
1	2	3	4	5	6	7	8	9
1	XI¹	Ph	(19)	89	0.21	215-226	10.42	4.75
2	XI²	Ph	(6)	88	0.07	83-88	10.41	2.99
3	XI²	Ph	(9)	85	0.09	91-96	10.42	4.49
4		Me	2	76 (53)	0.14 (0.06)	-43-	8.60	36.4
5		Me	4	79 (56)	0.16 (0.07)	-78-	-	-
6		Me	8	86	0.18	-93	-	-
7		Me	20	90	0.26	-118	-	-
8		Me	37	90	0.31	-123	7.24	144.1
9		Ph	2	78 (53)	0.13 (0.05)	+11	8.81	29.8
10		Ph	4	80 (57)	0.15 (0.06)	-40	-	-
11		Ph	8	82	0.18	-57	-	87.9
12		Ph	20	86	0.29	-87	-	-
13		Ph	37	90	0.39	-123	-	-

Note: Shown in brackets are parameters of copolymers derived in the absence of acceptor

Table 16. Physical and chemical properties of poly(organocarbosiloxane) copolymers of cyclolinear structure

No	Copolymer	R'	R''	[η], dl/g	T_{degr}* of 5% mass loss	Coke residue,(800°C), %	T_g, °C
1		Me	Me	0.08	240	52	-7
2		Me	Ph	0.06	320	45	26
3		Ph	Ph	0.04	370	41	13

Note: * TGA data on polymers processed by heptamethylvinylcyclotetrasiloxane
Copolymer structure was determined from ^{29}Si NMR spectral data

Thermomechanical studies have shown that the effect of disilaindane fragment on dimethylsiloxane backbone is negligible, and T_g of the copolymer, even in the presence of tetramethylsiloxane units, remain below zero at the methyl framing, whereas phenyl framing in copolymer 9 (Table 16) gives $T_g \approx +11°C$. The expansion of linear fragment length decreases T_g down to T_g of PDMS (~ -123°C). Figure 3.22 shows dependence of T_g of copolymers on content of disilaindane fragments in the backbone. Similar to the case of poly(dimethylphenylsiloxane) copolymers [91], T_g is linearly increased with concentration of non-dimethylsiloxane units in them. Note also that copolymers 4 – 8 (Table 16) T_g dependence of disilaindane fragment content in them close to that on content of diphenylsiloxy units in poly(dimethylphenylsiloxane). For copolymers 9 – 13 (Table 16), increase of mol% concentration of disilaindane fragments induces sharper raise of T_g.

X-ray diffraction studies have shown that diffraction patterns of all studied copolymers possessing disilaindane fragments in the backbone displays two diffraction maximums: $d_1 \approx$ 7.24 - 8.81 Å and $d_2 \approx 4.45$ Å, typical of amorphous polymers. Substitution of methyl framing by phenyl one in silaindane ring is accompanied by an increase of interchain dispance. Analysis of diffraction patterns obtained proves the increase of interchain distance in copolymers with concentration of disilaindane fragments in them.

4. POLY(ORGANOCARBOSILOXANES) WITH CYCLIC SILOXANE FRAGMENTS IN THE BACKBONE

Besides HFC reaction of preliminarily prepared cyclic organosiloxanes with functional groups and bifunctional organosilicon compounds, which give an opportunity to preserve cyclic groups in the polymeric backbone, hydride polyaddition is also widely used, which proceeds under soft conditions and does not involve cyclic structures, introduced into the backbone.

For the purpose of synthesis of poly(carbosiloxanes) with cyclic tetrasiloxane fragments in the dimethylsiloxane backbone, hydride polyaddition of divinylorganocyclosiloxane by dihydrodimethylsiloxane was studied [92]. Polymers were synthesized in argon at 1 : 1 molar ratio of initial reagents in the absence of diluter or in inert organic solvent (toluene) at 100 - 110°C. The reaction temperature was selected at the level causing no disclosure of organosiloxane rings. Platinum-hydrochloric acid, double added to the reaction mixture in amount $1 - 1.5 \times 10^{-5}$ g of $H_2PtCl_6 \cdot 6H_2O$ per 1 g of the initial mixture, was used as the catalyst. A half of this amount was added before the reaction start, and the second half 25 – 140 hours after heating beginning.

Platinum-hydrochloric acid was added in the form of 0.01 M solution in tetrahydrofuran. Isopropyl alcohol, used as diluter for $H_2PtCl_6 \cdot 6H_2O$, decreased relative viscosity of synthesized polymers, apparently, due to proceeding of side alkoxylation reaction:

$$\equiv Si\text{-}H + HOC_3H_7 \rightarrow \equiv Si\text{-}O\text{-}C_3H_7 + H_2 \qquad (14)$$

Linear poly(organocarbosiloxanes) with cyclic structures in the backbone were synthesized in accordance with the following scheme [92]:

$$x[Me_2SiO]_2[MeVinSiO]_2 + x\ H(SiMe_2O)_{n-1}SiMe_2H \xrightarrow{H_2PtCl_6}$$

$$\rightarrow \left[-CH_2-CH_2-Si\begin{matrix}Me\\|\\\end{matrix}\begin{matrix}&Me_2\\&Si\\O^{\nearrow}&\,^{\searrow}O\\&\,\\O_{\searrow}&\,^{\nearrow}O\\&Si\\&Me_2\end{matrix}\begin{matrix}Me\\|\\Si\end{matrix}-CH_2CH_2-(SiMe_2O)_{n-1}SiMe_2- \right]_x$$

(15)

where n = 0; 1; 4; 5; 6; 10; 20; 27; 34; 57; 94; 150; 200.

Synthesized polymers represent viscous and highly viscous colorless, transparent liquids, soluble in aromatic hydrocarbons and lower ethers.

The effect of reaction proceeding in inert organic solvent (for example, toluene) on inherent viscosity of derived polymer is negligible. Obtaining of high viscosity values in the presence of solvent requires just longer-term heating up of the reaction mixture.

To the authors' point of view, polymers of such structure, possessing cyclic fragments in the backbone, are of interest due to their high reactivity. For example, these polymers are capable of easy formation of cross-linked structures under the effect of anionic catalysts.

Initial divinylhexamethylcyclotetrasiloxane was synthesized by combined hydrolysis of dimethyldichlorosilane and methylvinyldichlorosilane. Despite the application of efficient rectification columns and analytical chromatograph with preparative add-on device, the attempts of the authors to separate isomeric 1,3- and 1,5-divinylhexamethylcyclotetrasiloxanes, which may be formed in cooperative hydrolysis, have failed. That is why isomeric structural groups as follows (XIII) may also be present in synthesized polymers:

$$\left[\begin{matrix} & Me & Me & \\ & | & | & \\ -CH_2-CH_2-Si & -O- & Si-CH_2CH_2-(SiMe_2O)_{n-1}SiMe_2 \\ & | & | & \\ & O & O & \\ & | & | & \\ & Me_2Si & -O-SiMe_2 & \end{matrix} \right]_x$$

Semiquantitative assessment of the ratio of isomeric 1,3- and 1,5-cyclic structures in synthesized polymers with the help of NMR spectra was performed, which was found 1 : 1.

In a series of processes variation of functional groups' content ($\equiv$Si-H due to IE-spectroscopy data) during reaction proceeding and type of increase of reaction mixture specific viscosity were studied. Maximal viscosities of polymers ($[\eta] \approx 0.17 - 0.97$ dl/g) are reached after 50 – 160 hours of heating and in majority of cases depend on the length of α,ω–dihydropolydimethylsiloxane chain and purity of initial compounds used.

Studies of IR spectra of synthesized poly(organocyclocarbosiloxanes) and preliminary experiments on long-term heating of the mixture of initial hexamethyldivinylcyclotetrasiloxane isomers under polyaddition conditions allow a suggestion that polymers are synthesized due to hydride polyaddition proceeding with preservation of structures of initial compounds, but not polymerization of cyclic hexamethyldivinylcyclotetrasiloxane. The presence of organocyclotetrasiloxane fragments in the structure of synthesized poly(organocyclocarbosiloxanes) may be proved by their transition into non-fusible, insoluble state due to polymerization of organosiloxane cycles existing in the polymer structure. As re-precipitated polymers are heated at 100 - 110°C in the presence of 0.001 – 0.01 wt. % of anionic polymerization catalysts, viscosity is considerably increased and gel is formed. Varying length of alkylenesiloxane bridge between organocyclotetrasiloxane fragments of poly(organocyclocarbosiloxane) backbone, one may change the average distance between cross-link points and, consequently, properties of cross-linked polymers formed.

Hydride polyaddition between 1,5-divinyl-1,5-dimethyl-3,3,7,7-tetraorganocyclotetrasiloxane and methylphenylsilane has been studied [93]. All attempts to separate initial divinylorganocyclotetrasiloxanes into cis- and trans-isomers have failed. Thus, according to NMR data, initial divinylorganocyclotetrasiloxanes represent mixturea of cis- and trans-isomers. The reaction proceeds as follows:

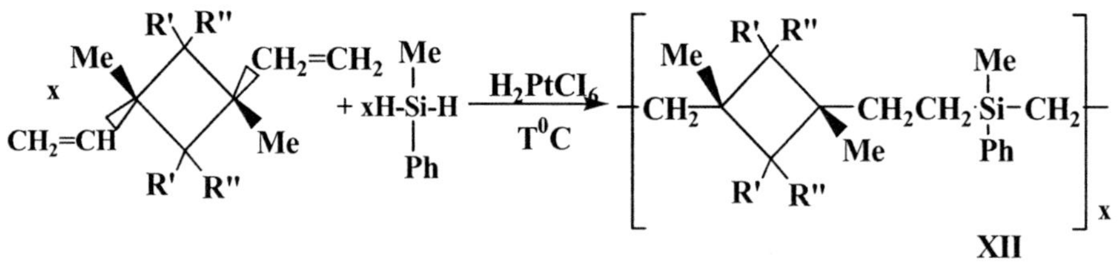

$$\text{(scheme 16)}$$

(16)

where $R' = R'' = Me = Ph$; $R' \neq R''$.

Polyaddition was carried out at 60 – 70°C, and at the final stage the mixture was heated up to 100°C. Catalyst in amount 5×10^{-4} Pt mol/mol was added to vinylcyclosiloxane, heated up to 50°C. Some parameters of synthesized copolymers are shown in Table 17.

For the purpose of synthesis of carbosiloxane copolymers with organocyclopentasiloxane fragments in dimethylsiloxane backbone, hydride polyaddition reaction of α,ω-dihydridedimethylsiloxanes to 1,5-divinyl-1,5-dimethylhexaphenylcyclopentasiloxane in the presence of platinum hydrochloric acid as a catalyst was studied at temperatures below 100°C: 75°C, 80°C and 85°C. Forasmuch as the initial 1,5-divinyl-1,5-dimethylhexaphenylcyclopentasiloxane represents a mixture of cis- and trans-isomers (with the ratio 52 : 48), which cannot be separated preparative methods, copolymers derived from them are atactic. Preliminary heating of initial compounds within the temperature range of 80 - 95°C in the presence of catalyst indicated that under these conditions organocyclopentasiloxane fragments are not polymerized.

The reaction proceeding was detected by a decrease of amount of active $\equiv$Si-H groups. It was observed that the rate and depth of the polyaddition reaction decrease with the increase of α,ω-dihydridedimethylsiloxane chain length. Hydride polyaddition proceeds in accordance with the following scheme [94, 95]:

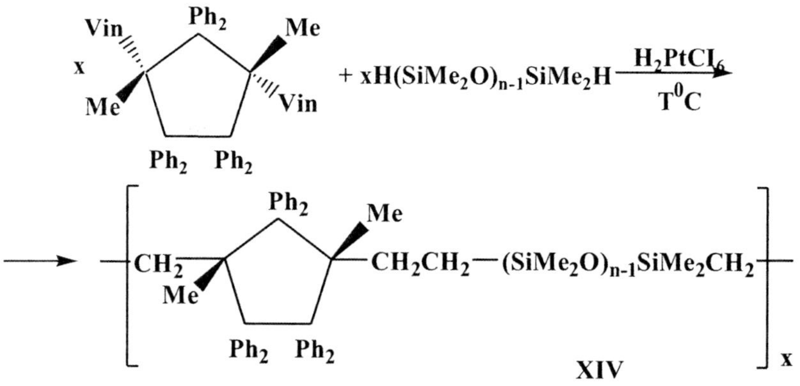

$$+ xH(SiMe_2O)_{n-1}SiMe_2H \xrightarrow{H_2PtCl_6}{T^0C}$$

(structure XIV) (17)

where $n = 2 - 23$.

As a result of the reaction, copolymers with $\eta_{spec} \approx 0.09 - 0.26$ are obtained, which are liquid or glassy light-yellow products, soluble in ordinary organic solvents. Some physical and chemical parameters and the yield of copolymers are listed in Table 18. As indicated by data in the Table, in the case of short lengths of dimethylsiloxane backbone, n, the yield of copolymers is low. This may be explained by the fact that besides intermolecular reaction, intramolecular cyclization proceeds forming a polycyclic structure. This conclusion is in agreement with data from the literature [12 - 17].

Table 17. Physical and chemical parameters of structure XIV carbosiloxane copolymers containing cyclopentasiloxane fragments

No	Copolymer	n_{SiO}	Yield %	T, °C	η^*_{spec}	T_g, °C	d_1, Å	5% mass loss	$\bar{M}_\omega \times 10^{-3}$
1	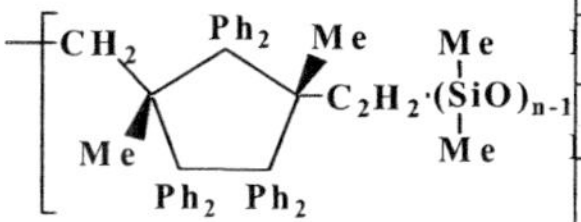	2	75	85	0.09	0÷-2	9.20	320	189
2		4	80	85	0.14	-22	-	-	-
3		6	92	75	0.15	-	-	-	-
3'		6	93	80	0.18	-	-	-	-
3''		6	95	85	0.20	-53	-	295	211
4		12	95	85	0.24	-82	-	-	-
5		23	96	85	0.31	-123	7.21	285	236

Note: * In toluene at 25^0C

The amount of active $\equiv$Si-H groups was decreased during proceeding of hydride polyaddition. Figure 22 shows that the rate of hydride polyaddition increases with temperature (at one and the same values of dimethylsiloxane units, n), but on the other hand, with an increase of the length of dimethylsiloxane links (n) at the same temperatures, the rate of hydride polyaddition decreases. Figure 22 shows that conversion of active $\equiv$Si-H groups is not complete and decreases from 20% (n=6) to 15% (n=12).

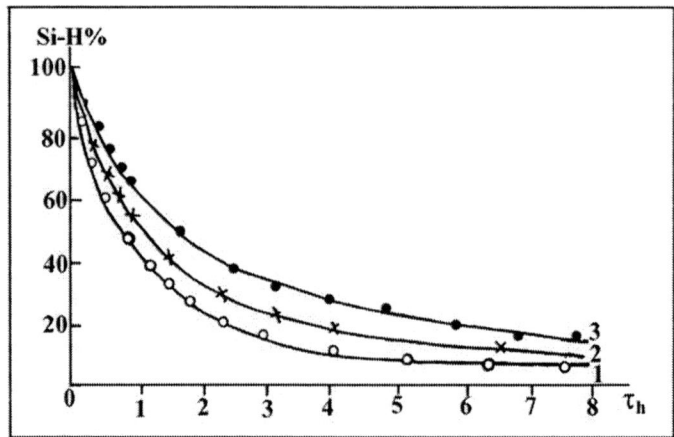

Figure 22. Time dependence of changes of active ≡Si-H% groups during polyaddition of α,ω−dihydridedimethylsiloxane (n=6) with 1,5-divinyl-1,5-dimethylhexaphenylcyclopentasiloxane: 1 - at 85°C, 2 - at 80°C, 3 - at 75°C

It was found, that polyaddition is the second order reaction. The reaction rate constants and the activation energy were calculated: $k_{75°C} \approx 1.4004 \times 10^{-2}$, $k_{80°C} \approx 1.965 \times 10^{-2}$, $k_{85°C} \approx 2.559 \times 10^{-2}$; $E_{act} \approx 62.1$ KJ/mol, respectively.

^{1}H NMR spectra of copolymers indicate that catalytic hydride polyaddition mainly proceeds by the Farmer rule with formation of dimethylenic bridges. In these spectra a reflex of $-CH_2-CH_2-$ group with chemical shift of $\delta \approx 0.34$ ppm is observed, it is indicated that the hydride polyaddition partly (about 6-7%) proceed by Markovnikov rule.

Cyclolinear carbosiloxane copolymers with 1,7- and 1,5-disposition of dimethyloctaphenylcyclohexasiloxane fragments in the dimethylsiloxane backbone were synthesized by hydride polyaddition of α,ω−dihydridedimethylsiloxane to 1,7-divinyl-1,7-dimethyloctaphenylcyclohexasiloxane and 1,5-divinyl-1,5-dimethyloctaphenylcyclohexasiloxane in the presence of a catalyst. Polyaddition reactions were studied below 100°C. It was also indicate that under these conditions polymerization or polycondensation of initial compounds does not take place. Polyaddition proceeds in accordance with the following scheme [95-97]:

$$x \underset{Vin}{\overset{Me}{Si}} \underset{O(SiPh_2O)_m}{\overset{O(SiPh_2O)_l}{}} \underset{Vin}{\overset{Me}{Si}} + xH(SiMe_2O)_{n-1}SiMe_2H \xrightarrow[T°C]{H_2PtCl_6}$$

$$\longrightarrow \left[-CH_2-\underset{O(SiPh_2O)_m}{\overset{Me \quad O(SiPh_2O)_l \quad Me}{Si}} Si-C_2H_4-(SiO)_{n-1}SiCH_2- \right]_x$$

XV, XVI

(18)

where $m = l = 2$ (XV); $n = 2 - 23$; $l = 1$, $m = 3$ (XVI); $n = 2 - 23$.

Forasmuch as 1,7- and 1,5-divinylcyclohexasiloxanes, used in polyaddition, represent mixtures of cis- and trans-isomers of the approximate 52 : 48 ratio, synthesized copolymers are atactic.

Re-precipitation of copolymers from toluene solution by methyl alcohol has given viscous or solid (with regard to the value of flexible junction) transparent products with $\eta_{spec} \approx 0.09 - 0.29$, well soluble in different organic solvents. It is found that at short length of dimethylsiloxane unit ($n \leq 4$), copolymer yields are slightly decreased, which may be explained by partial proceeding of hydride polyaddition by intramolecular cyclization mechanism (see Tables 18 and 19).

After removal of the mother solution of re-precipitated copolymer 1 (Table 18), a semi-crystalline compound with the molecular mass equal ~1,100 was obtained [96, 97]. The product of intramolecular cyclization of 1,7-divinyl-1,7-dimethyloctaphenylcyclohexasiloxane and 1,3-dihydridetetramethyldisiloxane of the following structure only may display current molecular mass:

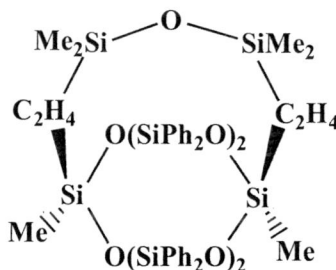

because divinylorganocyclohexasiloxane of the trans-structure participates in formation of macromolecular chain.

Structure and composition of synthesized cyclolinear carbosiloxane copolymers were determined by functional and ultimate analysis, IR and ESR spectral data. Some parameters of copolymers are shown in Tables 18 and 19.

A signal with chemical shift at $\delta \approx 0.35$ m.p. typical of $-CH_2-CH_2-$ group is observed in ^1H ESR spectrum of copolymer 1 (Table 18). This indicates that polyaddition proceeds pursuant to the Farmer rule. A duplet centered at $\delta \approx 1.06$ m.p., corresponded to methyl protons in $=CH-CH_3$ group, is also observed in the spectrum. Based on the ratio of intensities, it was concluded [96, 97] that polyaddition partly proceeds by the Markovnikov mechanism (6 – 8%). A complex multiplet with chemical shift at $\delta \approx 5.6 - 6.2$ m.p. typical of vinyl protons, not entered the polyaddition reaction, and a singlet for $\equiv Si-H$ protons with chemical shift at $\delta \approx 4.4$, not participated in the reaction, too, were observed in the spectra.

Hydride polyaddition proceeded at different temperatures. Figures 23 and 24 show variations of $\equiv Si-H$ bond concentration during polyaddition of α,ω-dihydride dimethylsiloxane ($n = 6$) to 1,7-divinyl-1,7-dimethyloctaphenylcyclohexasiloxane and 1,5-divinyl-1,5-dimethyloctaphenylcyclohexasiloxane. It is observed that hydride polyaddition depth is increased with the reaction temperature. Moreover, the effect of 1,7- or 1,5-disposition of vinyl groups in cyclohexasiloxane fragments is the negligible factor for their reactivity.

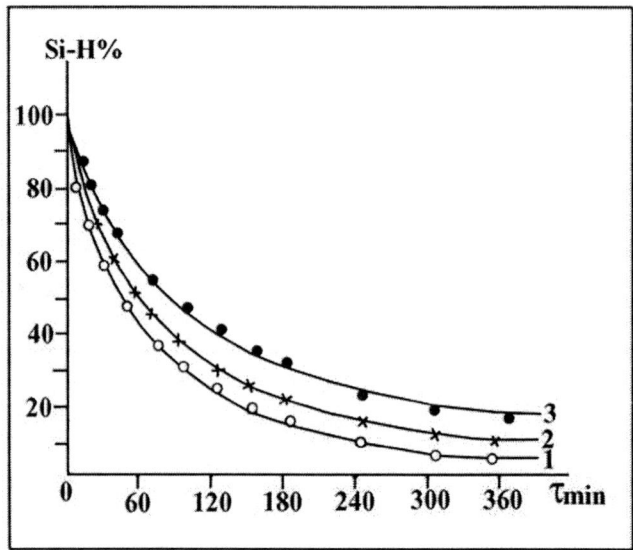

Figure 23. Decrease of ≡Si-H bond concentration during hydride polyaddition of α,ω-dihydride dimethylsiloxane (n = 6) to 1,7-divinyl-1,7-dimethyloctaphenylcyclohexasiloxane: 1 - 90°C; 2 - 85°C; 3 - 80°C

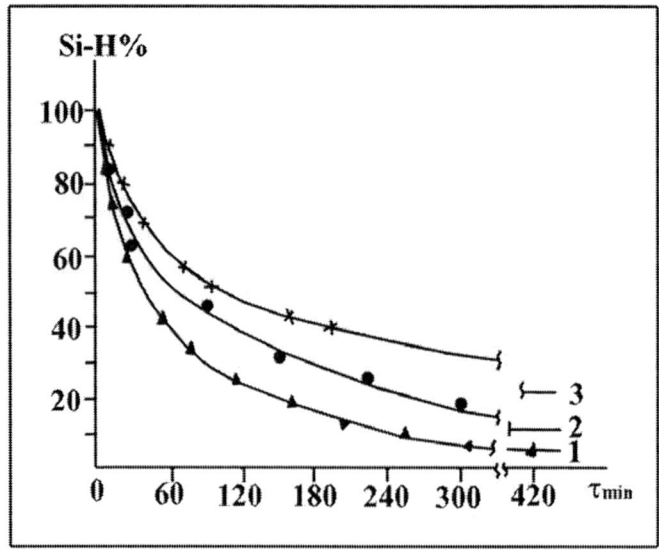

Figure 24. Decrease of ≡Si-H bond concentration during hydride polyaddition of α,ω-dihydride dimethylsiloxane (n = 6) to 1,5-divinyl-1,5-dimethyloctaphenylcyclohexasiloxane: 1 - 100°C; 2 - 90°C; 3 - 80°C

It is found that at the initial stages, polyaddition represents the second order reaction. In the case of 1,7-divinyl-1,7-dimethyloctaphenylcyclohexasiloxane, polyaddition rate constants for different temperatures were determined as follows: $k_{90°C} \approx 3.0797 \times 10^{-2}$; $k_{85°C} \approx$

2.3007×10^{-2}; $k_{80°C}$ ≈ 1.6781×10^{-2}. Activation energies for 1,7-divinyl-1,7-dimethyloctaphenylcyclohexasiloxane and 1,5-divinyl-1,5-dimethyloctaphenylcyclohexasiloxane were also calculated: E_a ≈ 66.7 and E_a ≈ 69.7, respectively. Obviously, these values are very close.

X-ray diffraction studies have indicated that copolymers are single-phase amorphous systems, and maximal interchain distance is observed for short dimethylsiloxane unit length (n = 2); hence, for copolymer 1 (Table 19), d_1 ≈ 9.60 Å. This value is slightly greater than the interchain distance of carbosiloxane copolymer 1 (Table 18) with 1,7-disposition of cyclohexasiloxane fragment in the dimethylsiloxane backbone (n = 2). As flexible junction length is increased, d_1 decreases and approaches the interchain distance of PDMS; it increases with the volume of cyclic fragment at the same lengths of flexible dimethylsiloxane unit, i.e. at transition from cyclopentasiloxane to cyclohexasiloxane fragment.

Thermogravimetric studies of carbosiloxane copolymers has indicated 5% mass loss of the compounds in the temperature range of 250 - 260°C. The main degradation process proceeds in the range of 380 - 630°C, and above 700°C the mass loss id not observed. It is found that thermal oxidative stability of copolymers is decreased with increase of the cyclic fragment volume, i.e. at the transition from cyclic pentasiloxane to hexasiloxane fragments in cyclolinear carbosiloxane copolymer. It is also found that carbosiloxane copolymers with 1,7- and 1,5-disposition of cyclic hexasiloxane fragments in the backbone are characterized by almost identical thermal oxidative stability. Thus, is has been concluded [96, 97] that the effect of 1,7- or 1,5-disposition of cyclic hexasiloxane fragment in carbosiloxane copolymer is negligible for thermal oxidative stability of copolymers (Figure 25). On the other hand, compared with pure siloxane analogues, thermal oxidative stability of carbosiloxane copolymers is lower.

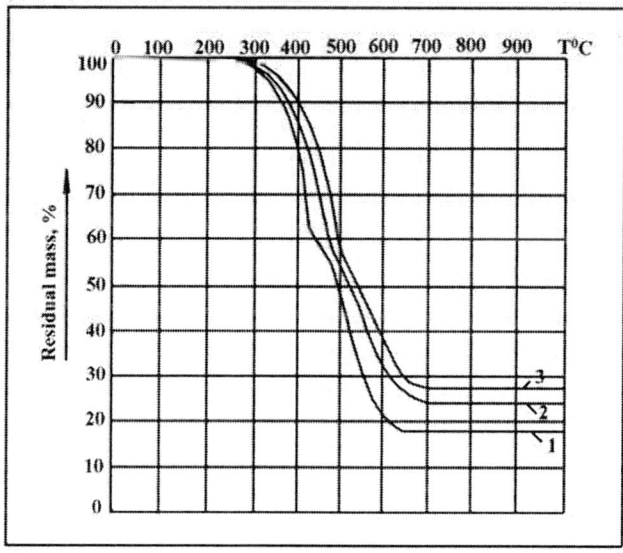

Figure 25. Thermogravimetric curves of carbosiloxane copolymers: 1 – copolymer 4 (Table 19) with 1,5-disposition of cyclic hexasiloxane fragment in the backbone; 2 – copolymer 1 (Table 18) with 1,7-disposition of cyclic hexasiloxane fragment in the backbone; 3 – copolymer 1 (Table 17) with cyclic pentasiloxane fragment in the backbone

Thermogravimetric studies have displayed that cyclic fragment causes a considerable effect on carbosiloxane copolymer at $n \approx 12$ only, and at $n \approx 23$ no effect of cyclic fragment on the glass transition temperature of the copolymer is observed.

Figure 26 shows dependence of T_g on the length of dimethylsiloxane unit for cyclolinear carbosiloxane copolymers.

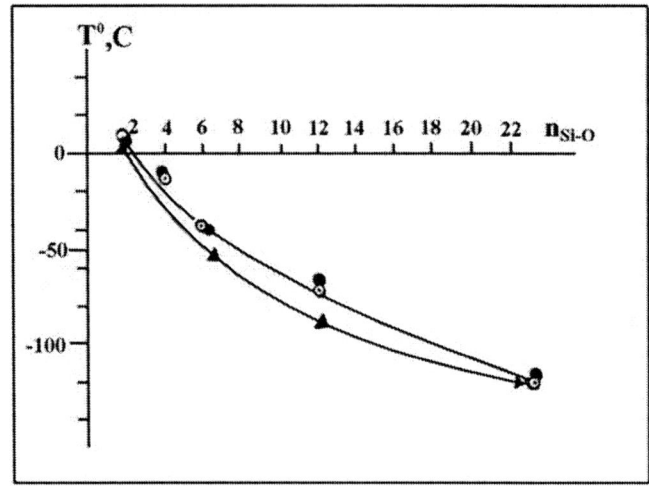

Figure 26. Dependence of T_g for cuclolinear carbosiloxane sopolymers on the length of dimethylsiloxane unit: 1 – copolymer with 1,7-position of cyclic hexasiloxane fragment; 2 – copolymer with 1,5-position of cyclic hexasiloxane fragment

It has been found that expansion of the cyclic fragment volume at the same length of dimethylsiloxane unit, i.e. introduction of a single diphenylsiloxane unit, T_g of the copolymer is increased by ~10°C. It is also shown that the effect of 1,7- or 1,5-disposition of cyclic hexasiloxane fragment on T_g of the copolymer is negligible, which conform to the previous results on pure siloxane copolymers.

Table 18. Physical and chemical parameters of carbosiloxane copolymers with 1,7-position of cyclic hexasiloxane fragments in the dimethylsiloxane backbone (structure XV)

No	Copolymer	n_{SiO}	Yield %	T, °C	η^*_{spec}	T_g, °C	d_1, Å	5% mass loss	$\bar{M}_\omega \times 10^{-3}$
1		2	74	90	0.10	+5	9.31	280	174
2		4	80	90	0.12	-10	-	-	-
3		6	94	80	0.17	-	-	-	-
3'		6	94	80	0.17	-	-	-	-
3''		6	95	85	0.18	-40	8.81	280	194
4		12	96	90	0.23	-68	8.40	-	-
5		23	95	90	0.29	-123	7.24	260	231

Note: * In toluene at 25°C

Table 19. Physical and chemical parameters of carbosiloxane copolymers with 1,5–position of cyclic hexasiloxane fragments in the dimethylsiloxane backbone (structure XVI)

No	Copolymer	n_{SiO}	Yield %	T, °C	η^*_{spc}	T_g, °C	d_1, Å	5% mass loss	$\bar{M}_\omega \times 10^{-3}$
1		2	72	10	0.09	+8	9.60	270	159
2	[Me, Ph₂, Me, Me, Me / C₂H₂·(SiO)ₙ₋₁SiCH₂ / Ph₂, Ph₂, Me, Me / Ph₂]	4	84	85	0.11	-12	-	-	-
3		6	86	80	0.15	-	-	-	-
3'		6	89	90	0.18	-	-	-	-
3''		6	94	100	0.15	-38	8.90	265	180
4		12	95	100	0.22	-72	8.34	-	210
5		23	95	100	0.28	-123	-	260	-

Note: * In toluene at 25°C

5. CYCLOLINEAR COPOLYMERS DERIVED BY POLYMERIZATION OF ORGANOBICYCLO-AND ORGANOTRICYCLOSILOXANES

The above-mentioned materials indicate that cyclolinear organosiloxane copolymers are usually synthesized by homopolycondensation and HFC reactions from difunctional organocyclosiloxanes or organopoly(cyclosiloxanes) with α,ω–dihydroxy(diamino, dichloro)diorganosiloxanes [99 – 102]. As known from the literature, polymerization and copolymerization of organopoly(cyclosiloxanes) not always gives soluble polymers and their yield is usually low [103].

5.1. Synthesis of Organodi- and Organotricyclosiloxanes with Labile Cyclic Fragments

To synthesize organodi- and organotricyclosiloxanes containing thermodynamically and sterically labile rings, step HFC reaction of cis-1,3,5,7-tetrahydroxy-1,3,5,7-tetraphenylcyclotetrasiloxane with dimethyldichlorosilane and trimethylchlorosilane at the ratio of initial components 1 : 1 : 2 was studied in the presence pyridine in 5% absolute ether. The reaction was performed in two stages. At the first stage, tetrole solution was added by dimethyldichlorosilane and a part pyridine, and at the second stage by trimethylchlorosilane and the rest of pyridine; at the final stage, the mixture was boiled and continuously mixed, simultaneously. The reaction preceeded in accordance with the scheme as follows:

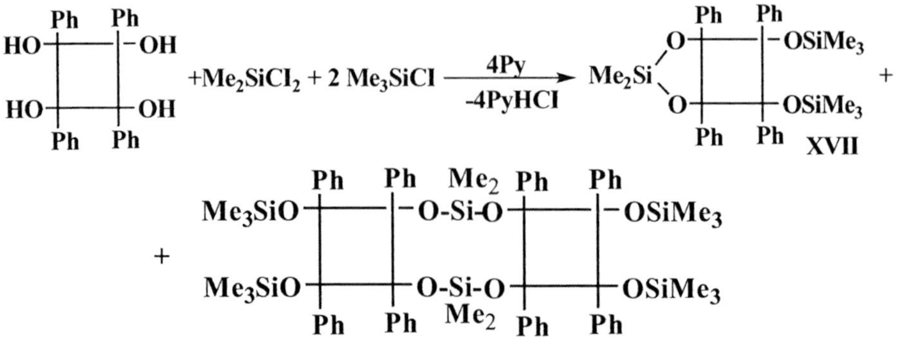

(19)

It was suggested that HFC reaction proceeds with formation of an intermediate complex, subject to both intramolecular cyclization forming bicyclic compound with structure XVII and intermolecular condensation forming bicyclic compound with structure XVIII.

Similarly, the step HFC reaction of tetrole with 1,3-dichlorotetramethylsiciloxane and trimethylchlorosilane (1: 1: 2 ratio of initial reagents) under the same reaction conditions was studied. In this case, the reaction is supposed to proceed via formation of a transitional complex, and as a result of intra- and intermolecular cyclization both bi- and tricyclic compounds are formed in accordance with the scheme as follows [104]:

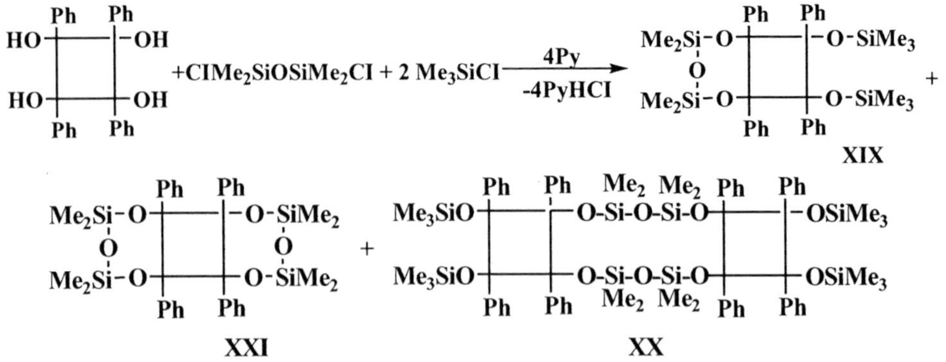

(20)

All synthesized organodi- and organotricyclic compounds represent dense transparent oil-like liquids, crystallized in calm. Tricyclic compound 5 with structure XXI was separated during re-crystallization of compound 3 with structure XIX (Table 20).

Composition and structure of obtained compounds was detected by the ultimate analysis, determination of the molecular mass and from IR and ^{1}H ESR spectral data. Yields and some physical and chemical parameters of synthesized bi- and tricyclic organosiloxanes are shown in Table 20. Figure 27 shows ^{1}H ESR spectrum of bicyclic compound 1 with structure XVII (Table 20) displaying a triplet with the chemical shifts at $\delta \approx$ 0.20, 0.33 and 0.46 m.p., and a complex multiplet in the range of 7.2 – 8.4 m.p.

Figure 28 shows ^{1}H ESR spectrum of bicyclic compound 3 with structure XIX (Table 20) containing a multiplet with chemical shifts at $\delta \approx$ 0.02, 0.12, 0.30, 0.33 and 0.46 m.p., corresponded to methyl protons.

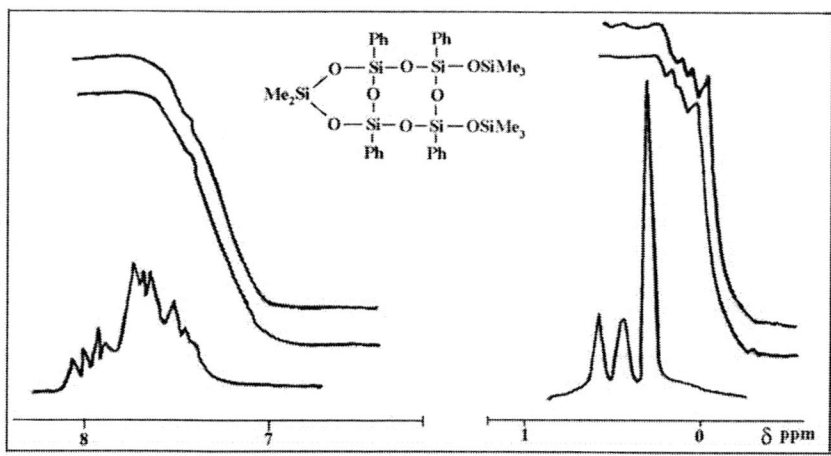

Figure 27. ¹H NMR spectrum of bicyclic compound 1 with structure XVII (Table 3.19)

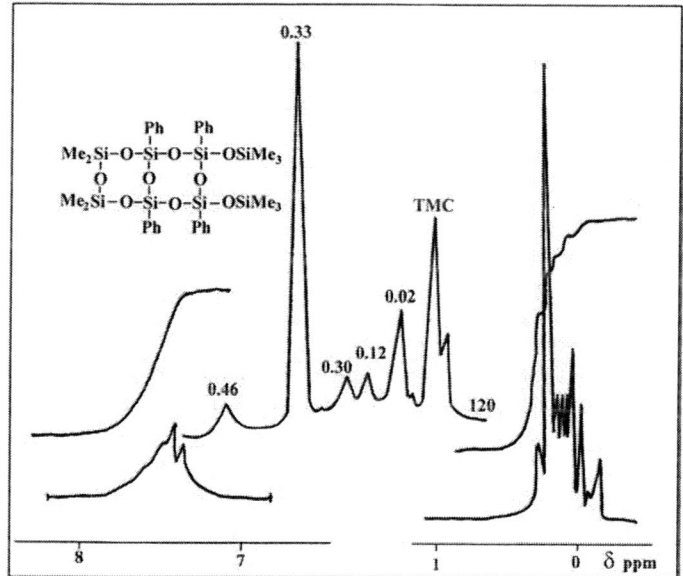

Figure 28. ¹H NMR spectrum of bicyclic compound 3 with structure XIX (Table 20)

Synthesized organodi- and organotricyclosiloxanes contain a single labile cycle each. A six-term cyclic fragment is more labile in bicyclic compound 1 (structure XVII), because it is more stressed than eight-term one, and the more so, the latter is framed by trimethylsiloxy- and phenyl groups, which naturally makes steric hindrances to polymerization. The compounds contain octa-, dodeca- and hexa-term cyclic fragments, framed by dimethyl radicals, which makes them accessible for impact of nucleophilic catalysts during polymerization.

Table 20. Physical and chemical parameters of bi- and tricyclic organosiloxanes

No	Structure	Yield, %	T_{boil}, °C (P, mm Hg)	T_{melt}, °C
1	XVII	43	260 – 268 (6×10^{-3})	100 - 102
2	XVIII	21	320 – 327 (6×10^{-3})	-
3	XIX	50	262 – 270 (6×10^{-3})	116 - 117
4	XX	24	330 – 335 (6×10^{-3})	-
5	XXI	5	-	128 - 129

5.2. Polymerization of Cyclolinear Organosiloxane Copolymers with Cyclic Fragments in the Backbone

Synthesized bi- and tricyclic organosiloxane compounds were used in the polymerization reaction, which was performed in the presence of α,ω–tetramethylammonium oxydimethylsiloxanolate ($n \approx 11$) and 0.1 H KOH alcohol solution in the temperature range of 50 - 200°C. It was found that the reaction proceeding at low temperature (< 100°C) gives low-molecular oligomers ($\eta_{spec} \approx 0.04$). The reaction temperature increase over 140°C induces decomposition of the catalyst mentioned. That is why future studies of polymerization of tricyclic organodi- and organotrisiloxanes were carried out in 0.1 H alcohol solution of potassium hydroxide.

Bi- and tricyclic compounds were polymerized with formation of low-molecular cyclolinear copolymers:

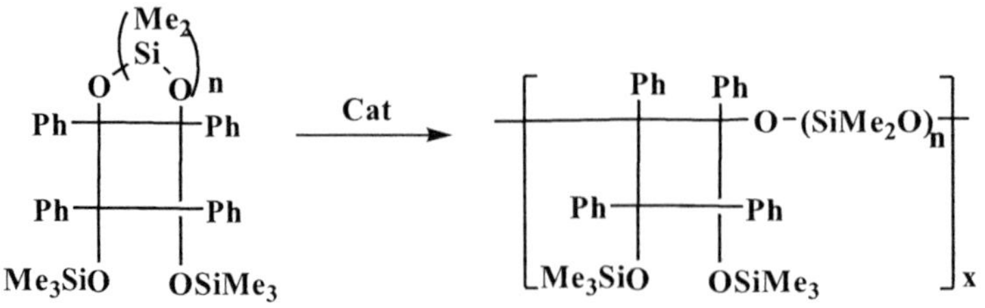

(21)

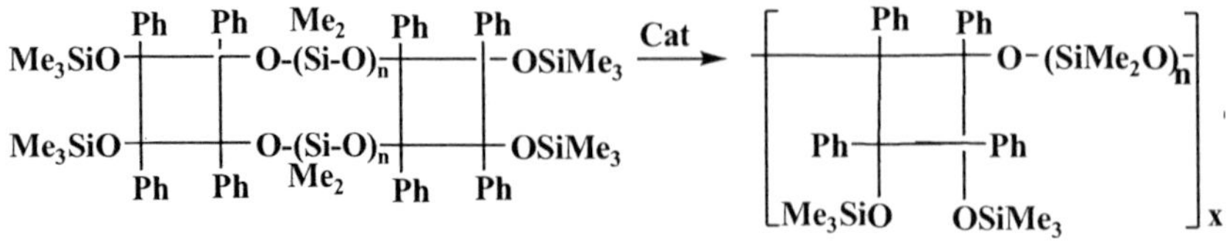

(22)

where $n = 1, 2$.

Synthesized regular copolymers represent light-yellow solids with $\eta_{spec} \approx 0.4 - 0.7$, well soluble in various organic solvents. It is found that the effect of increasing catalyst concentration (0.1 – 0.5 wt.% of initial compound) and reaction temperature on the molecular mass of low-molecular compounds obtained is negligible.

Thermogravimetric analysis has indicated mass losses of copolymers at 260 - 270°C not exceeding 2 – 3%. At 300 – 350°C thermal degradation rate is considerably increased, and at 550°C mass losses are maximal. The mass loss is regularly increased with the length of dimethylsiloxane backbone, i.e. at transition from dimethylsiloxane to disiloxane bridge.

Thus, synthesis of cyclolinear organosiloxane and carbosiloxane copolymers with monocyclic fragments in the backbone, obtained in HFC, hydride polyaddition and polymerization reactions, are considered in this Chapter. The above-shown data on high-molecular polymers with regular disposition of cyclic fragment in the backbone are mainly synthesized by HFC and hydride polyaddition reactions. As for polymerization of bicyclosiloxanes, they are almost useless for high polymer synthesis.

REFERENCES

[1] Andrianov K.A., *Doklady AN SSSR*, 1963, vol. 151(5), p. 1093. (Rus)
[2] Andrianov K.A., Talanov V.N., and Khananashvili L.M., *Izv. AN SSSR, Ser. Khim.*, 1969, No. 9, p. 2004. (Rus)
[3] Mileshkevich V.P., *Kauchuk i Resina*, 1978, No. 6, p. 4. (Rus)
[4] Thomas T.H. and Kendric T.C., *J. Polym. Sci.*, 1969, vol. 2A(7), p. 537.
[5] Alexander G. and Garzo G., *Chromatography*, 1974, vol. 7(4), p. 225.
[6] Andrianov K.A., Jenchelskaya S.M., and Petrashko Yu.K., *Zh. Obshch. Khim.*, 1958, vol. 28(3), p. 685. (Rus)
[7] Sokolov N.N., *Methods of Poly(organosiloxane) Synthesis*, Moscow – Leningrad, Gosnergoizdat, 1959, p. 103. (Rus)
[8] Andrianov K.A., *Polymers With Inorganic Backbones of Macromolecules*, Moscow, Izd. AN SSSR, 1962, p. 144. (Rus)
[9] Andrianov K.A., Jenchelskaya S.M., and Petrashko Yu.K., *Plast. Massy*, 1960, No. 3, p. 20. (Rus)
[10] Volkova L.M. and Andrianov K.A., *Vysokomol. Soedin.*, 1964, vol. 6(9), p. 1662. (Rus)
[11] Andrianov K.A., Khanashvili L.M., and Zavalny V.G., *Plast. Massy*, 1967, No. 6, p. 48. (Rus)
[12] Andrianov K.A., Khanashvili L.M., Makarova N.N., Mukbaniani O.V., and Meladze S.M., *A.C. No. 791,757* (USSR), *Bull. Izobr.*, 1981, No. 4. (Rus)
[13] Andrianov K.A., Nogaideli A.I., Slonimski G.L., Levin V.Yu., Makarova N.N., Kvachev Yu.P., and Mukbaniani O.V., *Vysokomol. Soedin.*, 1976, vol. 13B(1), p. 359. (Rus)
[14] Mukbaniani O.V., *Organosiloxane Copolymers and Block-copolymers With Different Cyclic Structure of Macromolecules*, Doctor's Dissertation on Chemistry, 1993, Tbilisi State University, Tbilisi, Georgia. (Rus)
[15] Koiava N.A., Mukbaniani O.V., and Khanashvili L.M., *Vysokomol. Soedin.*, 1985, vol. 27A(11), p. 2261. (Rus)

[16] Koiava N.A., Mukbaniani O.V., and Khananashvili L.M., *Abstr. Commun. 27th Intern. Symp. Macromol.*, Strasburg, 1981, p. 18.
[17] Mukbaniani O.V., Meladze S.M., and Khananashvili L.M., *Vysokomol. Soedin.*, 1984, vol. 24B(4), p. 250. (Rus)
[18] Andrianov K.A., Nogaideli A.I., Makarova N.N., and Mukbaniani O.V., *Izv. AN SSSR, Ser. Khim.*, 1977, No. 6, p. 1388. (Rus)
[19] Zhdanov A.A. and Astapova T.V., *Vysokomol. Soedin.*, 1981, vol. 23A(3), p. 626. (Rus)
[20] Yuzhelevski Yu.A., Kurlova T.V., Kogan B.G., and Suvorova M.V., *Zh. Obshch. Khim.*, 1972, vol. 42(9), p. 2006. (Rus)
[21] Mukbaniani O.V., *Synthesis of Polycyclic Organosilicon Compounds and Block-Copolymers on Their Basis*, Candidate's Dissertation on Chemistry, 1976, Tbilisi State University, Tbilisi, Georgia. (Rus)
[22] Bostick E.E., *Polymer Prep.*, 1969, vol. 10(2), p. 877.
[23] Tsetlin B.L., Gavrilov V.I., Velikovskaya N.A., and Kochkin V.V., *Zavodskaya Laboratoria*, 1956, vol. 22(3), p. 352. (Rus)
[24] Andrianov K.A., Slonimsky G.L., Levin V.Yu., Zhdanov A.A., and Godovsky Yu.K., *J. Polym. Sci.*, 1972, vol. 10(A-1), p. 1.
[25] Andrianov K.A., Slonimsky G.L., Levin V.Yu., Zhdanov A.A., and Godovsky Yu.K., *J. Polym. Sci.*, 1972, vol. 10(A-1), p. 23.
[26] Zhdanov A.A., Andrianov K.A., Malyikhin A.P., and Emelianov V.N., *Izv. AN SSSR, Ser. Khim.*, 1973, No. 10, p. 2299. (Rus)
[27] Andrianov K.A., Slonimski G.L., Zhdanov A.A., and Levin V.Yu., *Vysokomol. Soedin.*, 1968, vol. 10A(9), p. 2102. (Rus)
[28] Andrianov K.A., Slonimski G.L., Levin V.Yu., Godovski Yu.K., Kuznetsova N.K., Tsvankin D.Ya., Moskalenko V.A., and Kuteinikova L.N., *Vysokomol. Soedin.*, 1970, vol. 12A(6), p. 1268. (Rus)
[29] Khyinku E.S., Zhukov V.P., Gerasimov M.V., Il'ina M.N., Kvachev Yu.P., and Papkov V.S., *Thes. Proc. VII All-Union Conference on Chemistry, Production Technology and Application of Organosilicon Compounds*, 1990, November 20 – 23, Tbilisi, Part 1, p. 183. (Rus)
[30] Andrianov K.A., Zhdanov A.A., Zavin B.G., and Evdokimov A.M., *Doklady AN SSSR*, 1971, vol. 199(3), p. 597. (Rus)
[31] Andrianov K.A., Zhdanov A.A., Zavin B.G., Evdokimov A.M., and Biryukova T.V., *Vysokomol. Soedin.*, 1972, vol. 14B(1), p. 327. (Rus)
[32] Buch R., Klimisch H., and Johnson O., *J. Polym. Sci.*, 1970, vol. 8(A-2), No. 4, p. 541.
[33] Erenburg V.E., Kartasheva G.G., Eremina M.A., and Poddubni I.Ya., *Vysokomol. Soedin.*, 1967, vol. 9A(12), p. 2709. (Rus)
[34] Skazka V.S. and Shaltyiko L.G., *Vysokomol. Soedin.*, 1960, vol. 2(4), p. 572. (Rus)
[35] Andrianov K.A., Pavlova S.A., Tverdovhlebova I.I., Pertsova N.V., and Larina T.A., *Abstr. Commun., Prague IUPAC Microsimposium on Macromolecules. Thermodinamics of Interactions in Polymer Solutions*, 1971, September 6 - 9, p. 116.
[36] Tverdokhlebova I.M., Kurginyan P.A., Larina T.A., Makarova N.N., Ronova I.A., Pavlova S.S., and Mukbaniani O.V., *Vysokomol. Soedin.*, 1981, vol. 23A(5), p. 995. (Rus)

[37] Tverdokhlebova I.M., *Studies in the Sphere of Molecular and Conformational Transformations of Poly(organosiloxanes)*, Doctor's Dissertation on Chemistry, Moscow, 1978, INEOS, AS USSR. (Rus)
[38] Flory P.J., Fox T.G., and Shaefgen J.R., *J. Amer. Chem. Soc.*, 1951, vol. 73(5), p. 1904.
[39] Inagaki H., Suzuki H., and Kurata M., *J. Polym. Sci.*, 1966, vol. 15, p. 409.
[40] Stokmayer W.H. and Fixmann M., *J. Polym. Sci.*, 1963, vol. 1(1), p. 137.
[41] Flory P.J., Gescenzi V., and Mark J., *J. Amer. Chem. Soc.*, 1964, vol. 86(3), p. 146.
[42] Bohdanecky M., *Collect Czech. Chem. Commun.*, 1968, vol. 33(12), p. 4397.
[43] Kartasheva G.G., Erenburg V.E., and Poddubni I.Ya., *Vysokomol. Soedin.*, 1972, vol. 14B, p. 665. (Rus)
[44] Mark J.E. and Flory P.J., *J. Amer. Chem. Soc.*, 1964, vol. 86(1), p. 138.
[45] Tarasova G.I., *Fractionation of Aromatic Polyesters. Chemistry and Technology of High-molecular Compounds*, Coll., Comp. Teplyakov M.S., Ed. V.V. Korshak, Moscow, VINITI, 1973, vol. 4, p. 109. (Rus)
[46] Severni V.V., *Organosiloxanes' Reaction with Alkyl(aryl)chlorosilanes*, Doctor's Dissertation on Chemistry, 1962, Moscow, INEOS AS USSR. (Rus)
[47] Sobolevski M.V., Skorokhodov I.I., Ditsent V.E., Sobolevskaya A.V., and Efimov V.M., *Vysokomol. Soedin.*, 1969, vol. 11A(5), p. 1109. (Rus)
[48] Andrianov K.A., Pavlova S.-S.A., Zhuravleva I.V., Tolchinski Yu.I., and Astapova B.A., *Vysokomol. Soedin.*, 1977, vol. 19A(4), p. 896. (Rus)
[49] Andrianov K.A., Pavlova S.-S.A., Zhuravleva I.V., Tolchinski Yu.I., Makarova N.N., and Mukbaniani O.V., *Vysokomol. Soedin.*, 1977, vol. 19A(4), p. 1387. (Rus)
[50] Andrianov K.A., Papkov V.S., Zhdanov A.A., and Yakushina S.E., *Vysokomol. Soedin.*, 1969, vol. 11A(9), p. 2030. (Rus)
[51] Pathnode W. and Wilcock D.F., *J. Amer. Chem. Soc.*, 1946, vol. 46, p. 358.
[52] Andrianov K.A., Kononov A.M., and Tsvankin D.Ya., *Vysokomol. Soedin.*, 1968, vol. 10B(5), p. 320. (Rus)
[53] Makarova N.N., Petrova I.M., Godovski Yu.K., Lavrukhin B.D., and Zhdanov A.A., *Doklady AN SSSR*, 1983, vol. 209(6), p. 1368. (Rus)
[54] Godovsky Yu.K., Makarova N.N., and Mamaeva I.I., *Macromol. Chem. Rapid Commun.*, 1986, vol. 7(1), p. 32.
[55] Makarova N.N., Godovsky Yu.K., and Kuzmin N.N. Makromol. Chem., 1987, Bd. 188(1), S. 119.
[56] Mamaeva I.I., Makarova N.N., Petrova I.M., Tverdokhlebova I.I., and Pavlova S.A., *Vysokomol. Soedin.*, 1987, vol. 29A(7), p. 1507. (Rus)
[57] Makarova N.N., Kuzmin N.N., Lavrukhin B.D., Godovski Yu.K., Mamaeva I.I., Matukhina E.V., and Petrova I.I., *Vysokomol. Soedin.*, 1989, vol. 31B(9), p. 708. (Rus)
[58] Makarova N.N., Godovsky Yu.K., and Lavrukhin B.D., *Vysokomol. Soedin.*, 1995, vol. 37A(3), p. 1. (Rus)
[59] Matukhina E.V., Boda E.E., Timofeeva T.V., Godovski Yu.K., Makarova N.N., Petrova I.I., and Lavrukhin B.D., *Vysokomol. Soedin.*, 1996, vol. 38A(9), p. 1545. (Rus)
[60] Godovsky Yu.K., Makarova N.N., Petrova I.M., and Zhdanov A.A., *Macromol. Chem. Rapid Commun.*, 1984, vol. 5(3), p. 427.
[61] Timofeeva T.V., Boda E.E., Polishchuk A.P., Antipin M.Yu., Matukhina E.V., Petrova I.M., Makarova N.N., and Struchkov Yu.T., *Mol. Cryst. Liq. Cryst.*, 1994, vol. 248, p. 125.

[62] Makarova N.N., Petrova I.M., Matukhina E.V., Godovski Yu.K., and Lavrukhin B.D., *Vysokomol. Soedin.*, 1997, vol. 39A(10), p. 1616. (Rus)
[63] Destrade C., Tinh N.H., Gasparoux H., Malthete J., and Levelut A.M., *Mol. Cryst. Liq. Cryst.*, 1981, vol. 71, p. 111.
[64] Kuzmin N.N. and Matukhina E.V., *Vysokomol. Soedin.*, 1991, vol. 33B(7), p. 547. (Rus)
[65] Beatty C.L. and Karasz F.E., *Bull. Am. Phys. Soc.*, 1973, vol. 18(3), p. 461.
[66] Papkov V.S., Godovsky Yu.K., Svistunov V.S., Litvinov V.M., and Zhdanov A.A., *J. Polym. Sci., Polym. Chem. Ed.*, 1984, vol. 22(12), p. 3617.
[67] Godovsky Yu.K. and Papkov V.S., *Adv. Polym. Sci.*, 1989, vol. 88, p. 129.
[68] Tsvankin D.Ya., Papkov V.S., Zhukov V.P., Godovsky Yu.K., Svistunov V.S., and Zhdanov A.A., *J. Polym. Sci., Polym. Chem. Ed.*, 1985, vol. 23(4), p. 1043.
[69] Godovsky Yu.K., Makarova N.N., Papkov V.S., and Kuzmin N.N., *Macromol. Chem. Rapid Commun.*, 1985, vol. 6, p. 443.
[70] Out G., Turetski A., Moeller M., and Oelfin D., *Macromolecules*, 1994, vol. 27(12), p. 3310.
[71] Makarova N.N., Petrova I.M., Godovski Yu.K., Lavrukhin B.D., and Zhdanov A.A., *Doklady AN SSSR*, 1983, vol. 209(6), p. 1368. (Rus)
[72] Makarova N.N., Matukhina E.V., Godovski Yu.K., and Lavrukhin B.D.
[73] Beatty C.L. and Karasz F.E., *Bull. Amer. Phys. Soc.*, 1973, vol. 18(3), p. 461.
[74] Makarova N.N. and Godovski Yu.K., *Vysokomol. Soedin.*, 1986, vol. 28B(4), p. 243. (Rus)
[75] Ringsdorf H. and Scheller A., *British Polym. J.*, 1981, vol. 13, p. 43.
[76] Garzo G. and Perhsson K., *J. Organometal. Chem.*, 1971, vol. 30, p. 187.
[77] Garzo G., Tamas J., Szekely T., and Ujszaszi K., *Acta Chim., Acad. Sci. Hung.*, 1971, vol. 69, p. 273.
[78] Garzo G. and Alexander G., *Chromatogr.*, 1971, vol. 4, p. 554.
[79] Alexander G. and Garzo G., *Chromatogr.*, 1974, vol. 7, p. 190.
[80] Blazso M. and Garzo G., *J. Organomet. Chem.*, 1979, vol. 7, p. 273.
[81] Lengyel B., Garzo G., Fritz D., and Till F., *J. Chromatogr.*, 1966, vol. 24, p. 8.
[82] A.C. No. 842,572 (USSR), 1981; *Bull. Izobr.*, No. 24, 1981. (Rus)
[83] Mukbaniani O.V., Achelashvili V.A., Levin V.Yu., Meladze S.M., Inaridze I.A., and Khananashvili L.M., *Izv. AN GSSR, Ser. Khim.*, 1990, vol. 16(1), p. 20. (Rus)
[84] Inaridze I.A., *Synthesis and Investigation of Organocyclosiloxane and Organocarbocyclosiloxane Fragments Containing Polyorganosiloxanes*, Candidate Dissertation on Chemistry, 1993, Tbilisi State University, Tbilisi, Georgia. (Rus)
[85] Papkov V.S., Il'ina M.N., Makarova N.N., Zhdanov A.A., Slonimski G.L., and Andrianov K.A., *Vysokomol. Soedin.*, 1975, vol. 12A, p. 2700. (Rus)
[86] Achelashvili V.A., Mukbaniani O.V., Koiava N.A., and Komalenkova N.G., *Abstr. Commun. 31th IUPAC Macromolecular Symposium*, GDR, Merseburg, 1987, p. 155.
[87] Achelashvili V.A., Mukbaniani O.V., Khananashvili L.M., Levin V.Yu., Komalenkova N.G., and Chernyishev E.A., *Vysokomol. Soedin.*, 1990, vol. 32A(3), p. 480. (Rus)
[88] Chernyishev E.A., Komalenkova N.G., Klochkova T.A., Kuzmina G.M., and Bochkarev V.N., *Zh. Obshch. Khim.*, 1975, vol. 45(10), p. 2229. (Rus)
[89] Chernyishev E.A., Belkina T.V., Nikitin V.S., Komalenkova N.G., Shchapatin A.S., and Zhinkin D.Ya., *Zh. Obshch. Khim.*, 1978, vol. 48(3), p. 630. (Rus)

[90] Khananashvili L.M. and Mukbaniani O.V., *Intern. J. Polym. Mater.*, 1994, vol. 27, p. 31.
[91] Andrianov K.A., Slonimski G.L., Levin V.Yu., Podovski Yu.K., Kuteinikova N.K., Tsvankin D.Ya., Moskalenko V.A., and Kuteinikova L.I., *Vysokomol. Soedin.*, 1970, vol. 12A(6), p. 1268. (Rus)
[92] Zhdanov A.A., Andrianov K.A., and Malyikhin A.P., *Doklady AN SSSR*, 1973, vol. 211(5), p. 1104. (Rus)
[93] Zhdanov A.A., Pryakhina T.A., Strelkova T.V., Afonina R.I., and Kotov V.M., *Vysokomol. Soedin.*, 1993, vol. 35(5), p. 475. (Rus)
[94] Karchkhadze M.G., Mukbaniani N.O., Samsonia A.Sh., Tkeshelashvili R.Sh., Kvelashvili N.G., Chogovadze T.V., and Khananashvili L.M., *Bull. Georg. Acad. Sci.*, 1998, vol. 158(1), p. 75.
[95] Mukbaniani N.O., Synthesis and Investigation of Properties of Carbosiloxane Cyclolinear Copolymers, Candidate Dissertation on Chemistry, 2001, Tbilisi State University, Tbilisi, Georgia.
[96] Karchkhadze M.G., Mukbaniani N.O., Khananashvili L.M., Meladze S.M., Kvelashvili N.G., and Doksopulo T.P., *Intern. J. Polym. Mater.*, 1998, vol. 41, p. 89.
[97] Mukbaniani N.O., Karchkhadze M.G., Samsonia A.Sh., Tkeshelashvili R.Sh., and Khananashvili L.M., *Bull. Georg. Acad. Sci.*, 1999, vol. 160(1), p. 84.
[98] *Patent No. 3,297,632* (USA), 1967; Wu T.C., *C.A.*, 1967, vol. 66, 46790x.
[99] *Patent No. 3,264,259* (USA), 1966.
[100] A.C. No. 794,029 (USSR); *Bull. Izobr.*, No. 1, 1981. (Rus)
[101] A.C. No. 791,758 (USSR); *Bull. Izobr.*, No. 48, 1980. (Rus)
[102] A.C. No. 794,029 (USSR); *Bull. Izobr.*, No. 1, 1981. (Rus)
[103] Andrianov K.A. and Zachernyuk A.B., *Vysokomol. Soedin.*, 1974, vol. 16A(8), p. 1435. (Rus)
[104] Mukbaniani O.V., Meladze S.M., Makarova N.N., and Khananashvili L.M., *Soobshch. AN GSSR*, 1980, vol. 99(4), p. 109. (Rus)
[105] Meladze S.M., *Candidate Dissertation on Chemistry*, 1980, Tbilisi State University, Tbilisi, Georgia.

In: Chemistry as Art
Editors: L. Liu and G. E. Zaikov, pp. 139-146
ISBN 1-59454-585-5

Chapter 6

CHARACTERIZATION OF NITROXYL RADICALS PRODUCED FROM HINDERED AMINES DURING ACCELERATED AGING OF POLYMERS◊

Leonid Yu. Smoliak*, Wolf D. Habicher [1], Gabriele Theumer [1], Nikolay R. Prokopchuk, Sergey G. Mikhalyonok, Svetlana V. Nesterova and Andrey V. Evsey
Belarusian State Technological University, 13-A Sverdlova str., 220050 Minsk, Belarus
[1] Institut für Organische Chemie, Technische Universität Dresden, Mommsenstr.13 D-01062 Dresden, Germany

ABSTRACT

Polymers containing hindered amine stabilizers (HAS) were studied by Electron Spin Resonance 'in situ' in order to measure the nitroxyl radical (NR) concentration in processes of thermo and photooxidative degradation. The correlation links between the type of degradation, NR concentration and efficiency of stabilizers were investigated. The role of chemical and physical aspects that could affect the efficiency in polyolefins depending on the conditions of degradation was discussed. It was observed that HAS substances in Poly(ethylene terephthalate) are inactive after thermal processing but produce NR and, therefore, participate in stabilization process during posterior UV aging.

Keywords: Nitroxyl Radical, Hindered Amine, Electron Spin Resonance, Efficiency, Polyethylene, Polypropylene, Poly(ethylene terephthalate).

◊ Based on a paper presented at the 3rd International Conference on 'Polymer Modification, Degradation and Stabilization (MoDeSt 2004)', Lyon, France, 29 August – 2 September 2004.

* Corresponding author E-mail: L_Smoliak@Yahoo.com

INTRODUCTION

One of the most common stabilizer classes for polyolefins are derivatives of 2,2,6,6-tetramethylpiperidine also known as hindered amine stabilizers (HAS). The recent investigations have allowed to clarify generically the mechanism of stabilizing activity for this class [1,2]. Nevertheless many aspects of this mechanism, including linkage between the structure of stabilizer molecule and its efficiency, have no theoretical validation until now. The present conceptions about the factors, that affect the efficiency of stabilizers of HAS class, are summarized below.

Chemical aspects of HAS activity. The chemical (or 'primary') aspects of stabilization action include the initial amine's oxidation to a stable nitroxyl radical (NR) and reactions of this radical with alkyl (R•) and then regeneration with alkoxyl radicals (ROO•), which is the mechanism known as Denisov cycle of NR activity [1,2].

$$R\bullet + \text{\textbackslash NO}\bullet \longrightarrow \text{\textbackslash NOR}$$

$$\text{\textbackslash NOR} + ROO\bullet \longrightarrow ROOR + \text{\textbackslash NO}\bullet$$

Based on the properties for nitroxyls to be regenerated and repeatedly involved in the reactions of interruption of kinetic oxidizing chain, the high efficiency of HAS is a result of the stability, i.e. the longer lifetime, for these radicals [2]. The stability of 2,2,6,6-tetramethylpiperidineoxyls is due to localization of an unpaired electron on the sterically shielded nitroxyl group. The unpaired electron migration on the ordinary bonds of piperidine cycle is impossible, therefore the substituent of 2,2,6,6-tetramethylpiperidine cycle can not influence on stability of the radical. Regarding the main conclusions from this conception (more nytroxyl concentration or more stability of nytroxyls – more efficiency), the considerable difference in the efficiency for the substances of HAS class can not be explained completely by only chemical aspects.

Physical ('secondary') aspects of polymer stabilization. The efficiency of the stabilization against oxidative degradation depends on the consumption and loss of the additives from the polymer matrix. Some experiments showed a direct relationship between such physical parameters as the Solubility (S) and Diffusion coefficient (D) of the stabilizer and the service life of the polymer [3-6].

The physical aspects determine an opportunity of the physical loss for stabilizers, adequate S allows the stabilizer to be deposited in polymer for a long term period. The theoretical models of linkage between the physical properties and the stabilizer efficiency were proposed for the UV-absorbers and the phenolic antioxidants by Billingham [4] and Moisan [5] correspondingly and detailed for the HAS by Malik [6].

The S values of the antioxidant in the polymer are governed by polymer-antioxidant interactions. According to [3], the S of phenolic antioxidants in ethylene polymers is related to the Hildebrand solubility parameter (HP) and the D to the fractional free volume in the polymer.

Conception of secondary aspects does not explain completely causes of high or low S and D, the linkage to the efficiency for the substances with different structure is not clear. Now this is one of the directions to be developed, in particular, by computer simulation of the structure effect on the molecule properties [7]. Theoretical and experimental studies on physical parameters propose the following possible relations between the HAS molecular structure and the efficiency: polymer-like fragments – more S – more efficiency; oligomeric structure – less D – more efficiency [6].

Behavior of NR in polymer during the oxidation can be studied by spectroscopy of Electron Spin Resonance (ESR) and such an investigation might be helpful for understanding of some uncovered aspects of stabilization mechanism of HAS, moreover the concentration of NR ($[C]_{NR}$) can be measured by ESR in polymer 'in situ' [8, 9].

Hypothesis of study. The detection of NR by ESR characterizes the nitroxyl yield from a hindered amine and provides some correlation links between molecular structure of HAS and efficiency in Polyethylene (PE) and Polypropylene (PP). As long as HAS can be deactivated by reaction with acids during degradation of such polymer as Poly(ethylene terephthalate) (PET), NR are unlikely to be detected in PET. Substituents of 2,2,6,6-tetramethylpiperidine cycle in HAS molecule can determine the physical parameters (S), polymer – additive interactions and, through this, affect the HAS / HAS-NR efficiency.

The aim of this work is to study the nitroxyl radicals produced from a series of hindered amine stabilizers in order to estimate the influence of their concentration and yield on the stabilizing activity in polymers.

EXPERIMENTAL

Stabilizers. Commercially available HAS products Chimassorb®944, Tinuvin®622 (Ciba Specialty Chemicals), Nylostab®S-EED (Clariant) and hindered amines described in our previous work [10] were tested. 4-hydroxy 2,2,6,6-tetramethyl-1-piperidinyl-oxyl (4-hydroxy-TEMPO) provided by Fluka® was used as NR standard.

4-hydroxy-TEMPO

Polymer compositions. PE-LD (pure and additive-free, MFI=2.0 g/10min) and stabilizers were repeatedly roll milled (3 min at 160°C), then 0.2-0.3 mm thick films were prepared by compressing molding (1 min at 160°C).

PP (pure and additive-free, MFI=25.0 g/10min) and stabilizers were repeatedly roll milled (3 min at 190°C), then 0.1-0.2 mm thick films were prepared by compressing molding (1 min at 210°C).

PET (medium molar mass polymer, viscosity η_{inh}=0.81) and stabilizers were mixed and oxidized during mixing in laboratory plunger mixer (10 min at 290°C), then 3-4 mm in diameter strands were extruded.

Accelerated aging. The polymer samples (PE and PP films and PET strands) were two-side exposed (with equal time of irradiation for each side) to mercury-quartz lamp DRT-375 (375 W). PE film were then tested at various exposure intervals for elongation properties and Time to 50 % loss of elongation (T_{50}) was calculated.

Chemiluminescence. An Atlas ChemiLume CL400 apparatus was employed in isothermal mode at 180°C in pure oxygen. Oxidation Induction Time (OIT-CL) for PE films before UV aging was determined as a criterion of the thermooxidation stability during processing.

ESR measurements. ESR spectra were collected at 20 °C with an RE 1306 Spectrometer operating at 10 GHz with the following parameters: modulation amplitude 2.0 G, microwave power 10 mW. The spin concentration standard (reference) was solutions of freshly re-crystallized 4-hidroxy-TEMPO in chloroform at various concentrations, containing an exact quantity of radicals [9]. PE and PP cylindrical samples 3-4 mm in diameter were rolled from the films (30×10 mm) to fit the 5 mm in diameter of the ESR sample tube; PET samples were cut from extruded strand 3-4 mm in diameter, 10 mm in length. Samples were placed in the ESR resonator with the symmetry axis along the long axis of the resonator and parallel to the direction of the magnetic field gradient. The $[C]_{NR}$ in polymer was calculated from total NR presence and the sample weight [9]. Polymer samples were tested in 24 hours after processing or UV aging.

The influence of molecular structure on probable additive compatibility with polymer was estimated by calculation of a Hildebrand solubility parameter according to the Askadsky procedures [11].

RESULTS AND DISCUSSION

The ESR measurement data for PE films are presented in Table 1 as the $[C]_{NR}$ and NR yield ($[C]_{NR}$ / $[C]_{HAS}$ relation after processing and aging) values. The trends of efficiency evolution as a function of $[C]_{NR}$ are shown on Figure where OIT-CL and T_{50} are used as the melt thermostabilization (during processing) and photostabilization efficiency criteria correspondingly.

At HAS concentration ($[C]_{HAS}$) of 0.4 %wt. (the hindered amine group content about 1-2×10^{-2} mol/kg) after processing (pre-oxidation under thermal and mechanic shear degradation on rolls and in press) the hindered amines produce NR with yield from 0.13 to 1.30%. Degree of HAS-to-NR transformation (yield) can indicate the intensity of pre-oxidation process (inefficient heat stabilization) or the ability for a HAS to produce NR more easily than another. The HAS with highest values of NR yield before UV aging (0 hrs) at studied concentration are Chimassorb®944 and Nylostab®S-EED, the same stabilizers show the best OIT-CL after processing.

The value of increase of NR amount in polymer after UV exposure (I_{20}/I_0) is to be connected with intensity of photooxidation process. At the high initial yields of NR the additional NR amount produced after UV exposure is relatively lower than at the low ones which is also can determine the I_{20}/I_0 value (as for Chimassorb®944).

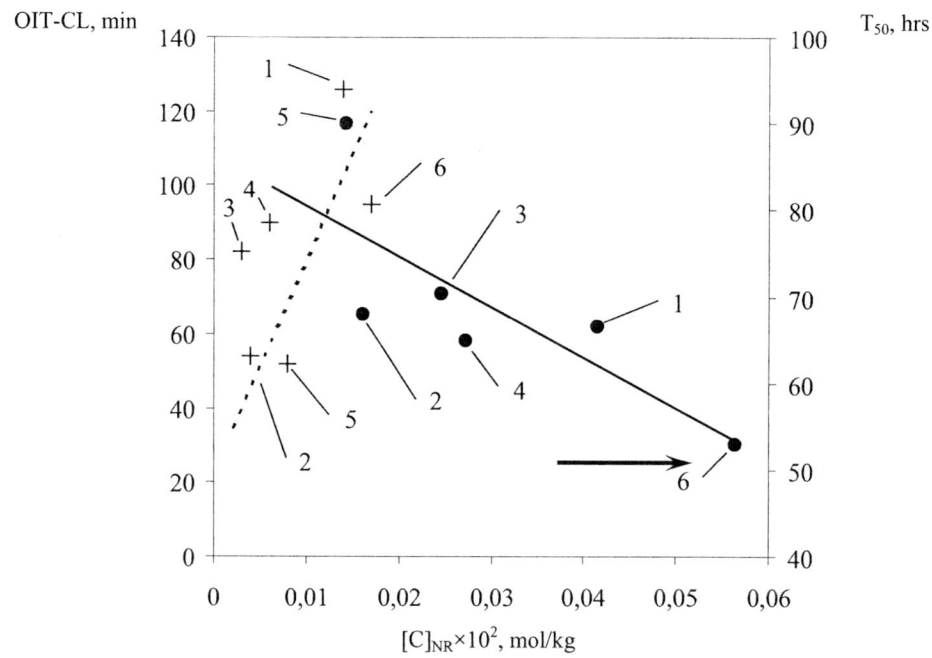

Figure 1. Oxidation Induction Time measured by CL in oxygen at 180° (crosses, dotted trend line) and Time to 50 % loss of elongation under UV aging (spots, solid trend line) for PE films versus Nitroxyl radical concentration (before and after aging respectively) at the concentration of 0.4%wt. for the HAS studied: 1 - Nylostab®S-EED, 2 - HA-2, 3 - HA-3, 4 - HA-4, 5 - Tinuvin®622, 6 - Chimassorb®944

Table 1. Characterization of nitroxyl radicals produced from hindered amine stabilizers in PE (according to ESR-data)

Stabilizer	HP, $(J/cm^3)^{0.5}$	$[C]_{HAS}$, %wt.	UV aging, hrs	$[C]_{NR} \times 10^2$, mol/kg	NR yield, %	I_{20}/I_0
HA-2 [10]	18.7	0.4	0	0.0034	0.36	
		0.4	20	0.0177	1.87	5.20
HA-4 [10]	19.8	0.4	0	0.0064	0.35	
		0.4	20	0.0272	1.51	4.25
Chimassorb®944	20.3	0.4	0	0.0174	1.30	
		0.4	20	0.0563	4.21	3.24
Tinuvin®622	20.3	0.4	0	0.0076	0.54	
		0.4	20	0.0143	1.01	1.88
HA-3 [10]	19.8	0.4	0	0.0023	0.13	
		0.4	20	0.0245	1.36	10.65
		0.8	0	0.0028	0.08	
		0.8	20	0.0282	0.78	10.07
Nylostab®S-EED	20.4	0.4	0	0.0144	0.80	
		0.4	20	0.0414	2.29	2.88
		0.8	0	0.0112	0.31	
		0.8	20	0.0522	1.44	4.65

At the same time, the ability to produce NR may be determined by physical properties of HAS, for example, high melting point or low S. HA-3 is characterized by the lowest NR yield at 0 hrs, the reason of such inactive transformation to NR during PE processing at 160°C could be the high melting point (Mp) - 205°C for this substance, taking into account that other HAS possess Mp below 140°C and produce 3-5 times more NR. However Nylostab®S-EED which is developed for polyamide stabilization and characterized by the highest Mp (about 270°C) gives close to the other HAS values of $[C]_{NR}$, yields and I_{20}/I_0 during PE processing and UV aging.

As said above, the values of NR yield before aging (0 hrs) can correspond to the degree of PE oxidation during processing, the high OIT-CL testifies good thermal stabilizing efficiency of HAS. Both values are correlated with $[C]_{NR}$ after processing (See 0 hrs lines in Table 1 and Figure) and allow us to presume that the rule 'more concentration – more efficiency' is true in this case. Any influence of the physical effects (S and D) seems to be almost negligible in conditions of melt oxidation therefore the chemical factors are dominant.

At the same time, the efficiency of HAS during UV aging corresponds to the less $[C]_{NR}$ detected in PE films (Figure). A higher nitroxyl content is probable a result of intensive polymer oxidation process which is the indication of less effective stabilization effect. In PE films (solid phase) the compatibility influence can play a more substantial role, one can see that the main trend, according to HP values in Table 1, is the following 'more polar substance – more nitroxyl concentration after UV aging'. Considering the fact that a less HP value corresponds to the more probability for substance to be compatible with non-polar PE, a higher nitroxyl content corresponds to the less compatible stabilizers.

Another point to be considered is the influence of HAS concentration in polymer on activity of NR. The chemical aspects presume direct correlation between $[C]_{HAS}$ and $[C]_{NR}$ while the physical ones may give more complicated dependence connected, for example, with S limit. It can be observed from Table 1 that the doubled amounts of HAS added to PE (Nylostab®S-EED and HA-3 at 0.4 and 0.8 %wt.) do not show the correspondent or any substantial increase of $[C]_{NR}$ detected during UV aging which is, from our point of view, an additional evidence that the HAS molecules act as deposits of NR. In this case, the long-term testing of NR content in HAS-stabilized samples seems more promising to prove the correlation of S and D values with efficiency.

Table 2 presents some results of $[C]_{NR}$ measurements in PP and PET samples. The main difference between PE and PP is more substantial transformation of HAS to NR during PP aging, that is the illustration of a more active radical process in PP. Final yield of NR after 20 hrs of UV amounts to 14.5 % from PP versus about 2.3 % from PE at the same stabilizer content (Nylostab®S-EED at 0.4% wt.). The more active NR yield from PP than from PE samples is probable connected with the more important role of alkoxyl radicals scavenging during PP aging.

Similar to PE, doubled amount of HAS in PP does not show direct increase of $[C]_{NR}$ before and after UV aging which can serve as an evidence of the S influence.

Weak ESR signals do not detect NR in PET samples after processing (the amounts are under sensitivity of our ESR instrument) which is probable due to deactivation of HAS by acid groups. The presence of NR in the same samples after UV aging (Table 2) indicates that nitroxyls participate in the PET photooxidative degradation reactions.

Table 2. Characterization of nitroxyl radicals produced from hindered amine stabilizers in PP and PET (according to ESR-data)

Polymer	Stabilizer	$[C]_{HA}$, %wt.	UV aging, hrs	$[C]_{NR} \times 10^2$, mol/kg (ESR)	NR yield, %
PP	Nylostab®S-EED	0.2	0	0.0831	9.19
		0.2	20	0.1773	19.59
		0.4	0	0.0458	2.53
		0.4	20	0.2624	14.50
PET	HA-4 [10]	0.2	0	undetected	<0.01
		0.2	40	0.0035	0.39
	Nylostab®S-EED	0.2	0	undetected	<0.01
		0.2	40	0.0023	0.26

It should be emphasized that the results are the only initial phase of the study which can not cover all aspects of the problem concerned. In this work we have not drawn conclusions on the correlation between molecular structure of HAS and its efficiency except the influence of compatibility (HP) on the NR concentration and yield. With more experimental data it will be able to study the influence of such parameters as the presence of mono- or bis-radicals (for bis-substituted HAS), distance and type of 'bridge' between groups on the NR activity and stabilization efficiency during polymer oxidation. There also should be more minutely considered the role of polymer-like structures, for example, long aliphatic chains in HA-2 and Chimassorb®944 (similar to polyolefins) and highly polar phthalate fragments of Nylostab®S-EED, HA-3 and HA-4 (similar to PET).

CONCLUSIONS

Thus, the ESR spectroscopy 'in situ' is useful method of HAS investigation in polymers, the measurements of nitroxyl concentration in different conditions of aging provide necessary insight into the mechanisms and roles of chemical and physical aspects in hindered amine stabilization action.

Concentration of nitroxyls produced from HAS during processing and UV aging depends on the stabilizer structure, polymer matrix and conditions of aging. After thermooxidation during processing, the yield of nitroxyls is much higher in PP than in PE, the amount of NR is increased in 2-5 times for both PP and PE samples during UV aging in similar conditions of treatment. The stabilizing efficiency of HAS in PE corresponds to the nitroxyl concentration for thermooxidation in melt and for accelerated UV aging in solid film.

In contrast to polyolefins, nitroxyl radicals are undetected in PET after processing but appear after UV aging with the yield about 0.3-0.4 % of initial amine concentration.

ACKNOWLEDGEMENTS

The authors wish to thank Dr. Jan Malik (Clariant) for his help and fruitful discussion. Leonid Smoliak is grateful to the DAAD (The German Academic Exchange Services) for financial support of his research stay in TU Dresden.

REFERENCES

[1] Gugumus F. Current trends in mode of action of hindered amine light stabilizers. *Polym. Degrad. Stab.* 1993; 40: 167-215.
[2] Scott Gerald Ed. Mechanism of Polymer Degradation and Stabilization. *Elsevier Applied Science.* London, 1990.
[3] Boersma A. Mobility and Solubility of Antioxidants and Oxygen in Glassy Polymers. *J. Appl. Polym. Sci.* 2003, 89, 2163 –2178.
[4] Calvert PD, Billingham NC. Loss of additives from polymers: a theoretical model. *J. Appl. Polymer Sci.* 1979; 24 (3): 357-370.
[5] Moisan JY. Diffusion des additifs du polyethylene-1. *Eur. Polym. J.* 1980; 16: 979-987.
[6] Malik J, Hrivik A, Tomova E. Diffusion of hindered amine light stabilizers in low density polyethylene and isotactic polypropylene, *Polym. Degrad. Stab.* 1992; 35: 61-66.
[7] Nath Sh.K., de Pablo J.J., DeBellis A.D. Molecular simulation of physical properties of hindered-amine light stabilizers in polyethylene. *J. Am. Chem. Soc.*,1999, 121, 4252-4261.
[8] Step EN, Turro N.J, Gande ME, Klemchuk PP. Mechanism of Polymer Stabilization by Hindered-Amine Light Stabilizers (HALS). *Macromolecules* 1994; 27: 2529-2539.
[9] Dudler V. Compatibility of HALS-nitroxides with polyolefins. *Polym. Degrad. Stab.* 1993, 42, 205-212.
[10] Smoliak L.Y., Prokopchuk N.R. Estimation of parameters that correlate molecular structure of hindered amines with their stabilizing efficiency. *Polym. Degrad. Stab.* 2003; 82, Issue 2, P. 169-172.
[11] Askadskii AA, Kondraschenko VI. Computer-Based Materiology of Polymers. Moscow: *Nauchnii mir* (Scientific World), 1999.

Chapter 7

PHYSICAL GROUNDS FOR CLUSTER MODEL OF POLYMER AMORPHOUS STATE STRUCTURE

G. V. Kozlov and G. E. Zaikov[1]

Kabardino-Balkarian State University, 173 Chernyshevskogo str.,
Kabardino-Balkaria, Nal'chik 360004, Russia
[1] N.M.Emanuel Institute of Biochemical Physics Russian Academy of Sciences,
4 Kosygin str., Moscow 119991, Russia

1. FUNDAMENTALS

The well-known experimental observations have become the pre-requisites for creation of the cluster model of the polymer amorphous state structure. As shown by Haward et al. [1, 2], at deformation of glassy polymers outside the yielding range (on the plateau of induced rubbery state) they obey the regularities of the rubbery elasticity theory. In this case, the polymer behavior under high deformations is described by either the Langevin equation [3, 4] or the Gaussian interpretation [5], when the polymer chain does not approach completely stretched state, and correlation between the real stress, σ^{real}, and the stretch degree, λ, for the axial tension is presented as follows [5]:

$$\sigma^{real} = G_p\left(\lambda^2 - \lambda^{-1}\right), \tag{1}$$

where G_p is the so-called strain hardening modulus.

Formally, G_p value allows determination of macromolecular entanglements network frequency, ν_{ent}, in accordance with the well-known formulae of the rubber elasticity [5]:

$$M_{ent} = \frac{\rho KE}{G_p}, \tag{2}$$

$$v_{ent} = \frac{\rho N_A}{M_{ent}}, \quad (3)$$

where ρ is the polymer density; R is the universal gas constant; T is the test temperature; M_{ent} is the molecular mass of chain segment between entanglements; N_A is the Avogadro number.

However, the attempts to assess M_{ent} (or v_{ent}) via G_p values, determined from the equation (1), have given incredibly low values of M_{ent} (or unrealistically high v_{ent}), which conflict with the requirements of the Gaussian statistics. This statistics suggests the presence of, at least, 13 units in the chain section between entanglements [6]. Possible reasons for such inconsistency under consideration of entanglements as traditional macromolecular "backlashes" (Figure 2a) are discussed in detail in the work [5]. The principally different solution of this problem, suggested in ref. [7], indicates that besides the network of the above-mentioned backlashes, there is another type of entanglements in the polymer glassy state, the structure of cross-link points in which is analogous to crystallites with extended chains (CEC), schematically depicted in Figure 2b. Such cross-link point possesses quite high functionality, F (by the cross-link point functionality the number of chains coming out of it is meant [8]). Hereinafter, this cross-link point will be named the cluster.

The cluster consists of segments of different macromolecules, and the length of each segment is postulated equal to the statistical segment length, l_{st}, of the chain "rigidity segment" [9]. In this case, the effective (real) molecular mass of the chain segment between clusters, M_{cl}^{eff}, can be calculated as follows [8]:

$$M_{cl}^{eff} = \frac{M_{cl} F}{2}, \quad (4)$$

where M_{cl} is the molecular mass of the chain segment between clusters, calculated by the equation (2).

Obviously, at quite high values of F one can obtain reasonable values of M_{cl}^{eff} meeting requirements of the Gaussian statistics. Further on, for the purpose of distinguishing parameters of the cluster entanglement network and macromolecular hooking network indices "cl" and "h" will be used, respectively. Thus, as assumed in the model suggested in the work [7], the structure of the polymer amorphous state represents areas consisting of collinear closely packed segments, composed of different macromolecules (clusters), submersed into a packless matrix. Simultaneously, the clusters play the role of multifunctional cross-link points of physical entanglements. The value of F can be estimated (back again in the framework of the rubber elasticity concept) as follows [10]:

$$F = \frac{2G_\infty}{kTv_{cl}} + 2, \quad (5)$$

where G_∞ is the equilibrium shear modulus; k is the Boltzmann constant.

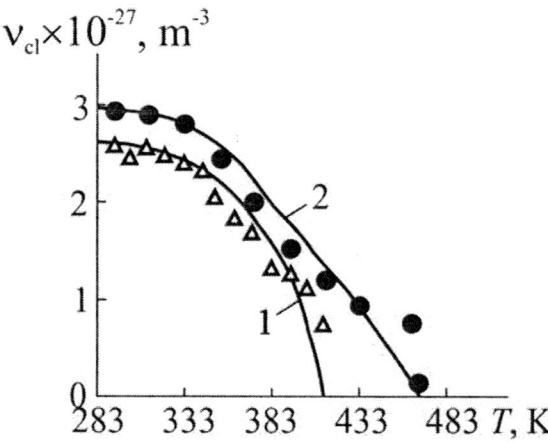

Figure 1. Dependencies of macromolecular entanglement cluster network frequency, v_{cl}, on temperature, T, for PC (1) and PAr (2) [7]

Figure 1 shows $v_{cl}(T)$ dependencies for polycarbonate, synthesized from bisphenol A (PC) and polyarylate (PAr). These dependencies indicate v_{cl} decrease with T increase that suggests thermofluctuational origin of clusters (the local order zones). Moreover, the mentioned dependencies display two characteristic temperatures. The first of them is the glass transition temperature of the polymer, T_g. It determined complete decay of clusters that corresponds to an inflection on $v_{cl}(T)$ curves and is located approximately by 50 K below T_g. Recently, in the framework of the local order concepts also, it has been shown [11, 12] that temperature T_g is associated with defrosting of the segmental mobility in packless zones of the polymer. This means that in the framework of the cluster model T_g can be associated with devitrification of the packless matrix. For the same polymers, $F(T)$ dependencies are of similar shape (Figure 2).

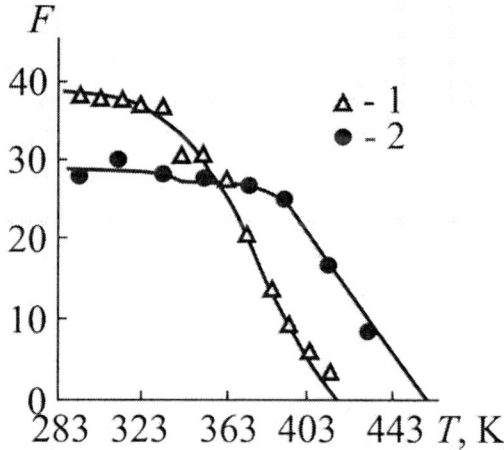

Figure 2. Dependencies of cluster functionality, F, on temperature, T, for PC (1) and PAr (2) [7]

Two basic models of the local order zones in polymers (with folded and extended chains, refer to Figure 1) possess the point of coincidence: they play the role of cross-link points of macromolecular entanglements physical network [7, 13]. However, their response to mechanical deformation should be significantly different: at high deformations the zones with folded chains ("packs") are capable of unfolding and forming stretched conformations, and clusters are incapable of doing that and polymers can be deformed by stretching "transitional" (cluster linking) chains, i.e. by their orienting in direction of applied stress. Turning back to the analogy with crystalline morphology of polymers, let us note that high deformations of amorphous-crystalline polymers (especially for polyethylenes, which give 1,000 – 2,000%) are realized due to unfolding of crystallites [14]. That is why using values of border deformations of polymers one can obtain arguments for the benefit of one or another type of the local order zones in the polymer amorphous state that was performed in the work [15].

It was suggested [33] that the "packs" are capable of delivering parts of chains to "interstices" (i.e. to inter-pack zones) by unfolding, thus acting similarly to crystallites with folded chains (CFC). On this evidence, of special attention is a great difference between maximum stretch degrees, λ_{max}, of amorphous elastic polymers and such amorphous-crystalline polymers as high density polyethylene (HDPE) and polypropylene (PP), for which unfolding of crystallite wrinkles at high deformations was proved in experiments ($\lambda_{max} \approx 1.6$ for PC, $\lambda_{max} \approx 6$ for PP, and $\lambda_{max} \approx 13$ for HDPE at room temperature [16]). Explanation of this difference requires quantitative assessments [15].

Gent and Madan [16] have suggested that stretching proceeds due to extension of crystalline and amorphous chain sequences. In this case, the value λ_{max} can be expressed via f, the number of times a molecule passes through the same crystallite (or "pack"):

$$\frac{1}{\lambda_{max}} = \frac{K}{f} + \frac{(1-K)^{1/2}}{n_{st}^{1/2}}, \qquad (6)$$

where K is the crystallinity degree; n_{st} is the number of equivalent statistical units between entanglements in the polymer melt. Usually, n_{st} varies within the range of 100 – 300; as this value is not sufficient for assessment results, for all polymers used in the work [15], it was accepted equal 225.

Clearly the value f determines the number of folds formed by a macromolecule in CFC (or the "pack"). Formation of folds requires meeting the condition $f \geq 2$. Assessment results of f for HDPE and PP are shown in Table 1. They have expectedly indicated that macromolecules of the former compound are folded over 50 times, and the latter ~5 times that correlates well with the known data [16].

For amorphous glassy polymers the equivalent of the crystallinity degree, K, is the relative part of clusters, φ_{cl}. The value φ_{cl} can be assessed as follows. Total length of macromolecules, L, per specific polymer volume under their dense packing equals [17]:

$$L = S^{-1}, \qquad (7)$$

where S is the macromolecule cross-section. The length of statistical segment, l_{st}, is assessed as follows [18]:

$$l_{st} = l_0 C_\infty, \qquad (8)$$

where l_0 is the skeletal bond length in the backbone; C_∞ is the characteristic relation, which is the statistical flexibility index of the macromolecule [19].

Total length of segments in the clusters, L_{cl}, per specific volume of the polymer is assessed as follows [20]:

$$L_{cl} = l_{st} v_{cl}, \qquad (9)$$

and φ_{cl} is assessed by the relation:

$$\varphi_{cl} = \frac{L_{cl}}{L}. \qquad (10)$$

As combined the equations (7) – (10) give the final formula for φ_{cl} calculation [20]:

$$\varphi_{cl} = S l_0 C_\infty v_{cl}. \qquad (11)$$

The values of φ_{cl} for PC and PAr, calculated by equation (11), are shown in Table 1. Using experimental values of λ_{max}, these results indicate that in all cases $f < 1$ (Table 1).

Table 1. Assessment of folding parameter f for amorphous and amorphous-crystalline polymers [15]

Polymer	T, K	λ_{max}	K	φ_{cl}	f
HDPE	293	13	0.69	-	53
PP	293	6	0.50	-	4.7
PC	333	1.91	-	0.33	0.70
	353	2.23	-	0.29	0.73
	373	2.15	-	0.24	0.57
	393	2.36	-	0.19	0.52
	413	2.75	-	0.11	0.35
PAr	293	1.66	-	0.50	0.89
	313	1.67	-	0.38	0.68
	333	1.66	-	0.37	0.66
	353	1.76	-	0.31	0.59
	373	1.66	-	0.25	0.45
	393	1.70	-	0.19	0.35
	413	1.75	-	0.16	0.30
	433	1.80	-	0.13	0.25
	453	1.86	-	0.11	0.22
	473	1.97	-	0.01	0.02

This proves the absence of macromolecule folding in the local order zones (macromolecular entanglement points of physical network) of these polymers. Note also that unfolding in crystallites of amorphous-crystalline polymers is initiated at deformations of about 50 – 100% [21]. Substituting $\lambda_{max} \sim 2$ and $\varphi_{cl} \approx 0.20$ for amorphous zones of HDPE into equation (6), we obtain $f < 1$. This means that in amorphous-crystalline polymers macromolecules are folded in the crystalline zones only.

Thus calculations executed in the work [15] have shown that the crystallite analogue with stretched chains – the cluster, is the most probable type of supersegmental structures in the polymer amorphous state (Figure 2b).

The well-known inhomogeneity of elastic deformation of amorphous glassy polymers [22, 23] allows their acceptance as heterogeneous systems. This statement is also true for the amorphous phase of amorphous-crystalline polymers [24, 25]. Nevertheless, the behavior of polymers of both classes is successfully described by both continual models (remind that primarily the known Dagdale model was developed for metals [26]) and molecular concepts. In this connection, the question about the scale which may be considered to be the lower border of continual models' applicability is raised.

One more problem is what the probability of an inhomogeneous molecular system (which unambiguously the amorphous glassy polymer is) consideration as a two-phase system is. If there is such a probability, this will allow application of the so-called "composite" models to description of the amorphous polymer behavior. These models are well developed and successfully used, for example, for description of artificial two-phase systems, which also include the filled ones. These two problems were discussed in the work [27].

Fellers and Huang [28] have applied the fluctuation statistical theory to description of crazing in amorphous polymers. They have deduced an expression for assessing the polymer volume, V_0, in which the fluctuation probability is of the unit value:

$$\frac{\sigma_c V_0^{1/2}}{(2kT_0 B)^{1/2}} = 3.87, \tag{12}$$

where σ_c is the crazing stress; T_0 is the equilibrium temperature, the lower border of which range is the glass transition temperature, T_g; B is the polymer volumetric modulus bound to the Jung modulus, E, by the following relation [29]:

$$B = \frac{E}{3(1-v)}, \tag{13}$$

where v is the Poisson index.

The distance between clusters, R_{cl}, can be assessed using a simple formula as follows [30]:

$$R_{cl} = 18\left(\frac{2v_{cl}}{A}\right)^{-1/3}, \text{Å}. \tag{14}$$

Table 2 shows comparison of R_{cl} and linear dimension L_0, at which the fluctuation probability is of the unit value ($L_0 = V_0^{1/3}$) for 5 amorphous glassy polymers. As indicated by the Table data, parameters R_{cl} and L_0 are close by both the absolute values and variation tendencies. This means that in the framework of characteristic dimensions of the cluster structure amorphous polymer (or amorphous phase of amorphous crystalline polymer) may be considered as the inhomogeneous system [27].

Table 2. Comparison of characteristic dimensions of the fluctuation theory, L_0, and the cluster model, R_{cl}, for amorphous glassy polymers [27]

Polymer	L_0, Å	R_{cl}, Å
Polystyrene	76.4	36.1
Poly(methyl methacrylate)	31.7	31.6
Polyvinylchloride	54.0	27.1
Polycarbonate	39.7	31.1
Polysulfone	36.4	25.0

Katsnelson [31] has suggested the following definition of the substance phases: they are "…states of the substance able, being in touch, to exist simultaneously in equilibrium with one another. Obviously different properties are corresponded to different phases. Hence, it should be taken into account that by different phases … parts of a body are meant, related to the solid phase, but possessing different structure and properties". Clusters and packless matrix which in accordance with the cluster model [7] are the main structural elements of the polymer amorphous state meet the above definition, at least, partly. It is known that these elements possess different mechanical properties [32] and different glass transition temperatures [33]. All these facts together give an opportunity to consider the amorphous state of a polymer to be a quasi-two-phase state, disclaiming full strength of the definition [27].

Let us now consider applicability of models, developed for two-phase filled polymers, for describing mechanical behavior of amorphous glassy and amorphous-crystalline polymers. By the analogy with the dispersion theory of strength the shear stress, τ_a, of the composite flow behavior (or for the case under consideration, the amorphous polymer structure) is presented by the equation as follows [34]:

$$\tau_a = \tau_m + \frac{Gb}{R_{cl}}, \qquad (15)$$

where τ_m is the shear flow stress of a packless matrix; b is the Burgers vector.

For polymers $\tau_m = 0$ [32, 35], and in the case of aggregation of the filler particles (i.e. association of segments to clusters) the equation (15) for the polymer amorphous state can be presented as follows [34]:

$$\tau_a = \frac{Gb}{k(d)R_{cl}}, \qquad (16)$$

where $k(d)$ is the aggregation parameter.

The equation (16) can be used for description of the temperature dependence of τ_a displaying one principal difference from filled polymers. As G value for amorphous and amorphous-crystalline polymers the macroscopic shear modulus should be used instead of its value for packless matrix. This is explained so that, contrary to τ_a value, the value of G for the polymer amorphous state is determined by the structure of both quasi-phases [32]. The plain truth is that application of G value to the equation (16) for packless matrix only would mean determination of the cluster property (τ_a) from properties of another component of the structure only, which is physically meaningless.

Figure 3 compares experimental and calculated (by the equation (16)) temperature dependencies of τ_a (under the condition $k(d)$ = const for every polymer). This comparison indicates good coincidence of the results thus proving consistency of equation (16) use for description of amorphous and amorphous-crystalline polymer properties.

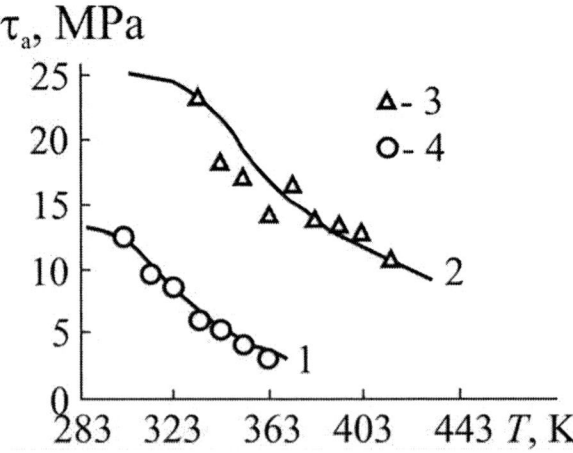

Figure 3. Experimental (1, 2) and calculated by equation (16) temperature dependencies of the shear flow stress, τ_a, for HDPE (1, 3) and PC (2, 4) [27]

The aggregation parameter $k(d)$ value was assessed by equation (16) for 8 polymers at T = 293 K. Obviously the physical meaning of it must be analogous to functionality, F, for the cluster structure of polymers. The relation between $k(d)$ and F values is shown in Figure 4, from which the expected correspondence follows. This observation and the fact that $k(d)$ is independent of temperature induce an assumption that for the polymer amorphous state $k(d)$ represents some qualitative measure of the polymer tendency to cluster formation at the segmental level [27].

For calculating filled polymer strength, σ_d, a series of empirical equations was deduced, for example [36]:

$$\sigma_d = a - c\varphi_{fl}, \tag{17}$$

where a and c are constants; φ_{fl} is the volumetric content of the filler.

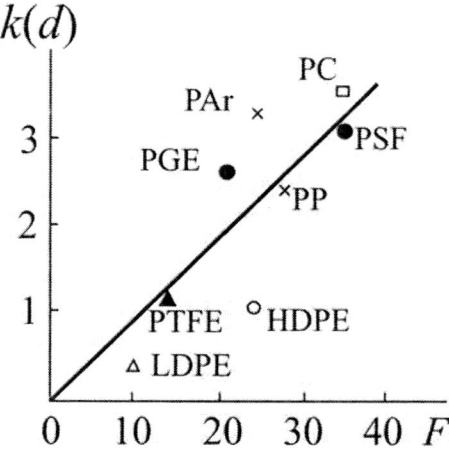

Figure 4. The relation between aggregation parameter, $k(d)$, and cluster functionality, F, for amorphous and amorphous-crystalline polymers [27]

Figure 5 shows dependencies of the degradation stress, σ_d, on φ_{cl} for three polymers (PC, PAr, and HDPE; for the latter results were obtained by the blow tests [37]). The Figure indicates that strength of the mentioned polymers is described by a simple equation as follows [27]:

$$\sigma_d \approx 119 \varphi_{cl} \text{ (MPa)}, \tag{18}$$

which at $a = 0$ and $c = -119$ MPa is analogous to the equation (17).

Thus, under some conditions, the "composite" models can be successfully applied to description of amorphous and amorphous-crystalline polymers' behavior in the framework of the quasi-two-phase cluster model.

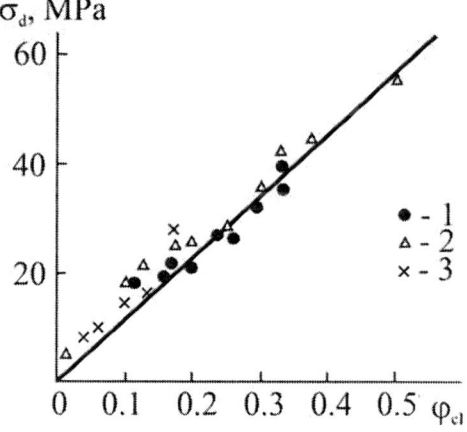

Figure 5. Destruction stress, σ_d, dependence on relative fracture of clusters, φ_{cl}, for PC (1), PAr (2) and HDPE (3) [27, 37]

In this case, obvious possibility for obtaining quantitative correlation between the theory and the experiment exists under the condition of heretophase structure of the amorphous state and consistency of the cluster model [38].

To finishing this Section, let us note that in the presence of the local order in the polymer amorphous state (irrelatively to particular model of its zones) there are the most general strict mathematical proofs. In accordance with the Ramsey theorem proved in the theory of numbers, any quite great amount of numbers, points or objects, $i > R(i, j)$, (in the case under consideration, statistical segments) necessarily contains highly oriented subsystem from $N_j \le R(i, j)$ such segments. That is why the absolute irregularity of large systems (structures) is impossible [39, 40].

2. THERMODYNAMICS OF THE LOCAL ORDER FORMATION

As shown in the present Section, cluster structure formation is the integral part of much more general concepts, for example, the theory of hierarchy systems evolution [41 – 46].

Correlation between the specific Gibbs function of intermolecular interactions, $\Delta \overline{\widetilde{G}}^{im}$ (im symbol means intermolecular or, in the present case, intersegmental type of interactions; symbol "-" indicates the specific type of the value; symbol "~" outlines heterogeneous type of the system; hereinafter, for the sake of simplicity it will be denoted as $\Delta \widetilde{G}^{im}$) and the melting temperature, T_m, [43, 45, 46] was chosen for experimental validation of the physicochemical theory of chemical systems evolution [41]. Selection of these parameters is stipulated in the works [43 – 46].

First of all, the fundamentals stated in ref. [78], necessary for better understanding of the material below should be reminded.

It is common knowledge that the Gibbs-Helmholtz equation is valid for processes proceeding in simple closed systems:

$$\left[\frac{\partial(\Delta G/T)}{\partial(1/T)} \right]_p = \Delta H, \tag{19}$$

where ΔG and ΔH are changes of the Gibbs function and enthalpy during the process, respectively; T and p are temperature and pressure, respectively.

If it is accepted that in a definite temperature range ΔH is independent of T, then for a non-equilibrium phase transition (self-assembly of an individual substance) at temperature T the following relation is valid:

$$\Delta G^{im} = \left(\Delta H_m^{im} / T_m \right) \left(T_m - T \right) = \left(\Delta H_m^{im} / T_m \right) \Delta T = \Delta S_m^{im} \Delta T, \tag{20}$$

where ΔG^{im} is the Gibbs function change during crystallization (self-assembling) of the studied substance from overcooled state at $T = T_m - \Delta T$, ΔH_m^{im} is the enthalpy change during crystallization (solidification); ΔS_m^{im} is the crystallization entropy (the change of entropy at the phase transition).

It has been suggested [41, 43, 45] to use the equation (20) for open systems, the composition and T_m values of which vary negligibly. Further on, the possibility of application of this equation was displayed for various chemical compounds melting at $T_m < 373$ K and condensing at constant standard temperature $T = T_0 = 298$ K [43 – 46]. At the extended approach, for these cases the equation (20) should be represented as follows:

$$\Delta G_i^{im} = \left(\Delta H_{mi}^{im}/T_{mi}\right)\left(T_{mi} - T_0\right) = \Delta S_{mi}^{im}\Delta T, \qquad (21)$$

where index $i = 1, 2, \ldots, n$ indicates different substances. In shape, the equation (21) represents the analogue of correlation (20). Simultaneously, these equations are principally different in the following. In equation (20) ΔG^{im} is a variable characterizing non-equilibrium transition of an individual substance in the system at any temperature $T < T_{mi}$. Values of ΔH_m^{im}, ΔS_m^{im} and T_m belong to this individual substance and are accepted as constant values. As a whole, the equation (20) represents the functional dependence $\Delta G^{im} = f(T)$.

In equation (21) ΔG^{im} is the variable related to non-equilibrium transitions of different compounds with different melting temperatures, T_{mi}, at a standard (constant) temperature, T_0. In this case, the equation (21) represents the function $\Delta G_i^{im} = f(T_{mi})$ [45]. The technique for calculating $\Delta \widetilde{G}^{im}$ for polymers is described in the work [47].

Figure 6 shows $\Delta \widetilde{G}^{im}$ dependence on $\Delta T = T_g - 293$ K for 15 amorphous glassy, amorphous-crystalline polymers and polymer networks [47]; $\Delta \widetilde{G}^{im}$ value is given in kcal/mol. As might be expected, linear decrease of $\Delta \widetilde{G}^{im}$ with ΔT (or, intrinsically, T_g) increase is observed. Yet more important fact is that the straight line plotted in Figure 6, which approximates well the results obtained, corresponds to the data of the works [43 – 46] shown in $\Delta \widetilde{G}^{im}$ - ΔT coordinates for absolutely different chemical compounds, but at 10:1 scale by $\Delta \widetilde{G}^{im}$ axis.

The latter circumstance is provided by, approximately, the order of magnitude difference between molar volumes of the segments (which are also kinetically independent fragments) in compounds, used in the work [47], which is proved by the plot in Figure 7 ($\Delta \widetilde{G}^{im}$ is measured in cal/g).

As follows from this graph, $\Delta \widetilde{G}^{im}(\Delta T)$ dependence for the mentioned polymers corresponds, both qualitatively and quantitatively, to the data shown in refs. [43, 45, 46]. Principally, this allows calculation of the segment dimension, which is different for different polymers. Deviation from the graph plotted for polymers with high T_g indicates the features of supersegmental structure of these substances [43].

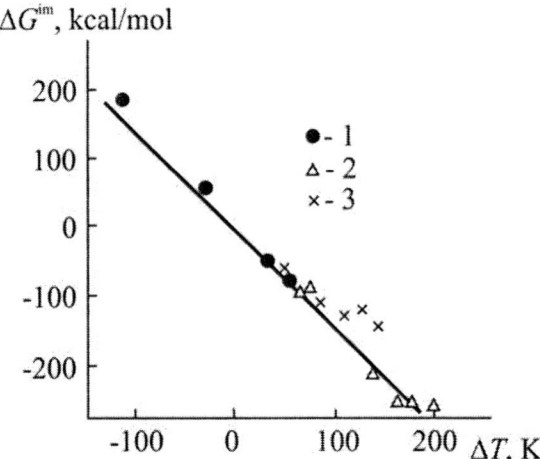

Figure 6. Dependence of the Gibbs specific function of non-equilibrium phase transition, $\Delta \widetilde{G}^{im}$ (kcal/mol), on $\Delta T = T_g - 293$ K for amorphous-crystalline (1) and amorphous glassy (2) polymers, and polymer networks (3). The straight line is plotted in accordance with ref. [43] at 10:1 scale by $\Delta \widetilde{G}^{im}$ axis [47]

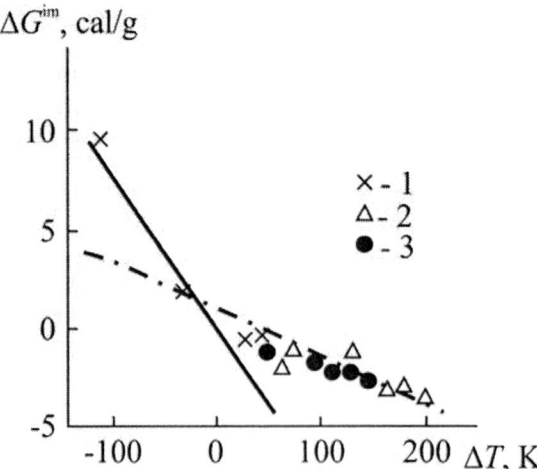

Figure 7. Temperature dependence of the Gibbs specific function $\Delta \widetilde{G}^{im}$ (cal/g) for amorphous-crystalline (1), amorphous (2) polymers and polymer networks (3) [47]. The straight line is plotted with respect to data from ref. [43]

Data from Figures 6 and 7 indicate that the cluster model postulated in ref. [7], based on the existence of the local order in the amorphous state of the polymers, qualitatively and quantitatively conforms with yet more general macrothermodynamic hierarchy model [41 – 46], occupying the corresponded energetic niche in the hierarchy of the real world structure. Graphs in Figures 6 and 7 demonstrate direction of the polymer structure evolution during its physical aging. Striving of the polymer structure to equilibrium means $\Delta \widetilde{G}^{im}$ striving to the minimum (i.e. $\Delta \widetilde{G}^{im}$ shift towards lower negative values) and increase of the local order degree, respectively, which is accompanied by T_g increase [48]. Polymer "rejuvenation", which is the process opposite by its thermodynamic directivity to the above-considered one, is also possible. In practice, this is realized by "injection" of energy (for example, mechanical) to the polymer [49].

Regularities (the Gibbs-Helmholtz-Gladyishev equation), shown in Figures 6 and 7, are true for both different polymers and a single polymer at varying temperature, T (the Gibbs-Helmholtz equation). Figure 8 shows the $\Delta \widetilde{G}^{im}(\Delta T)$ dependence for PC, which also quantitatively coincides with the one shown in Figure 6. Therefore, equations (20) and (21) are fulfilled for polymers simultaneously, i.e. formation of supersegmental structures represents a non-equilibrium transition resulting in formation of non-equilibrium structures. Typically, the beginning of their formation corresponds to glass transition, i.e. to the transition from equilibrium devitrified state to low non-equilibrium glassy-like one.

Finally, it should be noted that $\Delta \widetilde{G}^{im}$ values "regulating" formation of supersegmental structures in polymers are definitely associated with molecular characteristics of the latter. Since polymer is the solid consisting of long chain macromolecules, it should be expected that the most important (or at least one of the most important) property is flexibility of the polymer chain, which can be expressed with the help of characteristic relation C_∞ [19, 50]. That is why Figure 9 shows $\Delta \widetilde{G}^{im}(C_\infty)$ dependence, which clearly displays the tendency to $\Delta \widetilde{G}^{im}$ increase (and consequently, T_g decrease) with the chain flexibility.

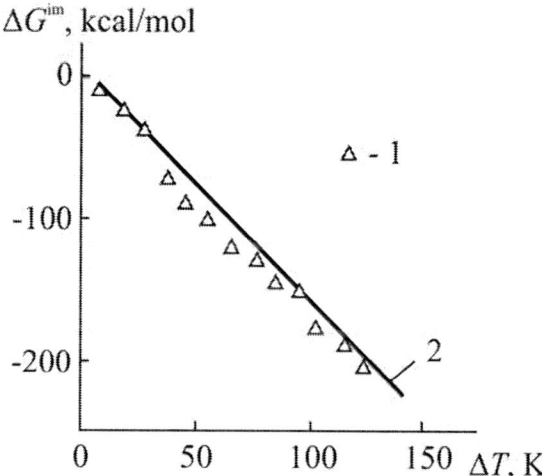

Figure 8. Dependence of the Gibbs specific function of non-equilibrium phase transition, $\Delta \widetilde{G}^{im}$, on $\Delta T = T_g - T$ for PC (1). Line (2) is plotted in accordance with Figure 6 [47]

The unique noticeable deviation, detected for polystyrene may be stipulated by the well-known specificity of its chemical structure [51]. Correlation of $\Delta \widetilde{G}^{im}(C_\infty)$ dependence fully conforms with the previously postulated [51] T_g increase with the polymer chain rigidity (which is now thermodynamically substantiated).

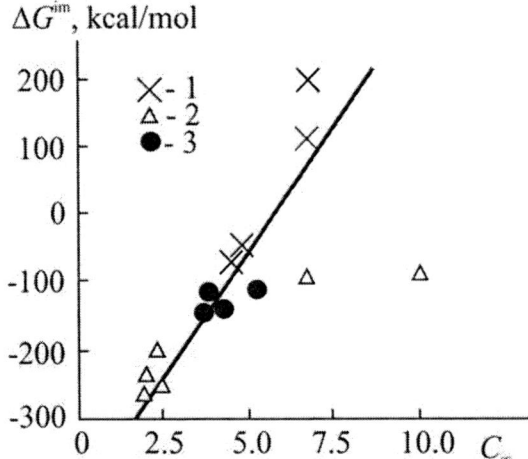

Figure 9. Dependence of the Gibbs specific function, $\Delta \widetilde{G}^{im}$, of non-equilibrium phase transition on characteristic relation C_∞ for amorphous-crystalline (1), amorphous glassy (2) polymers and polymer networks (3) [47]

Thus $\Delta \widetilde{G}^{im}$ (ΔT) dependencies for supersegmental structure of polymers, obtained in the framework of macrothermodynamic hierarchy model, qualitatively and quantitatively conform to previously obtained analogous correlations for a broad selection of substances [43, 45, 46, 52]. This proves the reality of these structures for the polymer amorphous state. Equations (20) and (21) are equally applicable to description of thermodynamic behavior of these structures and can be used for their quantitative simulation. If more rigid calculations are required, corrections for heat capacity change at phase transitions must be introduced [52].

3. POLYMER STRUCTURE REGULARITY AND CLUSTER MODEL

It should be expected that formation of the local order zones will affect the general regularity of the polymer amorphous state structure. At the present time, there are some methods allowing characterization of order (or disorder) of the polymer structure, and they will be compared below with parameters characterizing the cluster structure of these materials.

The possibility of using the Mooney-Rievlyn equation constants for characterization of the local order; therefore, the principal difference from the methods previously used for rubbers only is application of this approach to solid polymers [53]. The simplest form of the Mooney-Rievlyn empirical equation is as follows [55]:

$$f^* = 2C_1 + 2C_2\lambda^{-1}, \tag{22}$$

where f^* is the reduced stress; $2C_1$ and $2C_2$ are equation constants; λ is the stretch degree. The value of f^* is determined by the follow relation [56]:

$$f^* = \frac{\sigma}{\lambda - \lambda^{-2}}, \qquad (23)$$

where σ is the nominal stress, i.e. the one calculated using the initial cross-section of the sample.

The equation (22) is widely applied to the study of mechanical properties of rubbers. Basing on works by different authors, Boyer [56, 57] has made an assumption that the relation $2C_2/2C_1$ can be the measure of the short range order in cross-linked rubbers and has presented the summary Table for a series of polymers in the rubbery state, which proves this assumption.

Figure 10 shows typical dependencies corresponded to equations (22) and (23) for PC (test temperature $T = 403$ K) and HDPE ($T = 293$ K). Clearly the Mooney-Rievlyn equation is applicable to both amorphous glassy PC and amorphous crystalline HDPE and gives reasonable $2C_1$ and $2C_2$ values. The latter statement is based on the following observation. It is known [55] that $2C_1$ constant can be expressed as follows:

$$2C_1 = \frac{A\rho RT}{M_{ent}}, \qquad (24)$$

where A is a coefficient determined by functionality of the entanglement network cross-link points. The equation (24) gives a possibility to estimate M_{ent} with respect to the known $2C_1$ values. The estimation results displayed good correspondence of the values obtained to analogous ones, calculated by the equations (2) and (3). It is typical that for macromolecular overwhelms' network, M_{cl} but not M_{ent} values (which are by one or two orders of magnitude greater than M_{cl}) conform to $2C_1$ values [53].

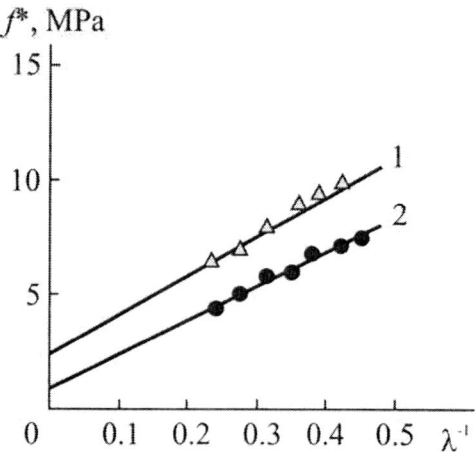

Figure 10. Dependencies of reduced stress, f^*, on the strain degree, λ, for PC (1) and HDPE (2) [53]

Figure 11 shows M_{cl} dependence on $2C_2/2C_1$ relation value, obtained from the Mooney-Rievlyn graphs. Data from this Figure display good linear correlation between the mentioned parameters that confirms the Boyer assumption about the possibility of $2C_2/2C_1$ relation use as the measure for the short range (local) order in polymers. However, Boyer has also assumed [56] that increase of the absolute $2C_2/2C_1$ value displays growth of the short range order degree in rubbers. For studied polymers [53, 54], M_{cl} increase with $2C_2/2C_1$ means decrease of the entanglement network frequency (equation (3)), decrease of the number of segments in clusters and, consequently, reduction of the local order degree (equation (11)). To put it differently, for amorphous glassy and amorphous-crystalline polymers, the increase of $2C_2/2C_1$ reflects the effect opposite to that observed in rubbers.

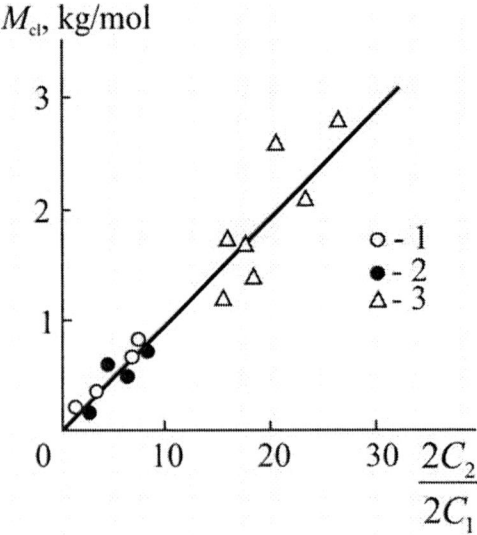

Figure 11. Dependence of the molecular mass, M_{cl}, of chain fracture between clusters on the relation of Mooney-Rievlyn equation constants for PC (1), poly(arylate sulfone) (PAS) (2), and HDPE (3) [53]

Such deviation is not accidental and reflects the difference in structures for these classes of polymers. If density of the cross-linked carcass and the possibility of chain segments packing between cross-link points in rubbers correspond to different structural elements and display opposite tendencies of variation [56], then for studied polymers both the entanglement network frequency and increase of the local order degree possess symbate tendencies that follows from the cluster model [38]. To put it differently, analogies between structural and mechanical properties of absolute cross-linked rubbers and linear polymers, studied in the work [53] are true to some extent only.

In the framework of phenomenological theory of the second-order transition by Landau [58], the order parameter ψ_0, unambiguously associated with one of the most important thermodynamic properties which is entropy change, ΔS, is determined as follows:

$$\psi_0 = (a/2C)^{1/2}(T_{tr} - T)^{1/2}, \qquad (25)$$

where a and C are parameters; T_{tr} is the transition temperature.

The experimental proof of the Landau theory applicability is correspondence of the temperature dependence of ψ_0 to $(T_{tr} - T)^{1/2}$ shape [58]. Graphs in Figure 12 indicate the same shape of ν_{cl} temperature dependence for amorphous-crystalline HDPE [59]. In this case, the glass transition temperature T_g (as usual for the Landau theory [58]), the melting temperature T_m and temperature of "liquid 1 – liquid 2" transition T_{ll} [60] were accepted for T_{tr}. Temperature of "liquid 1 - liquid 2" transition can be estimated as follows [9]:

$$T_{ll} = (1.20 \pm 0.05)T_m. \tag{26}$$

Data from Figure 12 give a possibility to assume that the cluster entanglement network frequency, ν_{cl}, is analogous to the order parameter ψ_0 and, consequently, characterizes local ordering degree in non-crystalline zones of polyethylenes.

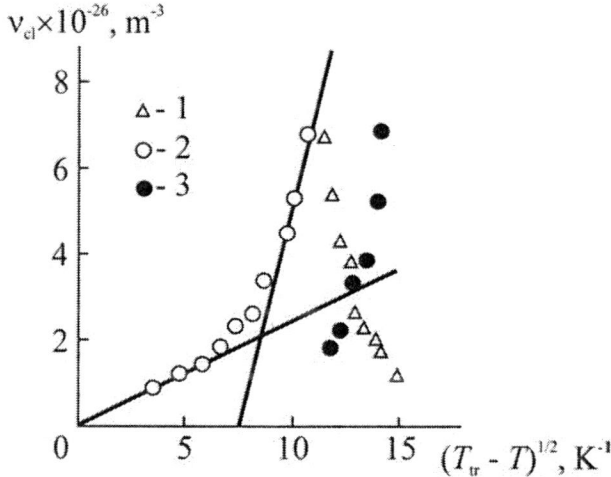

Figure 12. Dependencies of cluster entanglement network frequency vcl on (Ttr – T)1/2 parameter for HDPE, corresponded to the Landau equation for the second-order phase transitions. The transition temperature Ttr is as follows: Tg (1), Tm (2), and Tll (3) [59]

Two interesting features of dependencies shown in Figure 12 should be displayed. First of all, on the dependence $\nu_{cl}(T_{tr} - T)^{1/2}$ at T = 333 K an inflection is observed. As is known [61], a relaxation transition of polyethylenes, which Boyer named the "glass transition I", was detected at this temperature. Secondly, the condition $\psi_0 = 0$ at $T = T_{tr}$ is assumed in the equation (25). The graph in Figure 12 indicates that the identity $\nu_{cl} = 0$ is reached at $T_{tr} = T_m$, but not at $T_{tr} = T_g$. To put it differently, temperature T_m is corresponded not only to melting of crystallites, but also to "melting" of segments, and above T_m a network from macromolecular backlashes remains only [30]. This situation can be explained by the specificity of the local order formation in non-crystalline zones of amorphous-crystalline polymers similar to polyethylenes, for which this process is of "induced" type due to chain tensioning in the amorphous phase during crystallization [62, 63]. Note also that similarity of ψ_0 and ν_{cl} parameters eliminates consideration of entanglement cross points as traditional backlashes [30] and supposes that the local order zones (clusters) play their role [59].

Obviously the structure of amorphous polymers (or amorphous phase of amorphous-crystalline polymers) gives grounds to assume the presence of a definite chaos degree in it. It is also absolutely obvious that the chaotic character of the amorphous phase structure represents an important parameter determining structural characteristics and, consequently, properties of the polymer. That is why the question about interconnection of these indices is brought up. The second important problem is the physical origin of chaos in polymers: if it is random (and unpredictable) or deterministic chaos.

Among possible measures of the chaos degree in dynamics of a system (in this case, in polymer structure), the common one is Lyapunov's index, λ_L [64]. It represents the measure of exponential rate of neighboring trajectories divergence or convergence in the phase space along the current axis of coordinate. Chaotic processes are characterized by exponential divergence of neighboring trajectories, which gives, at least, one positive Lyapunov's index. The technique for λ_L assessment is described in the work [65]. Because the basic distinguishing characteristic of a polymer is its composition from long chain macromolecules, and the basic characteristic of the latter is their flexibility, the interconnection between Lyapunov's index λ_L and characteristic relation C_∞ has been studied [65]. There are two reasons for choosing C_∞ parameter for the flexibility measure. The first reason is that C_∞ is determined more precisely compared with other similar parameters [66]. The second one is that it can be estimated basing on chemical structure of the polymer macromolecule only [67]. Primarily, it should be noted that the presence of positive λ_L [65] indicates chaos existence of in the polymer structure. Further on, a systematic increase of λ_L (i.e. chaos intensification) with the chain flexibility is observed. Obviously C_∞ increase means growth of the chain mobility and, as a consequence, intensification of chaotic processes in the system (Figure 13).

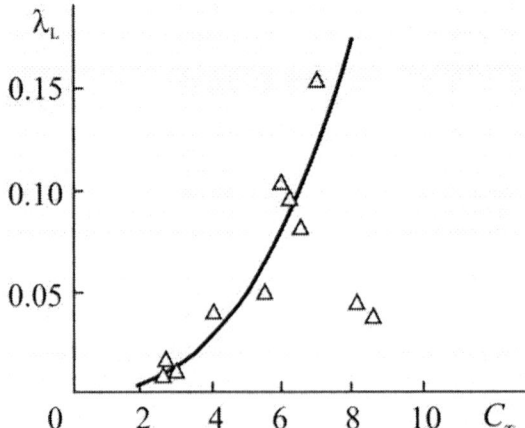

Figure 13. Dependence of Lyapunov's index, λ_L, on characteristic relation, C_∞, for amorphous glassy and amorphous-crystalline polymers [65]

Two polymers are outside of the general dependence: PS and PMMA, which was frequently observed before [66]. This suggests that C_∞ value is not always the unique molecular parameter that determines structural characteristics and properties of polymers, as suggested in ref. [67]. It is a matter of familiar experience [68] that both PS and PMMA have side groups, which are the reason for sharp increase of the cross-section, S, of their macromolecules. That is why one may suggest that applying S as the normalizing factor will improve $\lambda_L(C_\infty)$ correlation. Actually, the dependence $\lambda_L(C_\infty/S)$ does not give fall out results (Figure 14) and, despite a definite data scattering, can be approximated by a straight line. The most important result following from the above-mentioned data is regular variation of the chaos, λ_L, exponent for the structure of polymers possessing molecular characteristics C_∞ (or C_∞/S), which presumes deterministic (predictable) chaos [65].

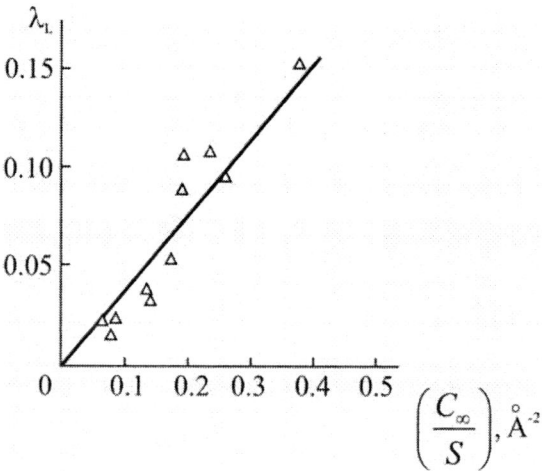

Figure 14. Dependence of Lyapunov's index, λ_L, on molecular parameter (C_∞/S) for amorphous glassy and amorphous-crystalline polymers [65]

Figure 15 shows dependence of macromolecular entanglement cluster network, v_{cl}, on λ_L. As might be expected, chaos intensification (λ_L increase) decreases v_{cl} value, i.e. the local ordering degree in the structure of polymer amorphous state [65]. More precise interpretation of the chaotic character of the polymer structure within the framework of multifractal formalism is given below.

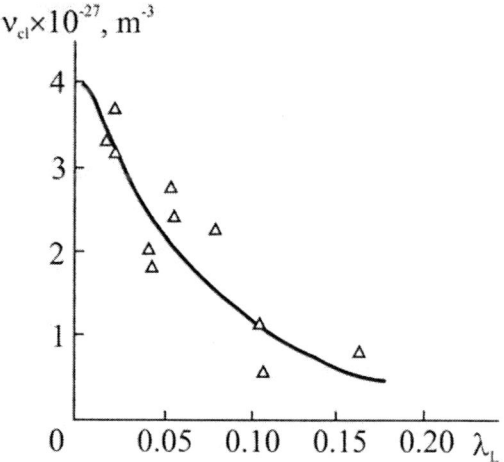

Figure 15. Dependence of macromolecular entanglement cluster network frequency, v_{cl}, on Lyapunov's index, λ_L, for amorphous glassy and amorphous-crystalline polymers [65]

Finishing this Section, let us discuss from positions of thermodynamics the interconnection between disorder degree of the polymer structure and the local regularity degree. One of possible characteristics of disorder can be the random free volume, f_f [69], which is associated with entropy variation, ΔS, as follows [70]:

$$\Delta S = (3 \div 5) R f_f \ln f_f. \tag{27}$$

One more quantitative interpretation of disorder is suggested in ref. [71], in which the disorder parameter, δ, associated with thermal mobility of macromolecules in liquid at the melting point T_m is determined as follows:

$$S = \frac{P_i(\widetilde{V}_m - 1)}{p^*}, \tag{28}$$

where P_i is the internal pressure; $\widetilde{V}_m$ is the reduced molar volume at the melting point; p^* is the characteristic pressure.

The value of P_i can be determined as follows [71]:

$$P_i = \frac{p^*}{\widetilde{V}_m^2}, \tag{29}$$

and finally, after substitution of equation (29) into (28), the equation for δ determination is obtained [72] as follows:

$$\delta = \frac{\widetilde{V}_m - 1}{\widetilde{V}_m^2}. \tag{30}$$

Taking into account successful application of liquid theories to description of amorphous polymer behavior [9], the range of equation (30) application was extended upon the glassy state of these polymers (by substituting $\widetilde{V}_m$ by $\widetilde{V}$) [72].

Obviously the problem of quantitative assessment of δ is reduced now to determination of $\widetilde{V}$ that for the polymer may be implemented by using the equation as follows [73]:

$$\frac{(\widetilde{V}^{1/3} - 1)}{(4 - 3\widetilde{V}^{1/3})} = \alpha T, \tag{31}$$

where α is the heat linear expansion coefficient.

Temperature dependence of α can be assessed by different methods. To the first approximation, α value may be accepted constant, and direct experimental measurements (which is intrinsically the most accurate method) or the known Barker equation can be applied [74]:

$$\alpha^2 E = C_1, \qquad (32)$$

where E is the elasticity modulus; C_1 is the coefficient varying in the range of 7.5 – 24 N/m^{2}K^2 giving, on average, 15 N/m^{2}K^2 [74].

The equation (32) gives an opportunity of quite simple assessment of α by mechanical test results. Variation of C_1 does not change the quality of δ assessment results, but gives a possibility of improving quantitative correspondence of parameters, calculated in ref. [72], by 15 – 20%, approximately. Preliminary assessments indicate the best correspondence at C_1 equal 7.5, 15 and 24 N/m^{2}K^2 for PAr, PC and PMMA, respectively [72].

Relative random free volume, f_f, represents the function of Poisson coefficient, v, expressed by the following equation [65]:

$$f_f = C_2\left(\frac{1+v}{1-2v}\right). \qquad (33)$$

Constant C_2 value may be varied, but this causes no effect on the quality of f_f assessment; thus it was accepted [29, 75, 76] that:

$$C_2 = 0.017. \qquad (34)$$

Figure 16 shows temperature dependencies of disorder parameter δ, assessed by the above-considered method, for three glassy polymers. Clearly dependencies $\delta(T)$ are symbate and indicate the increase of disorder degree with temperature. It should be noted that this result is not trivial. There are several concepts considering the disorder degree as the "frozen" one at transition of amorphous polymer to the glassy state. According to equation (27), the condition f_f = const also proves this point of view.

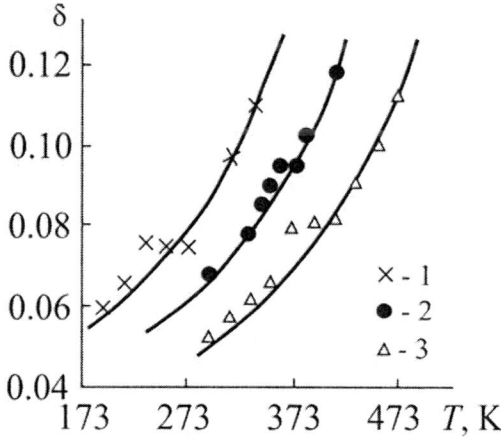

Figure 16. Temperature dependencies of disorder degree δ for PMMA (1), PC (2), and PAr (3) [72]

The symbate type of $\delta(T)$ curves (Figure 16) indicates the possibility of their superimposition by shifting along the temperature axis (i.e. applying the principle of temperature superimposition). Actually, the dependence $\delta(\Delta T)$, where ΔT represents the difference between the glass transition temperature T_g and the test one, T, indicated the reality of such superimposition (Figure 17). Practically, this means that the disorder degree in amorphous glassy polymers is independent of their chemical structure and is determined by temperature only (more specifically, by the difference between T_g and test temperature). This gives grounds to an assumption that disorder in the structure of amorphous polymers, determined by the above-mentioned method, is of thermofluctuation origin only (that follows from Sharma's determination of δ [71]).

Figure 18 shows comparison of δ and f_f values, calculated for PMMA, PC and PAr. This comparison displays approximate equality of the mentioned parameters, i.e. random free volume is also the disorder index for amorphous glassy polymer [72]. Thus as v_{cl} determines the number of closely packed segments in the local order zones and, consequently, is the index of this order, inversely proportional correlation between δ and v_{cl} must be observed. Obviously for the case of "backlash" network, none of such correlations is expected, because it is more probable that "backlashes" will open the structure of amorphous polymer [77], i.e. will intensify disorder.

Figure 19 shows the relation between δ and φ_{cl} (i.e. between disorder and order parameters, respectively). As might be expected, inversely proportional correlation approximated well by a straight line is observed for these parameters. The line is composed basing on the ideas as follows. Obviously at $\varphi_{cl} = 1.0$ the system possesses the ideal ordering and $\delta = 0$, whereas at $\varphi_{cl} = 0$, δ is accepted equal f_f at T_g with respect to the data in Figure 18 (in the present case, $f_f = 0.113$ is used [78]). The unambiguously indicated result favorably testifies the cluster interpretation of macromolecular entanglement cluster network.

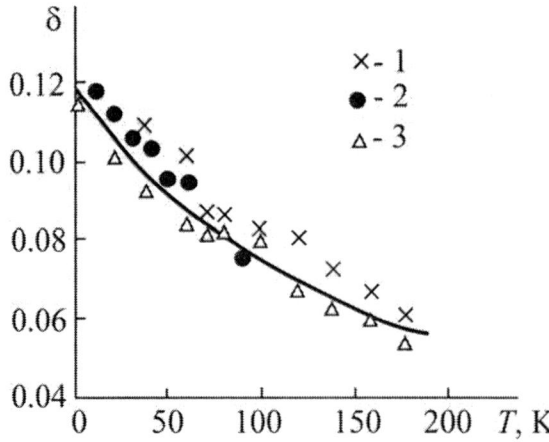

Figure 17. Dependence of disorder degree (δ) on the temperature difference ($\Delta T = T_g - T$) for PMMA (1), PC (2), and PAr (3) [72]

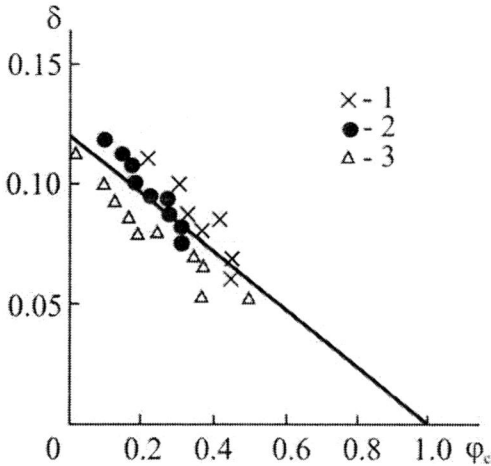

Figure 19. Correlation between the disorder parameter (δ) and the relative part of clusters (φ_{cl}) for PMMA (1), PC (2), and PAr (3) [72]

For linear polymers δ and f_f are approximately equal (Figure 18), and for glassy polymer networks $\delta > f_f$ [69]. This difference will be explained in Chapter 5.

Simultaneously, data from Figure 19 induce an assumption that there is a tight connection between the cluster model [7] and the random free volume theory [75, 76]: separation of a segment from a cluster means formation of the free volume microscopy, whereas addition of a segment to a cluster means its "implosion" or collapse. If it is true, the entropy variation, ΔS_f, due to random free volume formation (equation (27)) must be equal to entropy change, ΔS_{cl}, at partial thermofluctuational decay of clusters. The value of ΔS_{cl} can be assessed in the framework of the Forsman theory [79] by the equation as follows:

$$\Delta S_{cl} = \frac{Rc_3\varphi_{cl}}{M_0}\left[\frac{m-1}{m}\ln\left(\frac{c_3\rho\varphi_{cl}}{m}\right) - \frac{(m-1)+\ln m}{m}\right], \tag{35}$$

where c_3 is the polymer concentration in the system (in the case under consideration, $c_3 = 1$); M_0 is the molecular mass of the repeating unit; m is the number of repeating units in the cluster; ρ is the polymer density.

Figure 20 compares ΔS_f and ΔS_{cl} for PC and PAr and indicates an interesting detail. Let us note first that the change of $\Delta S_f(T)$ curve slopes correspond to devitrification temperature of the packless polymer matrix, T_g' [12]. Thus in the case of glassy packless matrix, the disorder increase due to partial cluster decay is transformed into disorder of random free volume microcavity formation. For devitrified packless matrix ΔS_{cl} is partly transformed into ΔS_f and partly increases disorder in accordance with other mechanisms (possibly, due to conformational changes [80]). As a consequence, strictly saying, δ value (and parameters f_f, v_{cl}, and φ_{cl} associated with it) characterize thermofluctuational "structural" disorder only, but not the entire spectrum of it [72].

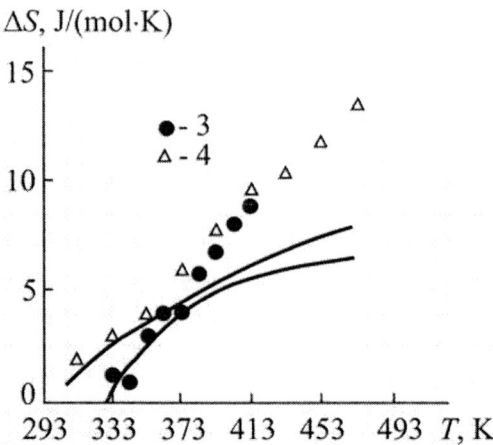

Figure 20. Temperature dependencies of entropy change stipulated by formation of the random free volume, ΔS_f (1, 2), and cluster decay, ΔS_{cl} (3, 4), for PC (1, 3) and PAr (2, 4) [72]

It should be noted that the study of disorder degree of amorphous polymer structure is not of the empirical significance only. Figure 21 shows dependence of the offset yield stress of three studied polymers on reverse δ value. Clearly this dependence is of the simplest shape: it is linear and passes through the origin of coordinates. That is why it is quite suitable for forecasting mechanical behavior of polymers. Similar dependencies were obtained for the yield strain, ε_y, and elasticity modulus, E, deformation [72].

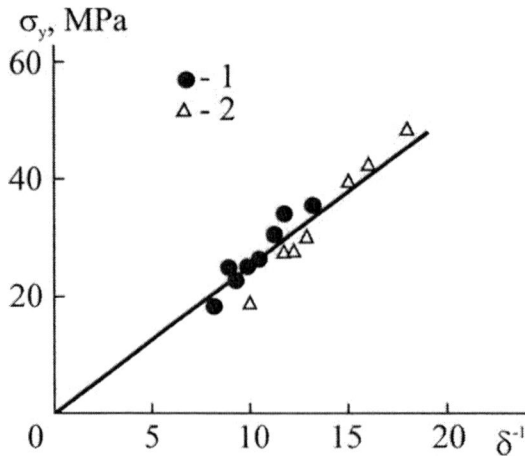

Figure 21. Dependence of offset yield stress σ_y on disorder parameter δ for PC (1), PAr (2), and PMMA (3) [72]

The above-shown results allow making the following conclusions. First of all for amorphous glassy polymers, the disorder degree is the function of temperature. Dependence of δ on temperature only suggests purely thermofluctuational origin of this disorder. Secondly, for amorphous polymers, the random free volume, f_f, is the quantitative parameter of the disorder degree. Thirdly, cross-link points of macromolecular entanglement cluster network represent the local ordering degree of the amorphous polymer structure, as postulated in the cluster model. Fourthly, disorder in the systems under consideration is not necessarily of the thermofluctuational origin, but the latter component of it defines mechanical behavior of the polymer (Figure 21) [72].

4. THERMOFLUCTUATIONAL ORIGIN OF CLUSTERS

A set of parameters indicating thermofluctuational origin of the local order zones (clusters) was mentioned above, for example, in equation (212), Figures 8 and 19. In the present Section, this point will be discussed in more detail. First of all, the interconnection between parameters of the entanglement cluster network and density fluctuations should be studied. Density fluctuations $\langle \Delta\rho/\rho \rangle^2$ represent the measure of disorder in polymers and are determined as follows [81, 82]:

$$\langle \Delta\rho/\rho \rangle^2 = \frac{\langle (N - \langle N \rangle)^2 \rangle}{\langle N \rangle}, \qquad (36)$$

where N is the number of electrons in randomly selected volume; $\langle N \rangle$ is the mean value of N.

For a liquid in the equilibrium state, statistical mechanics gives the following expression for $\langle \Delta\rho/\rho \rangle^2 (V)$, where V is the standard volume, in the limit $V \to \infty$ (i.e. in thermodynamic limit) [82]:

$$\langle \Delta\rho/\rho \rangle^2 (\infty) = \rho k T \chi_T, \qquad (37)$$

where χ_T is the isothermal compressibility.

Equation (37) shows that density fluctuations are stipulated by heat mobility of atoms with energy kT, but limited by volumetric rigidity $(\chi_T)^{-1}$.

Based on the data from literature for ρ and χ_T, assessment of $\langle \Delta\rho/\rho \rangle^2 (\infty)$ at T_{ll} or T_{ll}' temperatures (T_{ll}' is the analogue of T_{ll} for amorphous-crystalline polymers associated with T_m similar to equation (26)) has shown that the value mentioned is approximately constant at these temperatures. This observation suggests that as some critical value of $\langle \Delta\rho/\rho \rangle^2$ is reached, formation of the local order zones, i.e. thermofluctuational cluster network of macromolecular entanglements, is impossible due to high heat mobility of macromolecules. Vice versa, at $\langle \Delta\rho/\rho \rangle^2$ values below the critical ones the local order zones (clusters) in the polymer melt are formed, which according to the current interpretation are identified as cross-link points of macromolecular entanglement network, i.e. a state is formed, which was defined by Boyer as a "liquid with fixed structure" [56].

Basing on the above-discussed arguments, critical temperatures, T_{cr}, at which critical density fluctuations are reached, were calculated. For the latter, $\langle \Delta\rho/\rho \rangle^2$ value for PS at 433 K was accepted. Calculation results are shown in Table 3, from which it follows that T_{cr} and T_{ll} (T_{ll}') for a series of amorphous and amorphous-crystalline polymers are very close: maximal deviation of these temperatures does not exceed 6% [84, 85]. Note that in the case under consideration we are dealing with dynamic (short-living) local order, which is "frozen" below T_g (T_m) [9].

Table 3. Calculated temperatures T_{ll} (T_{ll}'), T_{cr} and their relative deviation Δ [85]

Polymer	T_{cr}, K	T_{ll} (T_{ll}'), K	Δ, %
Poly(methyl methacrylate)	398	415	4.1
Polystyrene	462	433	6.2
Polycarbonate	488	502	2.8
Low density polyethylene	488	480	1.7
High density polyethylene	496	480	3.3
Polypropylene	510	528	3.4
Polyamide-6	635	600	5.8

Sanditov and Bartenev [75] have shown that $\langle \Delta\rho/\rho \rangle^2$ value is associated with the relative random volume by the relation as follows:

$$\langle \Delta\rho/\rho \rangle^2 = f_f(V_h/V_{at}), \qquad (38)$$

where V_h is the microcavity volume of the random free volume; V_{at} is the atomic volume. The value of $\langle \Delta\rho/\rho \rangle^2$ can also be determined using the Poisson coefficient, ν [75]:

$$\langle \Delta\rho/\rho \rangle^2 \approx \frac{(1-2\nu)^3}{6(1+\nu)^2}. \qquad (39)$$

Figures 22 and 23 show temperature dependencies of $\langle \Delta\rho/\rho \rangle^2$ for high and low density polyethylene (HDPE and LDPE, respectively), Calculated by equations (38) and (39). Despite different absolute values of density fluctuations, calculated by the above-mentioned equations, shapes of their temperature dependencies correlate with the modern data [9, 82]. Typically, similar to Figure 12, dependencies $\langle \Delta\rho/\rho \rangle^2 (T)$ do again display bendings at ~333 K. Higher $\langle \Delta\rho/\rho \rangle^2$ values for LDPE are induced by higher V_h value for this polymer [87].

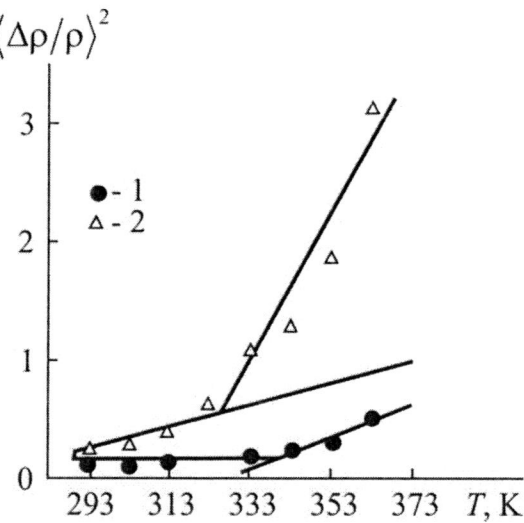

Figure 22. Temperature dependen-cies of density fluctuation $\langle \Delta\rho/\rho \rangle^2$, calculated by equation (38), for HDPE (1) and LDPE (2) [86]

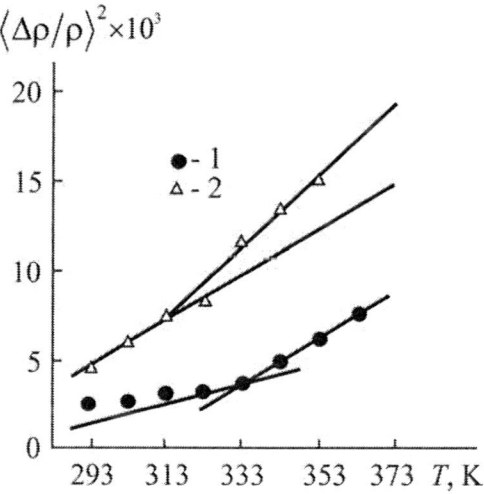

Figure 23. Temperature dependen-cies of density fluctuation $\langle \Delta\rho/\rho \rangle^2$, calculated by equation (39), for HDPE (1) and LDPE (2) [86]

Data in Figure 24 prove existence of inversely proportional correlation between order (ν_{cl}) and disorder $\left(\langle \Delta\rho/\rho \rangle^2 \right)$ indices for $\langle \Delta\rho/\rho \rangle^2$ values calculated by the equation (38). Correlation $\langle \Delta\rho/\rho \rangle^2 \left(\nu_{cl}^{-1} \right)$ is one more reason for the benefit of the thermofluctuational origin of local order zones (clusters) in the polymer amorphous state [86].

Thermofluctuational origin of entanglement cluster network allows theoretical assessment of temperature variation of its density v_{cl}. For this purpose, the authors of ref. [88] have used the model from [89], in which the following expression for relaxation time, τ_r, determining the speed of macromolecule fracture motion, N_s segments long, on the test temperature in the range of $T_s < T < T_m$ was used:

$$\tau_r = \tau(T)F(N_s), \quad (40)$$

where $\tau(T)$ dependence is described by the Frenkel-Eiring-Arrhenius formula [89]:

$$\tau(T) = \tau_0 \exp\left(\frac{\Delta U}{RT}\right), \quad (41)$$

where τ_0 is the constant; ΔU is the potential barrier of segment transition from one quasi-equilibrium state to another; the activation energy of viscous flow, U_{fl}, is suitable for estimating this value [89].

For function $F(N_s)$ the following scaling relation was accepted [89]:

$$F(N_s) = N_s. \quad (42)$$

Following the general way of deducing the model [89], it has been suggested [88] that at all temperatures the size of macromolecule segment between cross-link points of macromolecular entanglement network, N_{ent} (expressed by the number of statistical segments), equals N_s, at which $\tau_r(T, N_s)$ becomes equal some value, τ_{cl}. The latter parameter can be interpreted as time required for packing a segment of the chain possessing a definite axial collinearity with the ones already precipitated to the cross-link point (by analogy with crystallization), to an entanglement cross-link point. Clearly in this case, the cross-link point of thermofluctuational network of macromolecular entanglements is considered as a closely packed cluster of macromolecule segments with approximately parallel disposition. Solving the following equation for N_{ent}:

$$\tau_{cl} = \tau_r(T, N_{ent}), \quad (43)$$

we obtain that

$$N_{ent} = \left(\frac{\tau_{cl}}{\tau_0}\right)^{1/x} \exp\left(-\frac{\Delta U}{RTx}\right). \quad (44)$$

This dependence reflects the fact that at temperature decrease the mobility type becomes lower scale; in this case, N_{ent} can be considered the effective parameter characterizing the chain mobility dormancy or probability of its fluctuations [88].

Because at present the relation τ_{cl}/τ_0 can be hardly subject to reasonable theoretical and experimental assessment, the conditions mentioned in the work [89] were followed in further calculations. Taking $N_{ent} = 1$ at $T = T_g$ and $x = 3.3$, ΔU (or U_{fl}) was assessed in accordance with the data from ref. [90]. For PC, molecular mass of a segment, M_s, equals 726 g/mol, i.e. it is suggested that statistical segment of PC includes two monomeric units. Then $M_{ent} = N_{ent}M_s$. Figure 25 shows the comparison of M_{ent}, calculated by the above method, and experimental M_{cl} values of the chain segment molecular mass between entanglement cross-link points for PC as temperature function, where good coincidence is indicated. Such coincidence can be accepted to be the proof of the condition $N_{ent} = 1$ validity at $T = T_g$, shown in ref. [89].

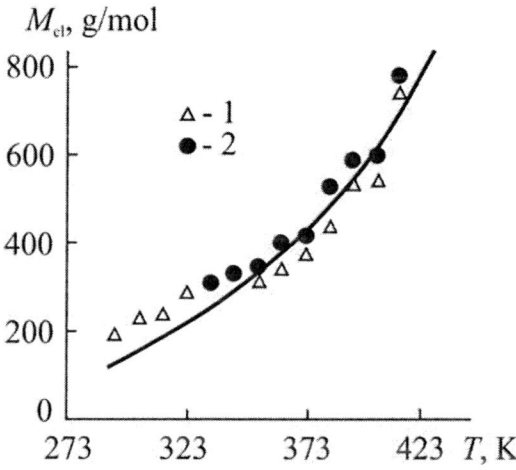

Figure 25. Theoretical (1) and experimental (2, 3) temperature dependencies of molecular mass M_{ent} (M_{cl}) of the chain segment between entanglement cross-link points for PC: initial (2) and annealed at 393 K (3) [88]

Fluctuation theory of glass transition allows assessment of the volume V_a of so-called cooperatively restructuring zones (CRZ) [8, 91]:

$$V_a = kT_g \frac{\Delta(C_V^{-1})}{\rho_g (\delta T)^2}, \qquad (45)$$

where $\Delta(C_V^{-1})$ is the reverse heat capacity jump at the glass transition; ρ_g is the polymer density at T_g; δT is the average temperature fluctuation per one CRA.

Bittrich [91] has assessed CRA characteristic length, ξ_A, which varies in the range of 0.2 – 2.5 nm in the temperature interval of 323 – 348 K and declines with temperature increase. Since both CRA [92] and clusters [38] are postulated to be of the thermofluctuational origin, consistency of two above-mentioned models should naturally be determined by comparing temperature dependencies of ξ_A and the size of clusters for considered polymers [93, 94]. As mentioned above, the cluster model suggests decrease of the number of segments in clusters with temperature increase [38]. As the cluster is considered the analogue to crystallite with stretched chains consisting of several collinear closely packed segments, decrease of the number of segments in the cluster with temperature increase induces the cross-section decrease, which is schematically shown in Figure 26 (insertion), and correspondingly, the cluster diameter, D_{cl}. Temperature dependencies of ξ_A and D_{cl} as characteristic parameters of microheterogeneity of the structure in both models were compared [93, 94].

Figure 26 shows comparison of the mentioned dimensions, where ξ_A values are taken from the work [91] and D_{cl} calculation technique from the work [94]. It is clear that temperature dependencies of these structural characteristics display good coincidence. This gives grounds for an assumption that CRA and clusters might be identical structural elements and proves thermofluctuational origin of clusters [93, 94].

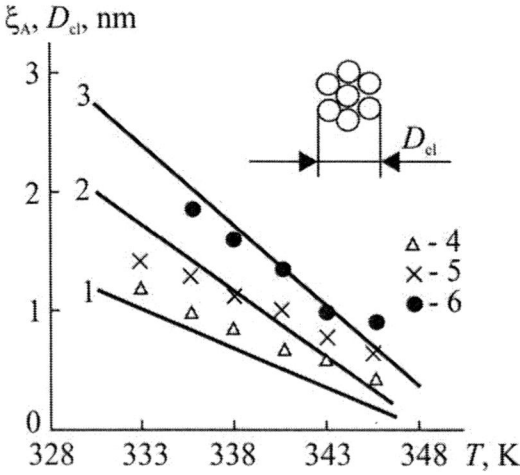

Figure 26. Temperature dependencies of CRA characteristic length, ξ_A, (1 – 3) and cluster diameter, D_{cl}, (4 – 6) for PMMA (1, 4), PVC (2, 5) and PS (3, 6). Insertion shows schematic image of the cluster cross-section [94]

To conclude this Section, let us discuss temperature stability of the cluster structure. A simple empirical approximation connecting v_{cl} to the Poisson coefficient, v, was deduced as follows [95]:

$$v_{cl} \approx 2.38 \times 10^{27}(1 + v)^2, \text{m}^{-3}. \tag{46}$$

In accordance with the Le Chatelier-Brown principle (the principle of least constraint) v value determining stability of a solid falls within the range as follows [39]:

$$0 \leq v \leq 0.5. \tag{47}$$

Combination of equations (46) and (47) gives the conditions of cluster structure stability for the polymer glassy state [95]:

$$2.38 \times 10^{27} \leq v_{cl} \leq 5.34 \times 10^{27}, \text{m}^{-3}. \tag{48}$$

Figure 27 shows temperature dependence of v_{cl} for polyhydroxyester (PHE), where horizontal dashed line marks the lower border $\left(v_{cl}^{min}\right)$ of the cluster structure stability. Good concordance of v_{cl}^{min} and v_{cl} values is observed, at which accelerated thermofluctuational decay of clusters is initiated. Similar dependencies were obtained for PC and polysulfone (PSF) [95].

As follows from the equation (46), v values become negative at $v_{cl} < v_{cl}^{min}$ (e.g. $T \geq T_g'$). The classic theory of elasticity assumes negative v values limiting the range of their variation as follows [39]:

$$-1 < v \leq 0.5. \tag{49}$$

It is absolutely obvious that according to the equation (46) at v approaching -1, v_{cl} approaches zero. Negative values of v observed for some substances near phase transitions testify about a change of their volumes. Making definite comparisons, note that the glass transition is also accompanied by discrete change of the polymer volume [75].

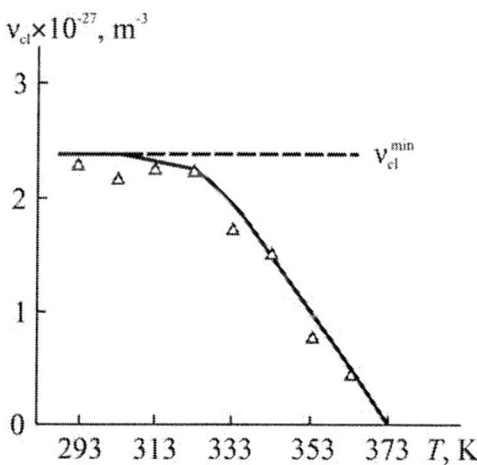

Figure 27. Temperature dependence of entanglement cluster network frequency, v_{cl}, for polyhydroxyester; horizontal dashed line marks the lower border $\left(v_{cl}^{min}\right)$ of the cluster stability [95]

Maximum possible value of v_{cl}, permitted by molecular structure of the polymer $\left(v_{cl}^{str}\right)$, can be determined by the following procedure. The length of macromolecules, L, in the specific volume of polymer is determined from the equation (7), and the segment length in cluster, taken equal to statistical segment length, l_{st}, according to the equation (8). Thus v_{cl}^{str} equals [95]:

$$v_{cl}^{str} = \frac{L}{l_{st}} = (SC_\infty l_0)^{-1}. \tag{50}$$

For PHE, PC and PSF, v_{cl}^{str} values estimated by the above method, approximately, equal 8.45×10^{27}, 10.78×10^{27} and 11.0×10^{27} m^{-3}, respectively. According to (48), these values are much higher than the upper border of v_{cl}. This means that the condition of the cluster structure stability does not allow complete packing of polymer chain segments in the local order zone (cluster). Other criteria of this effect are discussed below.

The results obtained [95] allow an assumption of the following variation of macromolecular entanglement cluster network frequency with temperature (Figure 27). As the lower border of the clusters' stability $\left(v_{cl}^{min} \right)$ is reached, their accelerated decay is initiated, which leads to much more abrupt decrease of v_{cl}. It is suggested [38] that a definite part of clusters with low functionality, F (e.g. with small number of segments in the cluster), is present in a packless matrix. These clusters preserve the matrix in the glassy state. This supposition is confirmed by experiments of polyarylate annealing, in which increase of v_{cl} and decrease of F were observed simultaneously. When T_g' (or v_{cl}^{min}) is reached, clusters with low F (thus thermodynamically unstable ones) decay completely, and the packless matrix devitrifies promoting an abrupt F decrease of stable clusters due to high molecular mobility of the environs. Abrupt decrease of v_{cl} is finished by complete thermofluctuational decay of the clusters "frozen" in the glassy state at $T = T_g$ [38].

5. FUNCTIONALITY OF CLUSTERS AND ASSESSMENT METHODS

As mentioned above, clusters consist of several collinear closely packed segments of different macromolecules and are considered as multifunctional cross-link points of macromolecular entanglement random network [38]. By functionality, F, of such cross-link point the number of chains in it is usually meant [8]. Thus full characterization of the polymer amorphous state requires obtaining of two parameters: macromolecular entanglement cluster network frequency, v_{cl}, and the number of segments, n_{seg}, in a single cluster (local characteristic). Because cluster represents an amorphous analogue of crystallite with extended chains, $n_{seg} = F/2$. Several possible techniques for F (or n_{seg}) value assessment were considered [96].

The first of possible methods of n_{seg} assessment is of a semi-quantitative type. It is supposed [9] that the volume of local order zones in polymers equals about 10 nm^3. Dividing this value by the statistical segment volume we obtain the following values: $n_{seg} \approx 88$ for PC and $n_{seg} \approx 40$ for HDPE. These values display the proper order (several ten segments in cluster) and can be the upper assessment border.

One more assessment method for F involves equation (5), suggested by Graessley [10].

Temperature dependencies of the cluster functionality F, calculated for HDPE and PC by the Graessley equation, are shown in Figures 28 and 29, respectively. As indicated by graphs in these Figures, the number of segments in a single cluster varies from 7 to 12 units for

HDPE and from 3 to 18 for PC. As should be expected [7], F value is decreased with temperature increase.

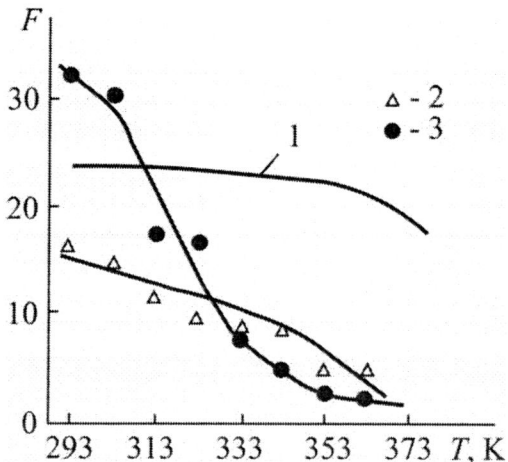

Figure 28. Temperature dependencies of the cluster functionality, F, for HDPE calculated as follows: 1 – by equation (5); 2 – by equation (51); 3 – by equation (52) [96]

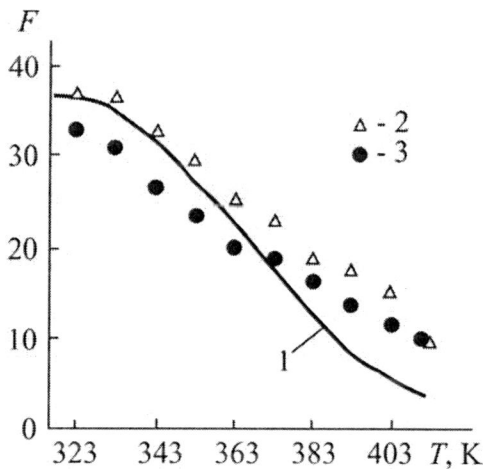

Figure 29. Temperature dependencies of the cluster functionality, F, for PC calculated as follows: 1 – by equation (5); 2 – by equation (51); 3 – by equation (52) [96]

There are two more approximate methods of F assessment. Several conditions, well approved at present, must be observed for application of any model of macromolecular entanglement network (cluster or "backlash" [18]). One of the conditions is fulfillment of the Gaussian statistics of polymer chains. Fulfillment of this condition requires the presence of, at least, 20 repeating units between cross-link points of the entanglement network [79]. Table 4 shows molecular masses of the chain segments between entanglement cross-link points in the "backlash" network alternative for HDPE and PC, accepted from references [17, 18, 30, 68,

97 – 99] or calculated in accordance with them. Though different authors have presented significantly M_{ent} values, on the whole, it is observed that the Gaussian statistics condition for chains is fulfilled. For multifunctional cross-link points, Flory [8] has displayed a relation between M_{ent} detected during experiment and effective molecular masses, M_{ent}^{eff}, of the mentioned chain fracture, expressed by equation (4). For entanglement cluster network M_{cl} is estimated in accordance with the equation (2), and M_{ent} is taken for M_{cl}^{eff}. Taking into account that the cross-link point of "backlash" network is of the four-functional type, it is finally obtained [96] that

$$F = \frac{4M_{eng}}{M_{cl}}. \tag{51}$$

Table 4. Molecular masses M_{ent} for chain segments between cross-link points of the "backlash" network for HDPE and PC in accordance with data from ref. [96]

Polymer	M_{ent}, g/mol			
HDPE	1,900 [17]	2,936 [68]	950÷2,140 [98]	737 [30]
PC	1,780 [18]	2,426 [99]	2,453 [98]	3,214 [97]

Figures 28 and 29 also show F values, estimated by the equation (51), and indicate their good coincidence with the values, calculated by the Greassley equation for PC and a bit poorer one for HDPE. Nevertheless, F magnitude and temperature dependence run conform in both cases.

It is common knowledge [8] that macromolecular coils in θ-solvent and in condensed state of polymers possess equal shapes, which is expressed by approximately equal mean-square distances between ends of the macromolecule. Boyer [56] has schematically indicated that this circumstance does not inhibit formation of local order zones (refer to Figure 1). More stricter this conclusion is proved in the work [96]. Obviously this experimentally proved condition should also be true for the cluster model. For assessment of n_{seg} in polymer clusters, Forsman [79] has suggested the following equation:

$$n_{seg}^{1/2} = \frac{\theta_2}{4A}\left\{1-\left[1-\frac{16A[1-\ln(c\rho\varphi_{cl})]}{\theta_2^2}\right]^{1/2}\right\}, \tag{52}$$

where θ_2 is a dimensionless (normalized by kT) value characterizing energy of the segments' interaction, the technique of determination of which is given in the next Chapter; c is the polymer concentration; ρ is the polymer density.

The value A describes variation of the macromolecule size before and after cluster formation and is determined from the following equation [79]:

$$A = \frac{\alpha^4 - \alpha^2 + 1}{\alpha^2}, \qquad (53)$$

where parameter α is given as follows [79]:

$$\alpha = \frac{\langle r^2 \rangle}{\langle r_0^2 \rangle}. \qquad (54)$$

Here r_0 and r are the mean-square distances between the ends of the macromolecule before and after cluster formation, respectively.

If the distance between ends of the macromolecule before and after cluster formation is the same ($r = r_0$), $\alpha = 1$ and assessment results by the equation (52) are also shown in Figures 28 and 29. Similar to the previous case, despite a definite quantitative inconsistency for HDPE, they give proper magnitude of F and expected shape of the temperature dependence.

Strictly saying, F value is not the exponent of the local regulation of the polymer structure, because clusters are formed from segments of different macromolecules, it can be the exponent of mutual penetration of macromolecular coils. As shown in refs. [19, 97], the same role can be played by the characteristic relation C_∞. If this assumption is correct, a definite correlation between F and C_∞ must be observed. Data from the Figure 30 show that such correlation is really observed for 9 amorphous and amorphous-crystalline polymers (F value is calculated for $T = 293$ K) [96].

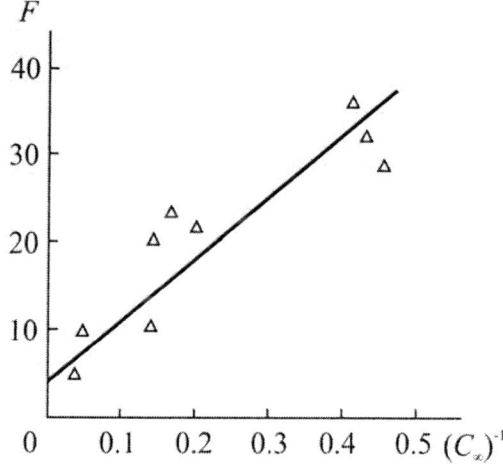

Figure 30. The relation between cluster functionality, F, and characteristic relation, C_∞, for 9 amorphous and amorphous-crystalline polymers [96]

As a consequence, the number of segments in the cluster with respect to characteristics of the polymer and temperature can vary from several ten segments. It is suggested that the most accurate assessment of F is given by the Greassley equation (equation 5). This conclusion follows from the fact, for example, that it describes decrease of v_{cl} at the glass transition temperature of packless matrix, T_g' (Figures 28 and 29). For equations (51) and (52), equal accuracy should not be expected, because the data of the former are highly scattered, and the latter represents an approximation.

These conclusions were proved [100] on the example of two series of cross-linked epoxidiane resin ED-20, hardened by 3,3'-dichloro-4,4'-diaminodiphenylmethane (EP-1 composite) and isomethyltetrahydrophthalic anhydride (EP-2) with variable curing agent : oligomer ratio. Figure 31 shows the comparison of clusters' functionalities, F_1 and F_2, calculated by equations (4) and (52), respectively, for epoxy-polymers EP-1 and EP-2. Clearly a good conformity between F_1 and F_2 values is obtained, which again indicates that invariability of sub-chains' statistics in epoxy-polymers during their cross-linking does not hinder formation of the local order zones (clusters) [100].

Of special attention is much worse conformity of F values, calculated by different methods for amorphous-crystalline polymers in contrast with amorphous ones (Figures 28, 2.29 and 31). This nonconformity is explained by different mechanisms of the cluster formation for the mentioned classes of polymers. Since clusters in amorphous glassy polymers display thermofluctuation origin and their relative part, φ_{cl}, is the function of temperature [38], for non-crystalline zones of such amorphous-crystalline polymers as polyethylenes and polypropylene the situation is somewhat more difficult.

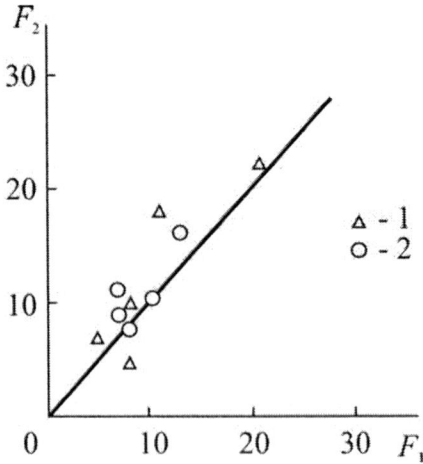

Figure 31. Correlation between cluster functionalities, F_1 and F_2, calculated by equations (4) and (52), respectively, for epoxy-polymers EP-1 (1) and EP-2 (2) [100]

As is known [48], clusters represent a "frozen" local order of the glassy state, and at temperatures above the glass transition temperature of the polymer their complete decay is observed. Nevertheless, for polyethylenes and polypropylene at about room temperature or higher, i.e. at $T > T_g$ of the amorphous phase, the presence of the local order in this phase is assumed [101 - 103]. It is postulated that in the current case, the mechanism of cluster formation is associated with the polymer chain tension during crystallization [29, 104]. The tension mentioned can be assessed with the help of parameter β [105], which is determined from the relation as follows [29]:

$$E = \frac{\rho RT}{M_{cl}} \cdot \frac{\beta^2}{5(1-K)^3}, \qquad (55)$$

where E is the elasticity modulus; K is the crystallinity degree.

Figure 32 shows the dependence of functionality F on parameter β for HDPE and PP. Clearly this dependence is approximated well by a straight line passing through the origin of coordinates, general for both polymers. As a consequence, $F(\beta)$ correlation proves the supposition made that straining of amorphous parts of the chains during crystallization is the basic mechanism of local order zones' formation in devitrified amorphous phase of amorphous-crystalline polymers [106].

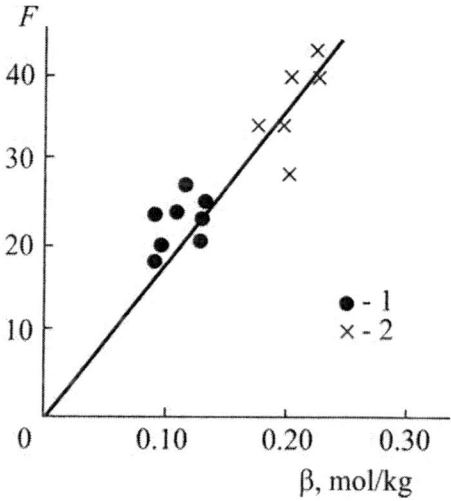

Figure 32. Dependence of the cluster functionality, F, on parameter β characterizing the chain tension during crystallization for HDPE (1) and PP (2) [106]

To complete this Section, let us consider one more principal point associated with formation of entanglement cluster network in amorphous-crystalline polymers. The equation (24) is valid for polymers, in which the entire volume of the compound is involved in formation of the carcass of macromolecular entanglements. In the case, if a part of polymer only with the volumetric fracture φ_{carc} is involved in the process, the equation (24) should be reduced to as follows [107]:

$$2C_1 = \frac{A\rho RT\varphi_{carc}^{1/2}}{M_{cl}}. \qquad (56)$$

At the present time, there are two points of view on the carcass structure of macromolecular entanglements in amorphous-crystalline polymers. One of them [98] suggests that the cross-link points of macromolecular entanglement carcass in polyethylene are crystalline zones, which are lamellar crystallites. Another point of view [108] presumes that the cross-link points are concentrated in non-crystalline zones of the amorphous-crystalline polymer. Obviously the condition $\varphi_{carc} = 1.0$ (i.e. the entire polymer represents the carcass) is corresponded to the former opinion, and φ_{carc} equal the volumetric part of non-crystalline zones of the polymer, e.g. about 0.3 for HDPE and 0.5 for LDPE [83] is corresponded to the latter one.

Prior to assessment of correctness of one or another φ_{carc} value, one should understand the physical meaning of the front-factor A in equations (24) and (56). One alternative [109] provides for A variation within the range of 0.5 – 1.0, where $A = 0.5$ is corresponded to the so-called "ghostable" or "phantom" carcass which displays the full freedom of fluctuation for cross-link points of macromolecular entanglements round their middle positions. Thus $A = 1.0$ is corresponded to an "affine" carcass, in which such freedom is completely suppressed. Another alternative [110] provides for quantitative relation between the factor A and functionality F of cross-link points of the macromolecular entanglement carcass:

$$A = 1 - \frac{2}{F}. \tag{57}$$

Obviously the first alternative allows assessment of tightness of the carcass entanglement cross-link points basing on A value, calculated using experimental $2C_1$ value and the equation (56). The second alternative provides for qualitative assessment of functionality F [54].

Table 5 shows calculation results for A value under conditions as follows: $\varphi_{carc} = 0.3$ and 1.0 for HDPE and $\varphi_{carc} = 0.55$ and 1.0 for LDPE, whence it follows that use of the condition $\varphi_{carc} = 1.0$ gives $A = 0.33 - 0.42$, i.e. physically meaningless values, whereas $\varphi_{carc} = 0.30$ and $\varphi_{carc} = 0.55$ provide for A values in the expected interval of 0.5 – 1.0. To put it differently, the calculation results indicate validity of the condition choice, in which the carcass is represented by the amorphous phase of amorphous-crystalline polymer and, consequently, there are no grounds for taking crystallites for cross-link points of the macromolecular entanglement carcass [54].

Table 5. Front-factor A and functionality F calculated for HDPE and LDPE [54]

Polymer	φ_{carc}	A	F
HDPE	0.30	0.75	8.0
	1.0	0.33	3.0
LDPE	0.55	0.64	5.6
	1.0	0.48	3.8

Figure 33 shows dependence of the front-factor A on the molecular mass M_{cl}. The graph shows that tightness of clusters' fluctuations is systematically varied from 1.0 at $M_{cl} = 0$ to 0.5 at $M_{cl} = 3,600$ g/mol. To put it differently, if the amorphous phase of amorphous-crystalline polymer represents an entire cluster (supercluster), fluctuations are completely suppressed, and its behavior corresponds to the "affine" model [107]. The value $M_{cl} = M_{ent} = 3,600$ g/mol corresponds to the melt of polyethylenes [17], where tightness applied by crystallites is absent. Thus in this case, behavior of the entanglement carcass of polyethylenes corresponds to the "phantom" alternative [110].

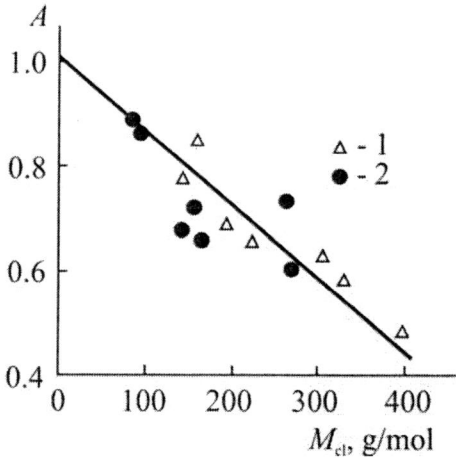

Figure 33. Dependence of front-factor A on molecular mass, M_{cl}, of a chain fracture between clusters for LDPE (1) and HDPE (2) [54]

Note also that in accordance with the equation (57) A increase is equivalent to increase of the cluster functionality F. As shown by Flory [111], the mean-square fluctuations of chain vectors, $\langle (\Delta r)^2 \rangle$, induced by joint fluctuations, is given by the following equation:

$$\langle (\Delta r)^2 \rangle = \frac{2\left(\langle r^2 \rangle_0\right)}{F}, \qquad (58)$$

where $\langle r^2 \rangle_0$ is the mean-square distance between ends of the chain.

As follows from equation (58), fluctuations in the system must be suppressed with F increase, which is supposed by data from Figure 33. Note also that the condition $F = 2$ (or $n_{seg} = 1$), which determines the cluster decomposition, occurs at $\langle (\Delta r)^2 \rangle = \langle r^2 \rangle_0$, e.g. in the case of equality of mean-square fluctuations of the chain vectors and the distance between end of the chain. Essentially, this condition is the thermofluctuation criterion of the cluster formation, another alternative of which is discussed in Section 4 (refer to Table 3).

Thus results obtained in the framework of the rubber elasticity theory suggest that the carcass of macromolecular entanglements in amorphous-crystalline polymers is limited by noncrystalline zones, and the cross-link points of this carcass represent clusters consisting of 2 – 9 closely packed collinear segments [54].

6. THE METHOD OF ARTIFICIAL CONTROL OF THE CLUSTER STRUCTURE (ON THE EXAMPLE OF HDPE)

Introduction of low amounts of highly dispersed Fe – FeO (Z) mixture into HDPE causes significant changes in the structure and properties of this polymer [101, 103, 112, 113]. Table 6 illustrates these changes in properties of HDPE + Z composites on the example of blow viscosity, A_p, cracking resistance in aggressive media, τ_{50}, and gas penetration coefficient by nitrogen, P_{N_2}.

Table 6. Blow viscosity (A_p), cracking resistance (τ_{50}) and gas penetration coefficient $\left(P_{N_2}\right)$ for HDPE + Z composites [103]

Concentration of Z in HDPE + Z composites, wt.%	A_p, kJ/m^2	τ_{50}, hour	$P_{N_2} \times 10^{-17}$, (mol/m)/(m^2·s·Pa)
0	12.0	10	2.70
0.01	17.3	36	-
0.05	37.4	250	0.16
0.10	12.0	38	-
0.50	13.0	-	-
1.0	19.5	39	1.70

Obviously so significant variations of mentioned (and some other [114]) properties suggest corresponded structural changes in the HDPE + Z composites compared with initial HDPE. Structural changes induced by introduction of Z into HDPE were studied [103] by classic methods for amorphous-crystalline polymers: differential scanning calorimetry (DSC), wide- and small-angle X-ray scattering, and IR-spectroscopy. Definite results were also obtained by mechanical tests, magnetic measurements, rheological studies, optic and electron microscopy, and gas penetration measurements.

Figure 34 shows dependencies of melting enthalpy, ΔH, and the melting temperature, T_m, determined by the DSC method, on Z concentration (C_Z) in HDPE + Z composites. As observed, minimal values of ΔH and T_m at C_Z = 0.05 wt.% correspond to the extreme variation of properties, shown in Table 6. This observation suggests decrease of the crystallite size at the mentioned concentration of Z [115]. Electron microscopy data prove this conclusion. Figure 35 shows electron micrographs of chemically etched initial HDPE samples and HDPE + Z composites. They visually illustrate changes in morphology of the samples studied. The initial HDPE possesses quite large lamellar crystallites with broad distribution by sizes. Moreover, the initial HDPE displays formation of larger morphological elements of the circular spherulite type. If just 0.1 wt.% of Z is injected to HDPE, circular structures disappear, and with further increase of Z concentration in the composites the size of crystallites is decreased and reaches its minimum in the HDPE + Z composite containing Z in concentration of 0.05 wt.% [116].

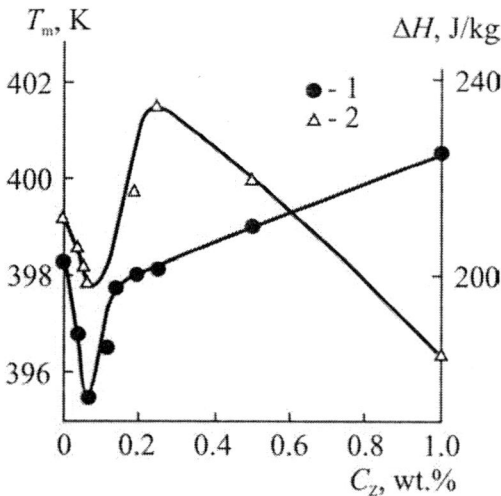

Figure 34. Dependencies of the melt enthalpy ΔH (1) and the melting temperature T_m (2) on Z concentration (C_Z) in HDPE + Z composite [103]

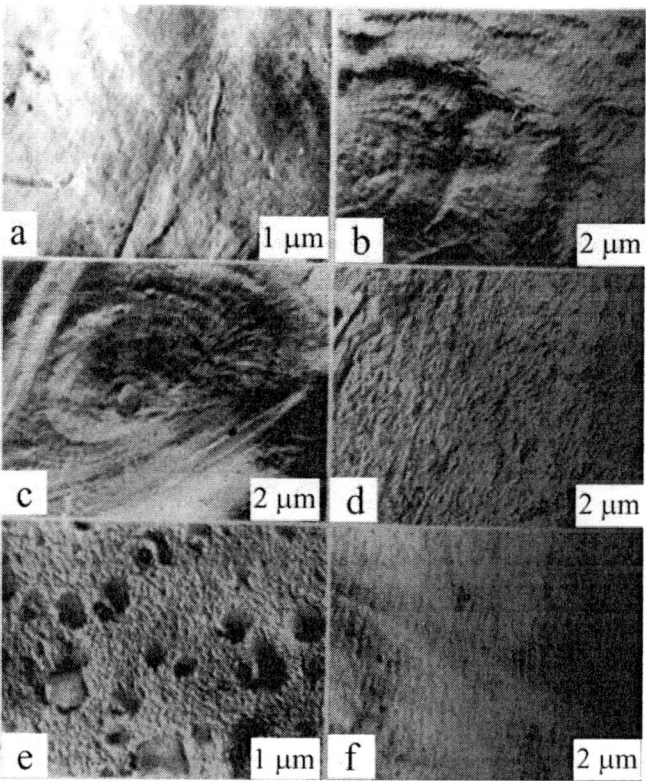

Figure 35. Electron micrographs of chemically treated surfaces of the initial HDPE (a – c) and HDPE + Z composites with Z concentration as follows: 0.01 (d), 0.05 (e) and 0.1 (f) wt.% [116]

Since properties of melts of HDPE and its composite with Z are interrelated with the material morphology formed during its crystallization, the melt flow indices (MFI) of the mentioned polymers were measured. Dependence of the flow index of melt of Z concentration is shown in Figure 36: an extreme change of the MFI function on C_Z for HDPE + Z composite containing 0.05 wt.% of Z is observed. As shown in the works [101, 102], this increase of melt viscosity for HDPE + Z composite compared with initial HDPE should be related to the account of increase of entanglement cluster network frequency, v_{cl}, induced by Z introduction. In accordance with currently existing concepts of crystallization of polyethylenes [117], the increase of the number of cross-link points of macromolecular entanglements, rejected from crystallizing zones during crystallization, induces a break in the process continuity and appropriate decrease of the size of crystallites.

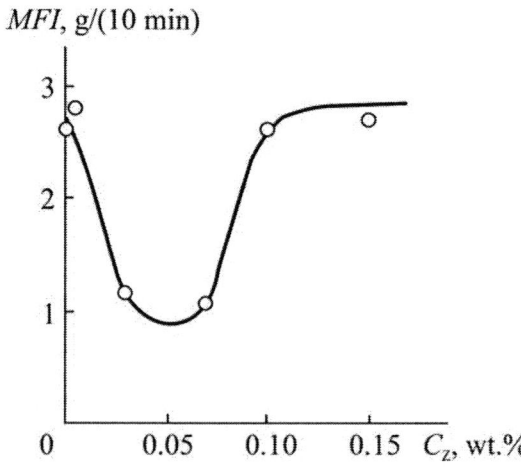

Figure 36. Dependence of the melt flow index (*MFI*) on Z concentration (C_Z) for HDPE + Z composites [102]

One more structural feature of HDPE + Z composite containing 0.05 wt.% of Z is detected by small-angle X-ray scattering method. Study of the dependence of the part of oriented crystallites L_{or} on C_Z indicated its broad variation with the latter (Table 7). Clearly this variation is of the extreme type also. For initial HDPE, $L_{or} \cong 25\%$, and for $C_Z = 0.05$ wt.% $L_{or} \cong 44\%$. It is assumed that L_{or} variation is induced by the melt viscosity increase (Figure 36), which determines an increase of shift stresses during polymer injection molding and orientation of crystallites [103].

Analysis of small-angle X-ray scattering patterns for the samples with different concentration of Z (Figure 37) indicates the broadest scattering in the samples with high concentration of oriented crystallites. For HDPE + Z composites with Z concentration of 0.05 wt.% the intensity of small-angle X-ray scattering reaches its maximum, and the pattern is maximum symmetrical and possesses the greatest half-width. All these observations indicate that samples of the current composite possess the most proper (regular) morphology with the minimum dispersion by sizes of both long period and crystalline and amorphous zones. These observations are in keeping with electron microscopy data (Figure 35). Such homogeneous structure results the oriented crystallization in HDPE + Z composite samples containing Z in concentration of 0.05 wt.%.

Table 7. The fracture of oriented crystallites (L_{or}) and normalized magnetic susceptibility $\left(\chi_m^n\right)$ for HDPE + Z composites [103]

Concentration of Z in HDPE + Z composites, wt.%	L_{or}, %	χ_m^n, Gs/g
0	25	-
0.01	27	0.30
0.05	44	0.25
0.10	36	-
0.15	25	-
0.20	24	0.18
0.50	27	-
1.0	25	0.14
12.0	-	0.17

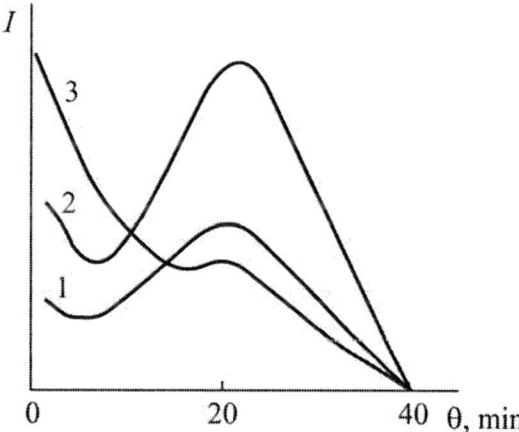

Figure 37. Small-angle X-ray scattering patterns recorded along the molding direction for initial HDPE (1) and HDPE + Z composite with Z concentration of 0.05 wt.% (2) and 1.0 wt.% (3) [101]

Completing the discussion on changes in crystalline morphology of HDPE + Z composites, one shall note that DSC, IR-spectroscopy, X-ray scattering and density measurement data indicated the absence of significant changes in the crystallinity degree, K, though its absolute values determined by DSC method were found minimal in the mentioned series of measurements [101].

To clear out differences in the structure of non-crystalline zones in HDPE + Z composites comparing with primary HDPE, it is suitable to search the relation of intensities (or extinction coefficients) of IR absorption bands at 1,303, 1,352 and 1,368 cm^{-1} using results obtained in the works [119, 120]. Figure 38 shows the change of extinction coefficient relations: $e_{1,303}/e_{1,352}$, $e_{1,303}/e_{1,368}$ and $e_{1,352}/e_{1,368}$, with respect to HDPE + Z composite density, ρ. This Figure also shows similar data from the work [119] for series of linear polyethylenes. Clearly good correspondence between these two series of data is observed, though for HDPE + Z composites the decrease of $e_{1,303}/e_{1,352}$ and $e_{1,303}/e_{1,368}$ relations with C_Z increase is much clearer [101, 121].

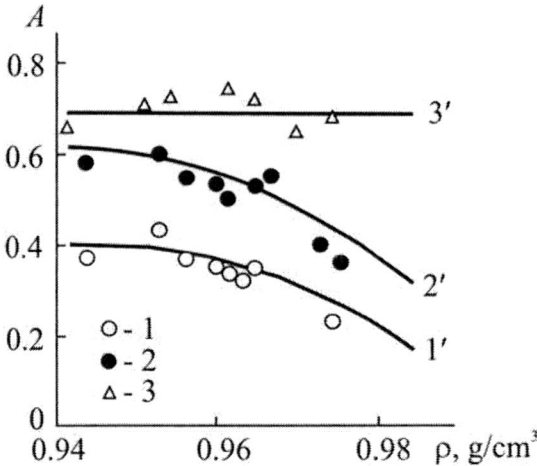

Figure 38. Dependencies of extinction coefficients' relations for IR-absorption bands, A, on density, ρ, for samples of HDPE + Z composites: 1, 1' - $e_{1,303}/e_{1,352}$; 2, 2' - $e_{1,303}/e_{1,368}$; 3, 3' - $e_{1,352}/e_{1,368}$; 1 - 3 – data from work [118], 1' - 3' - data from work [119]

Relations $e_{1,303}/e_{1,352}$ and $e_{1,303}/e_{1,368}$ were also estimated for paraffin melts [119]. They were found equal to the appropriate relations for polyethylene with the density below 0.960 g/cm³. This induced a conclusion that chain units in gauche-conformations of these samples (possessing $\rho \leq 0.960$ g/cm³) display the same structure (distribution of sequences) as in the primary melt at the same temperature. Decrease of the considered relations with ρ increase for $\rho > 0.960$ g/cm³ was assigned [119] to higher regularity of the distribution of units in gauche-conformations. Essentially, analogous explanation of dependencies from Figure 38 is given in the work [120]. It is supposed that increase of the crystallinity degree provides for more compact packing of crystallites and, as a consequence, limitation of the selection of possible conformations, which can be realized in non-crystalline zones. For example, as the distance between crystallites decreases, chains in non-crystalline zones have to orient more parallel to crystallite surfaces (the so-called barrier effect).

Thus the above shown results of experimental studies allow the conclusions as follows. Introduction of low amounts of Z (up to 0.10 wt.%) into amorphous-crystalline HDPE induces significant structural changes which imply corresponded variations of physicomechanical properties of HDPE + Z composites. Mechanical and rheological test results induce an assumption that the starting point of the structural changes is of macromolecular entanglement cluster network frequency, v_{cl}, in the HDPE melt. This phenomenon predetermined occurrence of two important structural effects. First, a significant increase of the melt viscosity leads to consequent increase of shear stresses during processing of HDPE + Z composites and oriented crystallization. In turn, this induces morphology of HDPE + Z composite with 0.05 wt.% concentration of Z representing a selection of lamellar crystallites with low dispersion by sizes. Secondly, variation of frequency v_{cl} leads to a change of lamellar crystallite sizes and thickness of non-crystalline interlayers between them. This is accompanied by the change of conformational state of chain sequences in non-crystalline zones. The second effect is the basic one, because qualitatively identical change of properties was also observed for HDPE + Z composite samples, obtained by press molding.

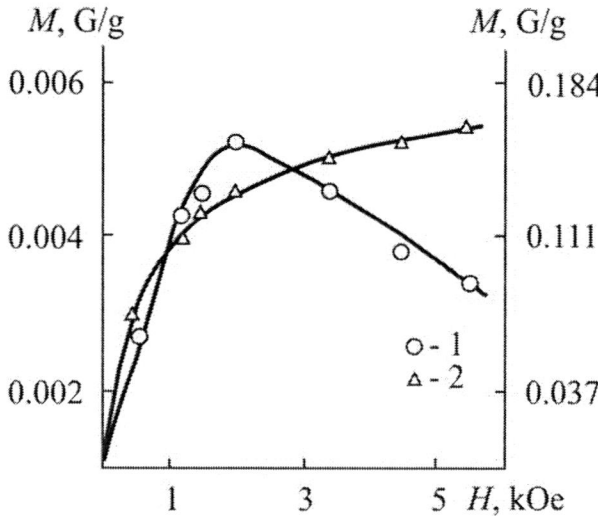

Figure 39. Dependencies of general magnetic moment M on external magnetic field intensity H for HDPE + Z composites containing 1 - 0.01 wt.% and 2 - 0.20 wt.% of Z

It may be assumed that the most probable origin of forces stipulating entanglement network frequency increase at Z injection to HDPE is either adhesion by short-range van der Waals forces or long-range magnetic interaction between Z and the polymer matrix. Since particles of injected modifier display ferromagnetic origin and possess very high specific surface, both possibilities are equally permissible. However, injection of disperse magnetic powder possessing high coercive force and low specific surface into HDPE has indicated M_{cl} decrease from (771 ± 45) g/mol for initial HDPE to (688 ± 27) g/mol for the composite. Statistically in the latter case, a noticeable decrease of M_{cl} bespeaks for the magnetic origin of forces, which increase the entanglement cluster network frequency [122].

Another argument for the benefit of this assumption is comparison of action of identical amounts of Z and highly dispersed copper, injected as before, on PS and HDPE. In contrast with Z, the effect of which is described above, injection of dispersed copper does not practically change physicomechanical properties of the material. Both modifiers (Z and Cu) possess large specific surface, but contrary to Z, Cu is diamagnetic substance.

For further proofing of this assumption two series of experiments were performed [103, 122]. The first series represented measurements of general magnetic moment, M, as the function of external magnetic field intensity, H, for a series of HDPE + Z composites. The measurements were performed on magnetic-field meter MAGNET B-E15 "Bruker" at 295 K [122]. Figure 39 show magnetization characteristics (general magnetic moment M as the function of H) for composites containing 0.01 and 0.20 wt.% of Z. These curves allow estimation of general magnetic susceptibility, χ_m, of the samples as follows [123]:

$$\chi_m = \frac{M}{H}. \tag{59}$$

Table 7 shows χ_m values normalized for Z concentration in HDPE + Z composites, χ_m^n. These results, as well as the data from Figure 39 provide for some conclusions. For example, HDPE + Z composite containing 0.05 wt.% of Z displays $M(H)$ dependence analogous by shape to curve 1 in Figure 39 and composites containing 1.0 and 12.0 wt.% of Z to curve 2. After the saturation point, dependencies $M(H)$ for HDPE + Z composites containing 0.01 and 0.05 wt.% of Z are characterized by negative gradient that assumes a noticeable increase of diamagnetic contribution (by the absolute value) to general magnetic moment for these composites compared with others [123]. As is known [124], one of the reasons for such increase can be growth of the number of conductivity electrons or, to put it differently, relative increase of the part of free iron in Fe/FeO mixture, injected to the polymeric matrix. The data shown indicate the following dynamics of χ_m^n change with Z concentration increase for HDPE + Z composites: it is significantly increased first and then (at $C_Z \geq 0.20$ wt.%) reaches asymptotic values [122].

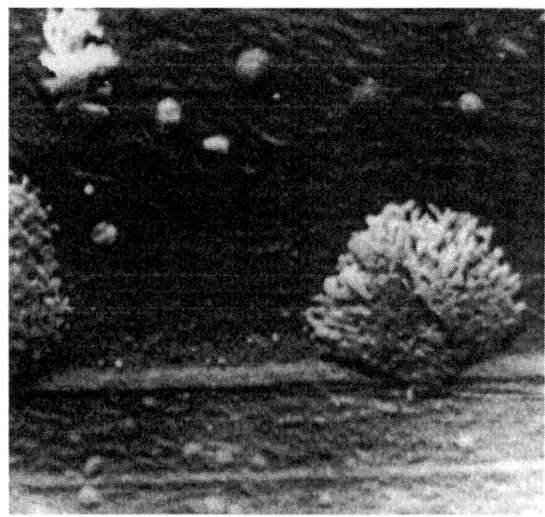

Figure 40. Scanning electron photomicrography of the degradation surface of HDPE + Z composite with 0.05 wt.% concentration of Z [125]

Figure 40 shows scanning electron photomicrograph illustrating the appearance of Z particles. Figure 41 shows the dependence of the average size of these particles, R_{av}, on C_Z value. Anomalous behavior of composites containing 0.01 and 0.05 wt.%, for which R_{av} are much lower than for the rest composites under consideration, is again observed. This suggests higher relative content of Fe particles in the mentioned composites, because they are smaller than FeO particles [122].

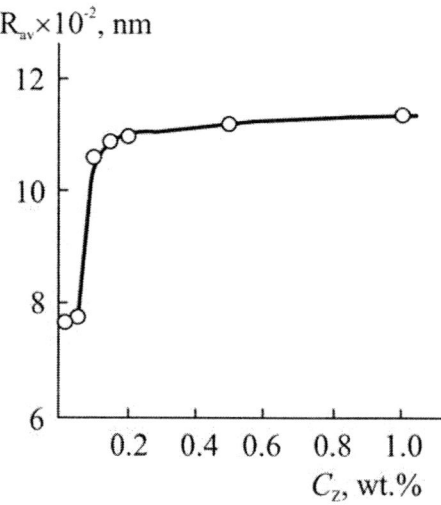

Figure 41. Dependence of the average size, R_{av}, of particles Z on C_Z in HDPE + Z composites [122]

The above stated results give an opportunity to suggest the following interpretation of experimentally observed extreme value of physicomechanical properties of HDPE + Z composites, which is associated with v_{cl} increase, analogous to the one suggested for single-phase metallopolymers [126]. As Z is injected into HDPE using the technique described [127], iron is reduced from oxides by the known formula [101, 122] as follows:

$$Fe^{+2} \underset{Red}{\overset{O_2}{\rightleftarrows}} Fe^{+3} \qquad (60)$$

where Red is the reducer, which can be HDPE oxidation products with hydroxyl and carbonyl groups, as well as mobile hydrogen. This assumption is proved by increase of diamagnetic contribution, decrease of Z particle size and higher χ_m^n values for composites with 0.01 and 0.05 wt.% concentration of Z. A part of Z fraction, due to high dispersion of Z particle size (refer to Figure 40), may fall within the range of sizes determining occurrence of super-paramagnetic domains [128] and, actually, the smaller is the average size of particles R_{av}, the greater part of them can become super-paramagnetic domains. The technique of metal injection to polymers used [126] suggests much smaller sizes of their particles and the absence of metal oxides. Nevertheless, general tendency in behavior of one-phase metallopolymers and HDPE + Z composites does really exist. In particular, global property of them is a significant decrease of the melt flow index [103].

Thus the following situation with formation of extreme value of properties for HDPE + Z composites can be suggested. In the case of low amounts of Z introduced into HDPE ($C_Z \leq$ 0.10 wt.%), intensive reduction of Fe from its oxides proceeds in accordance with the reaction (60) and relatively large part of small Z particles (super-paramagnetic domains) is formed. These particles increase frequency of macromolecular entanglement network. Below $C_Z =$

0.10 wt.%, v_{cl} is increased due to simple increase of Z amount. At $C_Z = 0.10$ wt.% reducing abilities of the polymer matrix are exhausted and dispersion composition of Z with dominating content of larger particles with greater part of FeO is practically preserved. Though possessing high coercive force, these particles are incapable of forming super-paramagnetic domains due to purely dimensional restrictions. That is why they do not modify the network of macromolecular entanglements and, consequently, do not affect properties of HDPE + Z composites [103].

Moreover, the above-described results also suggest that in non-crystalline zones of HDPE particles Z initiate entanglements of the type shown in Figure 2 (clusters), because magnetic field affecting polyethylene melt promotes ordering [129]. Thus it can be expected that local magnetic fields of super-paramagnetic particles will promote formation of the local order zones. From these positions, τ_{50} increase and P_{N_2} decrease for HDPE + Z composites (Table 6) are simply explained, because increase of the amount of macromolecular backlashes must induce the opposite effect due to structure pack-off [130].

Figure 42 shows temperature dependencies of the cross-link point functionality, F, of macromolecular entanglement network for HDPE and HDPE + Z composite containing 0.05 wt.% of Z. The functionality was calculated by the equation (5). For these polymers $F \geq 11$, whereas for macromolecular backlashes $F = 4$ [18] which also testifies for the benefit of the cluster formation in HDPE non-crystalline zones [131]. Note also that for initial HDPE, F declines with temperature, whereas for HDPE + Z composite it remains practically constant. This observation allows a supposition that magnetic field of a super-paramagnetic domain is more effective for hampering variation of the number of segments in the cluster with temperature than tension of the chain amorphous areas. To put it differently, despite the smaller number of segments, clusters in HDPE + Z composite with 0.05 wt.% of Z are more stable than the ones in the initial HDPE [131].

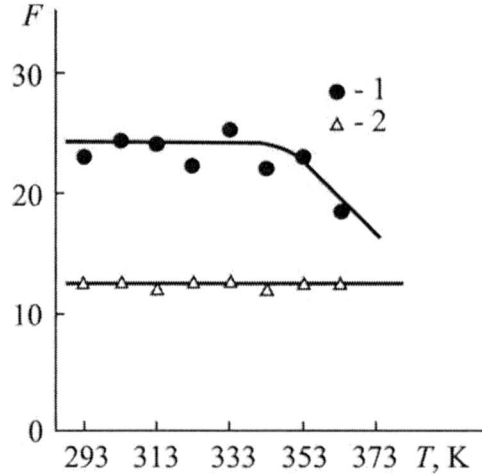

Figure 42. Temperature dependencies of cluster functionality F in the amorphous phase for HDPE (1) and HDPE + Z composite containing 0.05 wt.% of Z (2) [131]

Table 8. Comparison of parameters of macromolecular entanglement network obtained by different methods for PBTP + Z composites [134]

Z concentration, wt.%	M_{cl}^{m}, g/mol; equation (2.62)	$M_{cl}^{s.p.}$, g/mol; equation (2.2)
0	860	860
0.01	880	880
0.05	740	30
0.10	770	300
0.50	820	640
1.0	860	1,190

As is known [132], polymer melt viscosity is controlled by two parameters: molecular mass of the polymer, MM, and molecular mass of the chain segment between entanglements. General dependence of viscosity η of polymeric melt on the factors mentioned is of the following form [132]:

$$\eta \sim \frac{MM^2}{M_{cl}^3}. \tag{61}$$

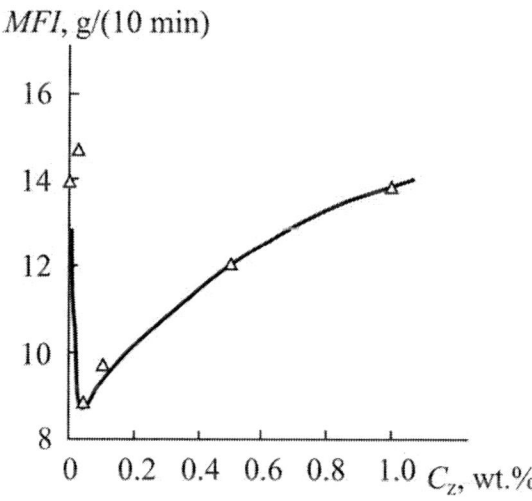

Figure 43. Dependence of the melt flow index (MFI) on Z concentration (C_Z) for PBTP + Z composite [134]

Correctness of the relation (61) was proved in the work [133], where for η the reverse melt flow index (MFI) was accepted for HDPE + Z composites. Figure 43 shows MFI dependence on Z concentration, C_Z, for poly(butylene terephthalate) composites with Z (PBTP + Z). Similar for HDPE + Z composites (Figure 36), the case under consideration displays a minimum of MFI (or a maximum of η) in the Z concentration range of $C_Z = 0.05 \div$

0.10 wt.%. Because there are no grounds for assuming changes of molecular mass of polymer at Z injection into PBTP [133], the relation (61) can be reduced to the form as follows [134]:

$$MFI = C_{ent} M_{cl}^r. \tag{62}$$

Constant C_{ent} in this equation can be determined using parameters of the initial PBTP (MFI = 14 g/10 min, M_{cl} = 860 g/mol): $C_{ent} \approx 22$. Table 8 shows comparison of $M_{cl}^{s.p.}$ (equation (2)) and M_{cl}^m (equation (62)) values, where index "s.p." indicates the solid phase and "m" means melt. Remind that for HDPE + Z composites this comparison has given: $M_{cl}^{s.p.} \approx M_{cl}^m$. However, for PBTP + Z composite at C_Z = 0 05 and 0.10 wt.% a significant (approximately, 2.5-fold) difference in the indicated parameters is observed. This means that for HDPE + Z composite the cluster network is formed already in the melt, and during crystallization its cross-link points (clusters) are rejected from crystallizing zones to non-crystalline (mainly, interphase) ones [117]. For initial PBTP and PBTP + Z composites containing 0.01, 0.5 and 1.0 wt.% of Z (i.e. in cases, for which super-paramagnetic domains of Z are absent) the process of cluster network formation is fully analogous to the above-described one for HDPE + Z. Thus the condition $M_{cl}^{s.p.} = M_{cl}^m$ (Table 8) is met for them. For PBTP + Z composites containing 0.05 and 0.10 wt.% of Z, a cluster network with cross-link points (clusters) formed by molecular interaction forces is added to the one formed by interaction between polymer chains and Z in the melt at $T < T_g$. The type of the former network formation is, probably, caused by additional chain straining by clusters and suppression of fluctuations. The relation of frequencies of these two types of the cluster network is 1:2.2. This difference is typical of amorphous-crystalline polymers with devitrified (HDPE) and glassy-like (PBTP) amorphous phase [134].

Thus the above results induce the following conclusions:

(1) cross-link points of macromolecular entanglement network in non-crystalline zones of amorphous-crystalline polymers represent the local order zones (clusters);
(2) variation of entanglement cluster network frequency causes a strong effect on structures and properties of this class of polymers;
(3) there is a possibility of purposeful control for v_{cl} and, consequently, properties of both amorphous-crystalline and amorphous polymers.

REFERENCES

[1] Haward R.N., *J. Polymer Sci.: Part B: Polymer Phys.*, 1995, vol. 33(8), pp. 1481 - 1494.
[2] Haward R.N., *Polymer*, 1987, vol. 28(8), pp. 1485 - 1488.
[3] Boyce M.C., Parks D.M., and Argon A.S., *Mech. Mater.*, 1988, vol. 7(1), pp. 15 - 33.
[4] Boyce M.C., Parks D.M., and Argon A.S., *Mech. Mater.*, 1988, vol. 7(1), pp. 35 - 47.
[5] Haward R.N., *Macromolecules*, 1993, vol. 26(22), pp. 5860 - 5869.

[6] Bartenev G.M. and Frenkel S.Ya., *Polymer Physics*, 1990, Leningrad, Khimia, 432 p. (Rus)
[7] Belousov V.N., Kozlov G.V., Mikitaev A.K., and Lipatov Yu.S., *Doklady AN SSSR*, 1990, vol. 313(3), pp. 630 - 633. (Rus)
[8] Flory P.Y., *Polymer J.*, 1985, vol. 17(1), pp. 1 - 12.
[9] Bernstein V.A. and Egorov V.M., *Differential Scanning Calorimetry in Physicochemistry of Polymers*, 1990, Leningrad, Khimia, 256 p. (Rus)
[10] Graessley W.W., *Macromolecules*, 1980, vol. 13(2), pp. 372 - 376.
[11] Perepechko I.I. and Startsev O.V., *Vysokomol. Soedin.*, 1973, vol. 15B(5), pp. 321 - 322. (Rus)
[12] Belousov V.N., Kotsev B.Kh., and Mikitaev A.K., *Doklady AN SSSR*, 1983, vol. 270(5), pp. 1145 – 1147. (Rus)
[13] Arzhakov S.A., Bakeev N.F., and Kabanov V.A., *Vysokomol. Soedin.*, 1973, vol. 15A(5), pp. 1154 - 1167. (Rus)
[14] Narisawa I., Polymer Material Strength, 1987, Moscow, Khimia, 400 p. (Rus)
[15] Kozlov G.V., Sanditov D.S., and Serdyuk V.D., *Vysokomol. Soedin.*, 1993, vol. 35B(12), pp. 2067 - 2069. (Rus)
[16] Gent A.N. and Madan S., *J. Polymer Sci.: Part B: Polymer Phys.*, 1989, vol. 27(7), pp. 1529 - 1542.
[17] Graessley W.W. and Edwards S.F., *Polymer*, 1981, vol. 22(10), pp. 1329 - 1334.
[18] Wu S.J., *J. Polymer Sci.: Part B: Polymer Phys.*, 1989, vol. 27(4), pp. 723 - 741.
[19] Budtov V.P., *Physical Chemistry of Polymer Solutions*, 1992, Saint-Petersburg, Khimia, 384 p. (Rus)
[20] Mashukov N.I., Vasnetsova O.A., Malamatov A.Kh., and Kozlov G.V., *Lakokrasochnye Materialy i Ikh Primenenie*, 1992, No. 1, pp. 16 – 17. (Rus)
[21] Barinov V.Yu., *Vysokomol. Soedin.*, 1981, vol. 23B(1), pp. 66 - 68. (Rus)
[22] Kambour R.P., *J. Polymer Sci.: Macromol. Rev.*, 1973, No. 7, pp. 1 - 54.
[23] Kramer E.J., *J. Polymer Sci.: Polymeer Phys. Ed.*, 1975, vol. 13(2), pp. 509 - 516.
[24] Lu X.-C. and Brown N., *J. Mater. Sci.*, 1986, vol. 21(11), pp. 4081 - 4092.
[25] Pertsev N.A., Romanov A.E., and Vladimirov V.I., *J. Mater. Sci.*, 1981, vol. 16(8), pp. 2084 - 2093.
[26] Dugdlave D., *J. Mech. Phys. Solids*, 1960, vol. 8(2), pp. 100 - 104.
[27] Kozlov G.V., Belousov V.N., and Mikitaev A.K., *Fizika i Tekhnika Vyisokikh Davleniy*, 1998, vol. 8(1), pp. 101 – 107. (Rus)
[28] Fellers J.F. and Huang D.C., *J. Appl. Polymer Sci.*, 1979, vol. 23(8), pp. 2315 - 2326.
[29] Kozlov G.V. and Sanditov D.S., *Anharmonic Effects and Physicomechanical Properties of Polymers*, 1994, Novosibirsk, Nauka, 261 p. (Rus)
[30] Lin Y.-H., *Macromolecules*, 1987, vol. 20(12), pp. 3080 - 3083.
[31] Katsnelson A.A., *Introduction to Physics of Solids*, 1984, Moscow, Izd. IGU, 293 p. (Rus)
[32] Shogenov V.N., Belousov V.N., Potapov V.V., Kozlov G.V., and Prut E.V., *Vysokomol. Soedin.*, 1991, vol. 33A(1), pp. 155 – 160. (Rus)
[33] Startsev O.V., Abeliov Ya.A., Kirillov V.N., and Voronkov M.G., *Doklady AN SSSR*, 1987, vol. 293(6), pp. 1419 – 1422. (Rus)
[34] Sumita M., Tsukumo Y., Miyasaka K., and Ishikawa K., *J. Mater. Sci.*, 1983, vol. 18(5), pp. 1758 - 1764.

[35] Mashukov N.I., Belousov V.N., Kozlov G.V., Ovcharenko E.N., and Gladyishev G.P., *Izv. AN SSSR, Ser. Khim.*, 1990, No. 9, pp. 2143 – 2146. (Rus)
[36] Schrager M., *J. Appl. Polymer Sci.*, 1978, vol. 22(8), pp. 2379 - 2381.
[37] Malamatov A.Kh. and Kozlov G.V., *Doklady Adyigeiskoi (Cherkesskoi) Mezhdunarodhoi AN*, 1998, vol. 3(2), pp. 78 – 81. (Rus)
[38] Kozlov G.V. and Novikov V.U., *Uspekhi Fizicheskikh Nauk*, 2001, vol. 171, pp. 717 – 764. (Rus)
[39] Balankin A.S., *Synergysm of Deformable Body*, 1991, Moscow, MO SSSR, 404 p. (Rus)
[40] Balankin A.S., Bugrimov A.L., Kozlov G.V., Mikitaev A.K., and Sanditov D.S., *Doklady RAN*, 1992, vol. 326(3), pp. 463 – 466. (Rus)
[41] Gladyshev G.P., Thermodynamics and Macrokinetics of Natural Hierarchic Processes, Moscow, Nauka, 1988, 290 p. (Rus)
[42] Gladyshev G.P., *J. Theor. Biol.*, 1978, vol. 75(4), pp. 425 - 444. (Rus)
[43] Gladyshed G.P. and Gladyshev D.P., *On Physicochemical Theory of Biological Evolution* (Preprint), 1993, Moscow, Olimp, 24 p. (Rus)
[44] Gladyshev G.P., *J. Biol. Systems*, 1993, vol. 1(2), pp. 115 - 129.
[45] Gladyshed G.P. and Gladyshev D.P., *Zh. Fiz. Khim.*, 1994, vol. 68(5), pp. 790 – 792. (Rus)
[46] Gladyshed G.P., *Izvestiya RAN., Ser. Biol.*, 1995, No. 1, pp. 5 – 14. (Rus)
[47] Kozlov G.V. and Zaikov G.E., In Book: *Fractal Analysis of Polymers: From Synthesis to Composites*, Ed. Kozlov G., Zaikov G., and Novikov V., New York, Nova Science Publishers, Inc., 2003, pp. 89 - 97.
[48] Beloshko V.A., Kozlov G.V., and Lipatov Yu.S., *Fizika Tverdogo Tela*, 1994, vol. 36(10), pp. 2903 – 2906. (Rus)
[49] Matsuoka S. and Bair H.E., *J. Appl. Phys.*, 1977. vol. 48(10), pp. 4058 - 4062.
[50] Wu S., *J. Appl. Polymer Sci.*, 1992, vol. 46(4), pp. 619 - 624.
[51] Privalko V.P. and Lipatov Yu.S., *Vysokomol. Soedin.*, 1971, vol. 13A(12), pp. 2733 – 2738. (Rus)
[52] Gladyshev G.P., *J. Biol. Phys.*, 1994, vol. 20(2), pp. 213 - 222.
[53] Belousov V.N., Kozlov G.V., and Lipatov Yu.S., *Doklady AN SSSR*, 1991, vol. 318(3), pp. 615 – 618. (Rus)
[54] Kozlov G.V., Aloev V.Z., and Lipatov Yu.S., *Ukrainski Khimicheski Zhurnal*, 2001, vol. 67(10), pp. 115 – 119. (Rus)
[55] Mark J.E. and Sullivan J.L., *J. Chem. Phys.*, 1977, vol. 66(3), pp. 1006 - 1011.
[56] Boyer R.F., *J. Macromol. Sci.-Phys.*, 1976, vol. B12(2), pp. 253 - 301.
[57] Boyer R.F., *Macromolecules*, 1992, vol. 25(20), pp. 5326 - 5330.
[58] White R.M. and Geballe T.H., *Long Range Order in Solids*, New York, Academic Press, 1979, 447 p.
[59] Kozlov G.V., Milman L.D., and Mikitaev A.K., *Local Order in Amorphous-Crystalline Polyethylene*, Manuscript deposited to VINITI RAS, Moscow, February 26, 1997, No. 622-V97. (Rus)
[60] Lobanov A.M. and Frenkel S.Ya., *Vysokomol. Soedin.*, 1980, vol. A22(5), pp. 1045 – 1057. (Rus)
[61] Boyer R.F., *Polymer Eng. Sci.*, 1968, vol. 8(3), pp. 161 - 185.

[62] Mashukov N.I., Serdyuk V.D., Kozlov G.V., Ovcharenko E.N., Gladyshev G.P., and Vodakhov A.B., *Stabilization and Modification of Polyethylene by Oxygen Acceptors* (Preprint), 1990, Moscow, IKhF AN SSSR, 64 p. (Rus)
[63] Belousov V.N., Kozlov G.V., and Mashukov N.I., *Doklady Adyigeiskoi (Cherkesskoi) Mezhdunarodnoi AN*, 1996, vol. 2(1), pp. 74 - 82. (Rus)
[64] Zhenyi M., Langford S.C., Dickinson J.T., Engelhard M.H., and Boyer D.R., *J. Mater. Res.*, 1991, vol. 6(1), pp. 183 - 195.
[65] Novikov V.U. and Kozlov G.V., *Materialovedenie*, 1999, No. 12, pp. 8 – 12. (Rus)
[66] Boyer R.F. and Miller R.L., *Macromolecules*, 1977, vol. 10(5), pp. 1167 - 1169.
[67] Miller R.L. and Boyer R.F., *J. Polymer Sci.: Polymer Phys. Ed.*, 1984, vol. 22(12), pp. 2043 - 2050.
[68] Aharoni S.M., *Macromolecules*, 1985, vol. 18(12), pp. 2624 - 2630.
[69] Kozlov G.V., Beloshenko V.A., and Lipskaya V.A., *Ukrainsky Fizichesky Zhurnal*, 1996, vol. 41(2), pp. 222 – 225. (Rus)
[70] Matsuoka S., Aloisio C.J., and Bair H.E., *J. Appl. Phys.*, 1973, vol. 44(10), pp. 4265 - 4268.
[71] Sharma B.K., *Acoust. Lett.*, 1980, vol. 4(2), pp. 19 - 23.
[72] Kozlov G.V. and Zaikov G.E., *Izv. KBHTs RAN*, 2003, No. 9, pp. 132 – 140. (Rus)
[73] Sharma B.K., *Acustica*, 1981, vol. 48(2), pp. 121 - 122.
[74] Barker R.E., *J. Appl. Phys.*, 1963, vol. 34(1), pp. 107 - 116.
[75] Sanditov D.S. and Bartenev G.M., *Physical Properties of Irregular Structures*, 1982, Novosibirsk, Nauka, 256 p. (Rus)
[76] Sanditov D.S., In Coll.: *Nonlinear Effects in Degradation Kinetics*, 1988, Leningrad, FTI AS USSR, pp. 140 – 149. (Rus)
[77] Privalko V.P. and Lipatov Yu.S., *Vysokomol. Soedin.*, 1974, vol. 16A(7), pp. 1562 – 1568. (Rus)
[78] Boyer R.F., *J. Macromol. Sci.-Phys.*, 1973, vol. 7B(3), pp. 487 - 501.
[79] Forsman W.C, *Macromolecules*, 1982, vol. 15(6), pp. 1032 - 1040.
[80] Hornboden E., *Int. Mater. Rev.*, 1989, vol. 34(6), pp. 277 - 296.
[81] Curro J.J. and Roe R.-J., *Polymer*, 1984, vol. 25(10), pp. 1424 - 1430.
[82] Rathje J. and Ruland W., *Colloid Polymer Sci.*, 1976, vol. 254(3), pp. 358 - 370.
[83] Kalinchev E.L. and Sakovtseva M.B., *Properties and Processing of Thermoplasts*, 1983, Leningrad, Khimia, 288 p. (Rus)
[84] Mashukov N.I., Serdyuk V.D., Belousov V.N., Kozlov G.V., and Khatsukova M.A., *Fluctuation Network of Molecular Entanglements As The Percolation System*, Manuscript deposited to VINITI RAS, Moscow, June 20, 1994, No. 1537-B94. (Rus)
[85] Belousov V.N., Kozlov G.V., and Mashukov N.I., *Doklady Adyigeiskoi (Cherkesskoi) Mezhdunarodnoi AN*, 1995, vol. 2(2), pp. 76 – 80. (Rus)
[86] Serdyuk V.D., Sanditov D.S., Mashukov N.I., Kozlov G.V., Novikov V.U., and Zaikov G.E., *Int. J. Polym. Mater.*, 1998, vol. 42(1), pp. 65 - 73.
[87] Afaunov V.V., Mashukov N.I., Kozlov G.V., and Sanditov D.S., *Izv. VUZov. Severo-Kavkazskii Region, Estestvennye Nauki*, 1999, No. 4(108), pp. 69 – 71. (Rus)
[88] Mashukov N.I., Serdyuk V.D., Gladyshev G.P., and Kozlov G.V., *Voprosy Oboronnoi Tekhniki*, Ser. 15, 1991, Iss. 3, No. 97, pp. 11 – 13. (Rus)
[89] Eliashevich G.K., *Vysokomol. Soedin.*, 1988, vol. 30A(8), pp. 1700 – 1705. (Rus)

[90] Kozlov G.V., Shogenov V.N., Kharaev A.M., and Mikitaev A.K., *Vysokomol. Soedin.*, 1987, vol. 29B(4), pp. 311 – 314. (Rus)

[91] Bittrich H.-J., *Acta Polymerica*, 1982, vol. 33(12), pp. 741.

[92] Adam G. and Gibbs J.H., *J. Chem. Phys.*, 1965, vol. 43(1), pp. 139 - 146.

[93] Kozlov G.V. and Lipatov Yu.S., *Vysokomol. Soedin.*, 2003, vol. 45B, pp. 660 - 664. (Rus)

[94] Kozlov G.V. and Lipatov Yu.S., In Coll.: *Perspectives of Chemical and Biochemical Physics*, Ed. Zaikov G., New York, Nova Science Publishers, Inc., 2002, p. 205-212.

[95] Kozlov G.V., Belousov V.N., Serdyuk V.D., Mikitaev A.K., and Mashukov N.I., *Doklady Adyigeiskoi (Cherkesskoi) Mezhdynarodnoi AN*, 1997, vol. 2(2), pp. 88 – 93. (Rus)

[96] Kozlov G.V., Belousov V.N., Mikitaev A.K., and Mashukov N.I., *Doklady Adyigeiskoi (Cherkesskoi) Mezhdynarodnoi AN*, 1997, vol. 2(2), pp. 94 – 98. (Rus)

[97] Aharoni S.M., *Macromolecules*, 1983, vol. 16(9), pp. 1722 - 1728.

[98] Mills P.J., Hay J.N., and Haward R.N., *J. Mater. Sci.*, 1985, vol. 20(2), pp. 501 - 507.

[99] Prevorsek D.C. and De Bona B.T., *J. Macromol. Sci.-Phys.*, 1981, vol. B19(4), pp. 605 - 622.

[100] Kozlov G.V., Novikov V.U., and Zaikov G.E., *Plasticheskie Massy*, 2002, No. 5, pp. 33 – 34. (Rus)

[101] Kozlov G.V., Temiraev K.B., MalamatovA.Kh., and Shustov G.B., Izv. KBNTs RAS, 1999, No. 2, pp. 95 – 99. (Rus)

[102] Mashukov N.I., Serdyuk V.D., Belousov V.N., Kozlov G.V., Ovcharenko E.N., and Gladyshev G.P., *Izv. AN SSSR, Ser. Khim.*, 1990, No. 8, pp. 1815 – 1817. (Rus)

[103] Mashukov N.I., Gladyishev G.P., and Kozlov G.V., *Vysokomol. Soedin.*, 1991, vol. 33A(12), pp. 2538 – 2546. (Rus)

[104] Serdyuk V.D., Kozlov G.V., Mashukov N.I., and Mikitaev A.K., *J. Mater. Sci. Techn.*, 1997, vol. 5(2), pp. 55 - 60.

[105] Krigbaum W.R., Roe R.-J., and Smith K.J., *Polymer*, 1964, vol. 5(3), pp. 533 - 542.

[106] Kozlov G.V. and Zaikov G.E., *Izv. KBNTs RAI*, 2003, No. 9, pp. 126 – 131. (Rus)

[107] Sanjuan J. and Lorence M.A., *J. Polymer Sci.: Polymer Phys. Ed.*, 1988, vol. 26(2), pp. 235 - 244.

[108] Popli R. and Mandelkern L., *J. Polymer Sci.: Part B: Polymer Phys.*, 1990, vol. 28(11), pp. 1917 - 1941.

[109] Jiang C.-Y., Carrido L., and Mark J.E., *J. Polymer Sci.: Polymer Phys. Ed.*, 1984, vol. 22(12), pp. 2281 - 2284.

[110] Falender J.R., Yeh G.S.Y., and Mark J.E., *J. Chem. Phys.*, 1979, vol. 70(11), pp. 5324 - 5325.

[111] Milagin M.F. and Shishkin N.I., *Vysokomol. Soedin.*, 1988, vol. 30A(11), pp. 2249 – 2254. (Rus)

[112] Ozden S., Hatsukova M.A., Mashukov N.I., and Kozlov G.V., *Int. Polymer Processing*, 1988, vol. 13(1), pp. 23 - 26.

[113] Ozden S., Hatsukova M.A., Mashukov N.I., and Kozlov G.V., *Plast., Rubber and Composites*, 2000, vol. 29(5), pp. 212 - 215.

[114] Mashukov N.I., Vasnetsova O.A., Kozlov G.V., and Kesheva A.B., *Lakokrasochnye Materialy i Ikh Primenenie*, 1990, No. 5, pp. 38 – 41. (Rus)

[115] Vunderlich B., *Physics of Macromolecules*, Vol. 3, 1984, Moscow, Mir, 484 p. (Rus)

[116] Afaunov V.V., Kozlov G.V., and Mashukov N.I., *Doklady Adyigeiskoi (Cherkesskoi) Mezhdunarodnoi AN*, 2001, vol. 5(2), pp. 114 – 119. (Rus)

[117] Seguela R. and Rietsch F., *Polymer*, 1986, vol. 27(5), pp. 703 - 708.

[118] Mashukov N.I., Gladyshev G.P., Kozlov G.V., and Mikitaev A.K., In Coll.: *Proc. VI All-Union Coordinat. Conference on Spectroscopy of Polymers*, 1989, Minsk, BGU, p. 81. (Rus)

[119] Okada T. and Mandelkern L., *J. Polymer Sci. Part A-2*, 1967, vol. 5(2), pp. 239 - 244.

[120] Wedgewood A.R. and Seferis J.C., *Pure and Appl. Chem.*, 1983, vol. 55(5), pp. 873 - 892.

[121] Malamatov A.Kh., Mashukov N.I., and Kozlov G.V., *Izv. KBNTs RAI*, 1999, No. 3, pp. 65 – 68. (Rus)

[122] Mashukov N.I., Kozlov G.V., Mikitaev A.K., and Vodakhov A.B., In Coll.: *Theory and Practice of Catalytic Reactions and Polymer Chemistry*, 1990, Cheboksary, ChGU, pp. 104 – 108. (Rus)

[123] Jones T.E., Butler W.F., Ogden T.R., Gottfredson D.M., and Gullikson E.M., *J. Chem. Phys.*, 1988, vol. 88(5), pp. 3338 - 3348.

[124] Yavorsky B.M. and Detlaf A.A., *Reference Book on Physics*, 1974, Moscow, Nauka, 847 p. (Rus)

[125] Shogenov V.N. and Kozlov G.V., *Fractal Clusters in Physicochemistry of Polymers*, 2002, Nalchik, Polygrafservis and T, 270 p. (Rus)

[126] Kosobudsky I.D., Kashkina L.V., Gubin S.P., Petrakovsky G.A., Piskorsky V.P., and Svirsky N.M., *Vysokomol. Soedin.*, 1985, vol. 27A(4), pp. 689 – 695. (Rus)

[127] Gladyshev G.P., Mashukov N.I., Eltsin S.A., Mikitaev A.K., Vasnetsova O.A., and Ovcharenko E.N., *Physicochemical Properties of Polyethylene Stabilized by "Non-chain" Inhibitors*, (Preprint), 1985, Chernogolovka, OIKhF AS USSR, 1985, 14 p. (Rus)

[128] Piskorsky V.P., Lipanov A.M., and Balusov V.A., *Zh. Vsesouzn. Khim. Obshch.*, 1987, vol. 32(1), pp. 47 – 50. (Rus)

[129] Belyi V.A., Snezhkov V.V., Bezrukov S.V., Voronezhtsev Yu.I., Goldade V.A., and Pinchuk L.S., *Doklady AN SSSR*, 1988, vol. 302(2), pp. 355 – 357.

[130] Privalko V.P., Andrianova G.P., Besklubenko Yu.D., Narozhnaya E.P., and Lipatov Yu.S., *Vysokomol. Soedin.*, 1978, vol. 20(12), pp. 2777 – 2783. (Rus)

[131] Malamatov A.Kh. and Kozlov G.V., *Doklady Adyigeiskoi (Cherkesskoi) Mezhdunarodnoi AN*, 1998, vol. 3(2), pp. 78 – 81. (Rus)

[132] Kavassalis T.A. and Noolandi J., *Macromolecules*, 1988, vol. 21(9), pp. 2869 - 2879.

[133] Mashukov N.I., Mikitaev A.K., Gladyishev G.P., Belousov V.N., and Kozlov G.V., *Plastmassy*, 1990, No. 11, pp. 21 – 23. (Rus)

[134] Kitieva L.I., *Candidate's Thesis on Chemistry*, 2000, Nalchik, KBGU, 138 p. (Rus)

Chapter 8

COMPLEX-RADICAL POLYMERIZATION OF STYRENE IN THE PRESENCE OF METALLOCENE INITIATING SYSTEMS

Yu. B. Monakov, N. N. Sigaeva, S. V. Kolesov, A. U. Abdulgalimova and E. M. Prokudina*

Institute of Organic Chemistry, Ufa Scientific Center,
Russian Academy of Sciences, Pr. Oktyabrya 71,
Ufa, 450054 Bashkortostan, Russia

ABSTRACT

For the radical polymerization of styrene initiated by metallocene systems, kinetic nonuniformity distributions of active centers have been determined based on experimentally derived molecular mass distribution curves using the Tikhonov regularization method. The polymodal pattern of the kinetic nonuniformity distribution curves indicates the existence of two types of active centers changing their kinetic activity during polymerization.

Metallocene catalytic systems actively employed in the past years in ion-coordination polymerization were applied as initiating systems components in radical polymerization [1, 2]. The conducted research [2] resulted in the fact that metallocenes combined with benzoyl peroxide provides higher polymerization rate and greater polymer yield. In addition, metallocene in the initiating system contributed to the increase of syndiotactic sequences in polymethylmetacrylate [2], i.e. it influenced the polymer streoregularity. Spectral studies [3] revealed that ferrocene (Cp_2Fe) was able to form complexes with benzoyl peroxide. It is evident that the complex being formed [4], as was noted in [5, 6], can be involved in coordination interaction with a monomer and initiate polymerization along with a benzoyl peroxide radical PhCOO'. Thus, several types of centers leading to chain propagation can

---

* Yu. B. Monakov: E-mail: monakov@anrb.ru

occur in the initiation; that is, the kinetic non-uniformity of the process can manifest itself. Consequently, polymers initiated by such systems can be characterized by a complicated molecular mass distribution (MMD) curves and their molecular properties depend on the polymerization conditions.

The objective of this work was to study molecular characteristics of polystyrene (PS) synthesized in the presence of initiating systems, viz., ferrocene-benzoyl peroxide (ferrocene system), titanocene-benzoyl peroxide (titanocene system), and zirconocene-benzoyl peroxide (zirconocene system), and to estimate the kinetic nonuniformity of the complex-radical polymerization process by solving the inverse molecular mass distribution problem [4, 7 - 10].

The polymerization of styrene was conducted in bulk at 45° and 60°C; the benzoyl peroxide concentration was $1 \cdot 10^{-3}$ mol/l, and the metallocene-to-benzoyl peroxide molar ratio was 1:1.

The molecular mass characteristics of PS were measured by gel-permeation chromatography. The measurements were made using a *DuPont Instruments* liquid chromatograph with four columns packed with a *Shimadzu* microgel with pore sizes of $1 \times 10^3 \div 10^6$ Å; toluene being used as an eluent.

Figure 1 demonstrates the values of M_w and M_w/M_n plotted as a function of the monomer conversion for PS samples prepared using metallocene initiating systems.

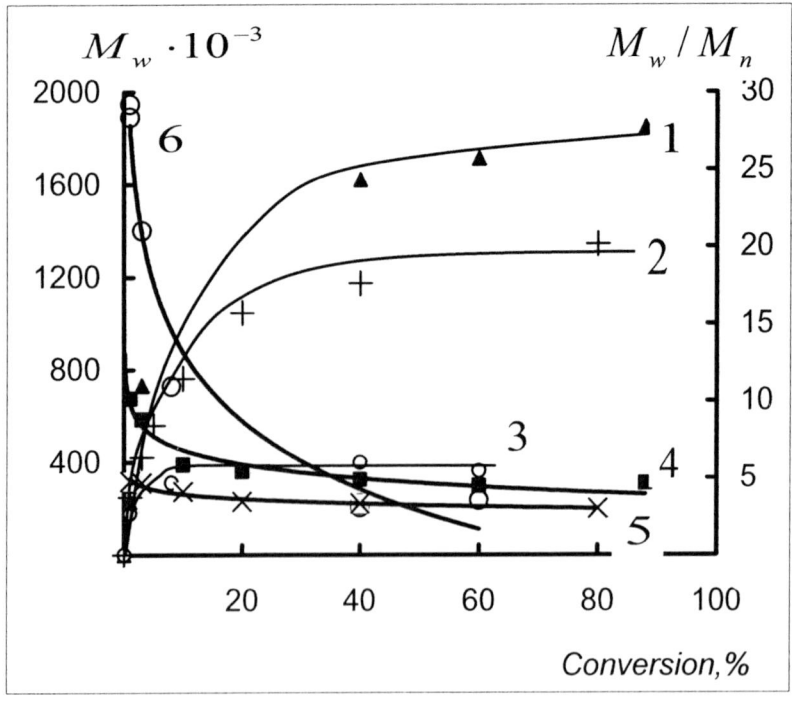

Figure 1. The plots of *(1-3)* M_w and *(4,5)* M_w/M_n of polystyrene vs. monomer conversion. *(1, 4)* Cp_2ZrCl_2-benzoyl peroxide, *(2, 5)* Cp_2Fe-benzoyl peroxide, and *(3, 6)* Cp_2TiCl_2-benzoyl peroxide

It is seen that the values of molecular masses (MM) sharply grow at the beginning of polymerization; however, as conversion rates of ~20-30% are achieved, this growth decelerates. The highest values of M_w are exhibited by PS samples prepared with the zirconocene system, while the lowest ones - that of titanocene. The polydispersity of PS (M_w/M_n) drops in the course of polymerization (Fig. 1). For example, for PS prepared with the Cp_2TiCl_2-benzoyl peroxide system, the polydispersity index falls from 29 to 2, while for the Cp_2Fe-benzoyl peroxide initiating system - from 5 to 2 respectively.

The nature of the metal in the metallocene initiating system affects not only the molecular mass of the resulting polymer, but also the pattern of molecular mass distribution curves (Figure 2). At the onset of polymerization (up to a conversion rate of ~20%), the molecular mass distribution curves are seen to show a bimodal pattern. One can see two well-resolution peaks: the first peak corresponds to a lower molecular mass fraction of PS, and the second one belonging to the higher molecular mass fraction.

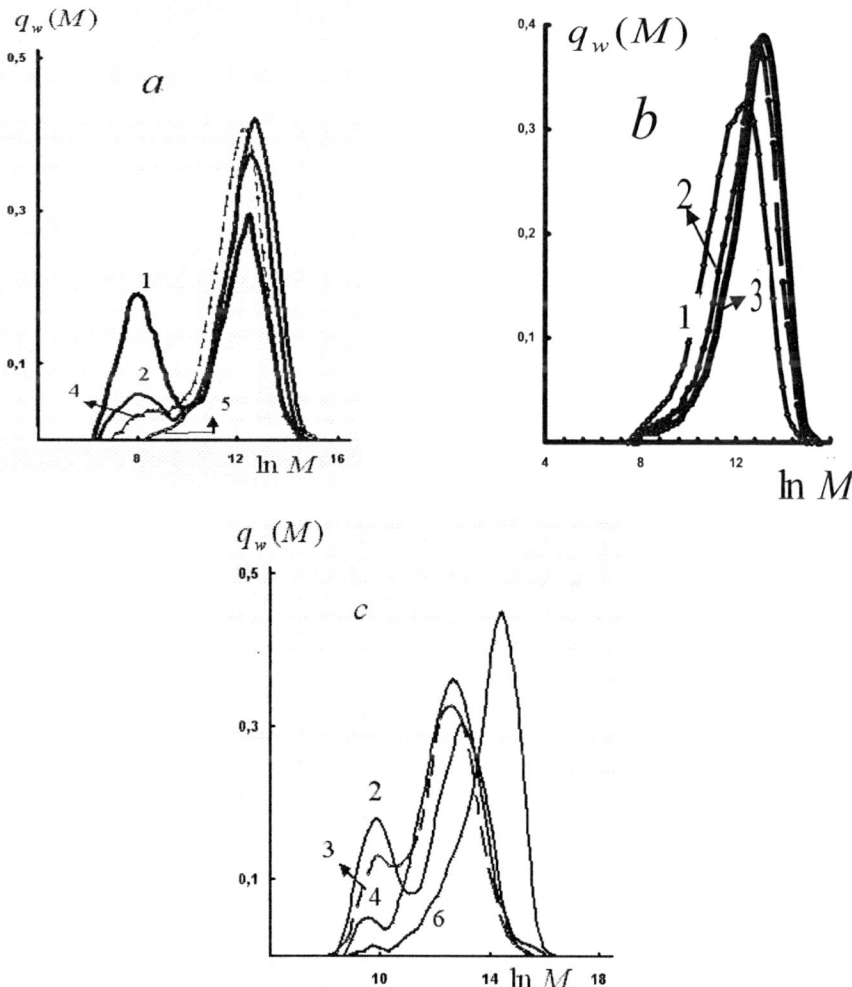

Figure 2. Molecular mass distribution curves of polystyrene synthesized using (a) titanocene-benzoyl peroxide, (b) ferrocene-benzoyl peroxide, and (c) zirconocene-benzoyl peroxide initiating systems. The monomer conversion: *(1)* 1, *(2)* 3, *(3)* 5, *(4)* 10, *(5)* 40, and *(6)* 60%

For PS prepared using the ferrocene system, the bimodal pattern of molecular mass distribution curves is slightly pronounced and can be detected only at small conversion rates of the monomer. As the latter goes up, the "low-molecular-mass" peak nearly disappears; that resulting in the polydispersity index drop to two, that is, to the value typical of free-radical polymerization.

One may suggest that the above-described dynamics of molecular characteristics is probably accounted for by the participation of two types of growing centers in polymerization. These types of active centers differ in the kinetic parameters of chain propagation and chain termination reactions (the kinetic nonuniformity) in which they are involved. The use of the mathematical analysis of molecular mass distribution curves developed for ion-coordination polymerization, based on the Tikhonov regularization method [9, 10], made it possible to calculate the kinetic nonuniformity distribution of chain-growth centers according to techniques described in [7, 8]. The mathematical model applied is based on the following assumptions.

(1) Each type of chain-growth centers produces polymer fractions whose molecular mass distribution curves are described by the Flory distribution, that is, they obey the law $\beta_i e^{-\beta_i M}$, where β is the reciprocal of the average molecular mass characterizing the probability of chain termination. The numerical value of β is equal to the ratio between the summary rate of chain termination reactions W_{oi} and the rate of chain propagation reaction W_p ($\beta(M) = \dfrac{\sum W_{ti}}{W_p m_o}$, where m_0 is the molecular mass of the monomer).

The employment of the Flory function for description of this polymerization process is substantiated by the fact that there is its closer resemblance in part of changes in the molecular mass characteristics to ion-coordination processes rather than the radical ones.

(2) There is a certain distribution $\varphi(\beta)$ of centers involved in chain propagation with respect to the statistical parameter of polymerization β.

(3) Diffusion limitations do not play a decisive role in polymerization.

The minimum of the Tikhonov functional was achieved at the optimum regularization parameter $\alpha = 9.4 \times 10^{-5}$ and a root-mean-square experimental error of $\delta^2 = 0.0001$; the discrepancy was 1.07×10^{-4}. The task was solved graphically in the coordinates $\psi \ln(\beta) - \ln M$. The curves of kinetic nonuniformity distribution thus obtained show a bimodal pattern (Figure 3), indicating the presence of at least two types of growing centers involved in free-radical polymerization.

The $\psi \ln(\beta) - \ln M$ curves were divided into individual functions (this division is shown in Figure 3) [7, 8]. During the polymerization, the positions of the Gaussian peak maxima remain unchanged. Each type of polymerization centers produces macromolecules with a certain average degree of polymerization; that is, the ratio between the rate constants of chain propagation and chain termination reactions remains unchanged for individual types of active centers.

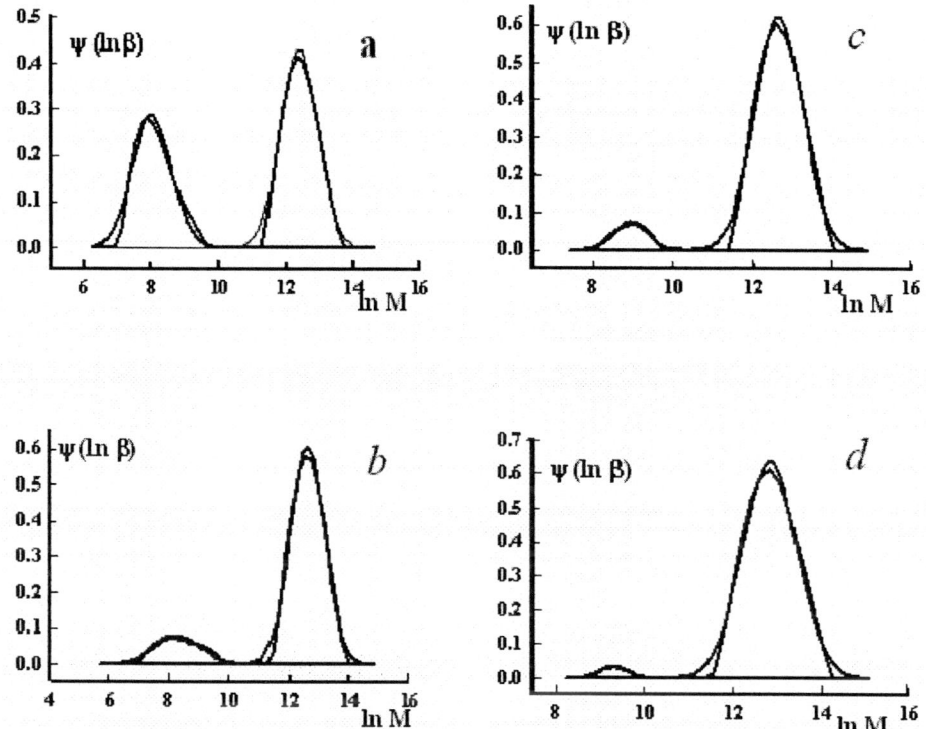

Figure 3. Variation in the kinetic nonuniformity of the titanocene initiating system in the polymerization of slyrene. The monomer conversion: (a) 1, (b) 3, (c) 10, and (d) 40%. The division into Gaussian functions is shown

The area under each Gaussian curve of kinetic non-uniformity distribution S_i is adequate to the kinetic activity exhibited by a given type of active centers in chain propagation reactions.

The areas under the Gaussian curves change in the course of polymerization; in other words, the kinetic activity of individual types of the centers changes (Figure 4). It is evident that active centers generating a lower molecular mass fraction and corresponding to the first maximum on the kinetic nonuniformity distribution curves are sufficiently active at the beginning of polymerization when zirconocene and titanocene initiating systems are used (Figure 4a). As the monomer conversion rate increases to ~10-20%, the activity of this type of centers declines markedly. On the contrary, the activity of centers giving rise to the high-molecular-mass fraction increases; within the monomer conversion range from 10 to 20%, it achieves its maximum value and then remains almost invariable throughout the polymerization. For the polymerization of styrene initiated by the ferrocene—benzoyl peroxide-system, the first type of centers generating a lower molecular mass fraction of PS exhibits a low activity; at conversions above 8%, these centers disappear. Since, as was noted above, the ratios between the rate constants of chain propagation and chain termination for each type of active centers (the positions of Gaussian peaks) remain invariable during polymerization, changes in the kinetic activity of individual types of active centers may be connected with changes in their concentrations.

The above experimental data show that the nature of the metal in metallocene initiating systems affects the molecular characteristics of PS and the kinetic nonuniformity of active centers.

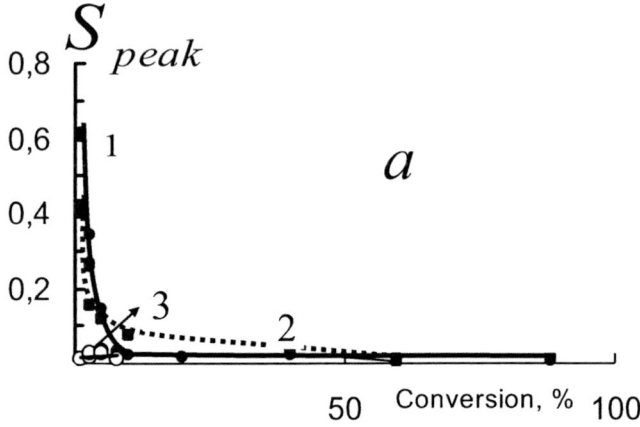

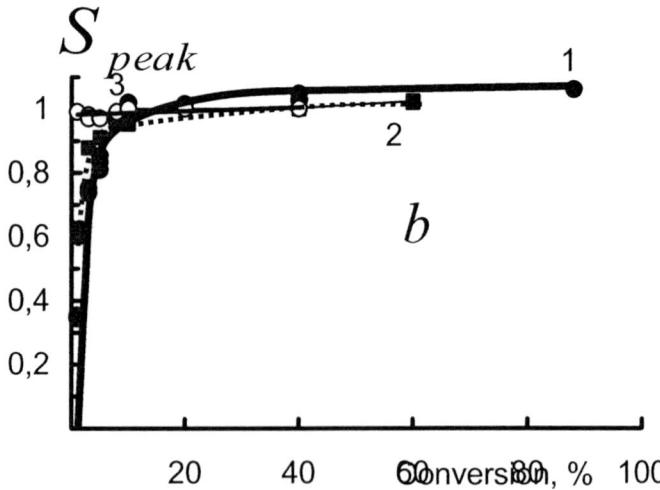

Figure 4. Variation in the kinetic activity of polymerization centers with styrene conversion. Initiating systems: (1) zirconocene-benzoyl peroxide, (2) titanocene-ben-zoyl peroxide, and (3) ferrocene-benzoyl peroxide. Panel (a) "low-molecular-mass" peak, and panel (b) "high-molecular-mass" peak

Based on the molecular mass distribution curves of PS, which were normalized to the monomer conversion by estimating a difference between the curve corresponding to the higher monomer conversion rate (5%) and the curve that corresponds to a lower conversion rate (3%), difference molecular mass distribution curves were obtained (Figyre 5). These curves refer to that part of the polymer that was synthesized during the time interval between two monomer conversion rates. The profile of difference chromatograms indicates which types of active centers are involved in polymerization over a specific period of time.

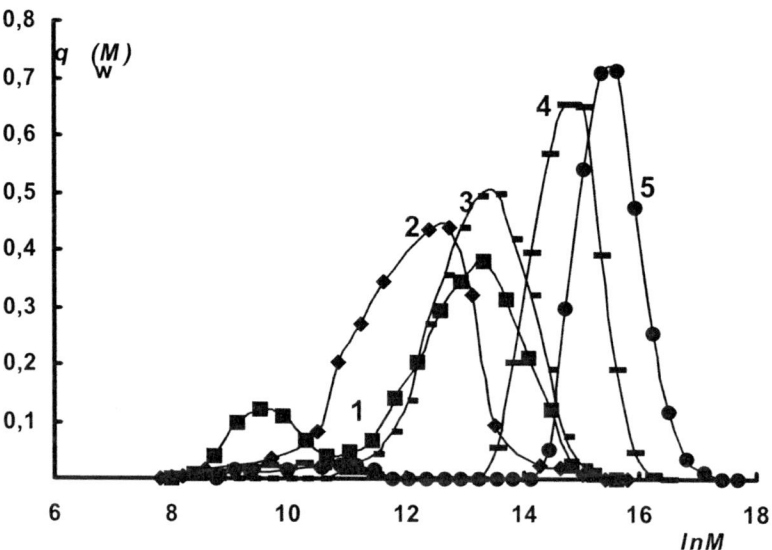

Figure 5. Molecular mass distribution curves for polystyrene derived from difference eliminate grams. Zirconocene-benzoyl peroxide initiating system and the monomer conversion: *(1)* 3-1, (2) 10-8, (3) 20-10, (4) 40-20, and (5) 88-60%

Since areas under the "low-molecular-mass" peak decrease with the increasing monomer conversion, whereas the areas under the peaks attributed to the higher molecular mass fraction of PS increase, the molecular mass growth of the polymer at conversions over 10-20% primarily proceeds at the expense of polymerization involving "high-molecular-mass" active centers.

Figure 6 presents the plots of M_w of polystyrene corresponding to the "high-molecular-mass" peak of the "difference" chromatograms as a function of the monomer conversion. It is seen that, for all three initiating systems, the molecular mass of the polymer goes up. The molecular masses of PS corresponding to the lower molecular mass peak remain constant. This implies that, in the course of polymerization, it is not only the weight of the polymeric product that grows, but the molecular mass also increases at the expense of polymerization involving the "high-molecular-mass" type of active centers.

The observations outlined above may be interpreted within the framework of the coordination-radical polymerization mechanism, that is, participation of both free and complex monomer (M) molecules and chain-growth radicals [11] arising from interaction with metallocene X in the elementary events of chain propagation and termination:

$$M + X \leftrightarrow (M\text{---}X) \qquad (1)$$

$$R^{\bullet} + X \leftrightarrow (R^{\bullet}\text{---}X) \qquad (2)$$

$$R^{\bullet} + nM \leftrightarrow R_n^{\bullet} \quad (3)$$

$$(M{-}{-}{-}X) + R^{\bullet} \leftrightarrow M + (R^{\bullet}{-}{-}{-}X) \quad (4)$$

$$(R_n^{\bullet}{-}{-}{-}X) + M \leftrightarrow (R_{n+1}^{\bullet}{-}{-}{-}X) \quad (5)$$

It can be proposed that a radical involved in chain propagation interacts with the metal atom of metallocene to yield a radical complex (2). Chain propagation may proceed both through the participation of chain-growth radicals $R^{\bullet}$ and as a result of insertion of the monomer via a labile bond $R^{\bullet}{-}{-}{-}X$ to give a new coordination-bound macroradical $R_{n+1}^{\bullet}{-}{-}{-}X$. Being coordinated with the growing radical according to scheme (2), metallocene is directly involved in the reaction of chain propagation and possibly hampers the bimolecular chain termination via recombination of growing macroradicals, as in the polymerization of vinyl monomers in the presence of some organoelemental compounds. The observed molecular masses considerably exceed the corresponding values for polystyrene prepared by the common radical polymerization initiated by benzoyl peroxide.

In the course of polymerization, the molecular mass of polystyrene increases at the expense of the process involving "high-molecular-mass" active centers (peak № 2). Thus, it may be suggested that this type of active centers corresponds to complex-bound radicals. As an increase in the monomer conversion leads to a relative rise in the fraction of the polymer produced on "high-molecular-mass" centers, one may assume that the transition of free to complex radicals according to reaction (2) takes place in the course of polymerization.

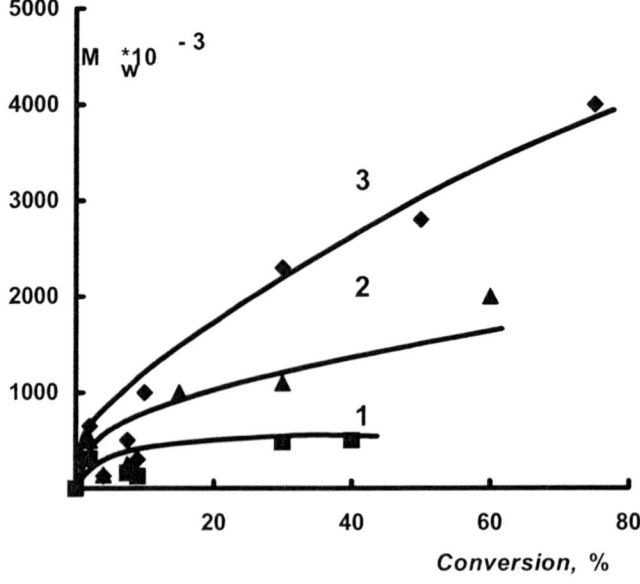

Figure 6. The plot of M_w of polystyrene corresponding to the high-molecular-mass peak of the difference chromatogram vs. monomer conversion for polymerization with (1) titanocene, (2) ferrocene, and (3) zirconocene initiating systems

This is consistent with the observed transition of the peak № 1 on the curves of kinetic nonuniformity distribution into peak № 2, considering that the former refers to the process involving free radicals, and the latter – complex-bound ones. In this case, the nature of the metal in the initiating system affects the fraction of both free and complex radical active centers. When the ferrocene initiating system is used, the quantity of free-radical centers being formed is small and they virtually disappear when a 20% monomer conversion rate is achieved (peak № 1 on Figures. 2, 4 disappears). In the case of the zirconocene system, the fraction of free-radical centers is rather high at the beginning of polymerization (Figures. 2, 4). However, this case also sees these centers virtually disappear as the monomer conversion rate goes up.

The probability of chain propagation involving complex radicals must increase due to the free radical generation rate drop resulting from the consumption polymerization initiator. The free radical generation rate is obviously affected by such factors as polymerization temperature and the ratio of initiating system components.

Figure 7 shows molecular mass distribution curves for monomer conversion rate of 5 %, those testifying to the fact of a significant impact of polymerization temperature on the polystyrene curves. Thus, for zirconocene initiating system at a lower polymerization temperature (45° C) the MMD curves are bimodal. There are two maxima: the dominating one in the high-molecular area and the second, small peak corresponding to lower-molecular fraction. At the polymerization temperature of 60° C the proportion of the peak accounting for lower-molecular fraction considerably goes up. This fact is consistent with our assumptions that second peak corresponds to free radical polymerization.

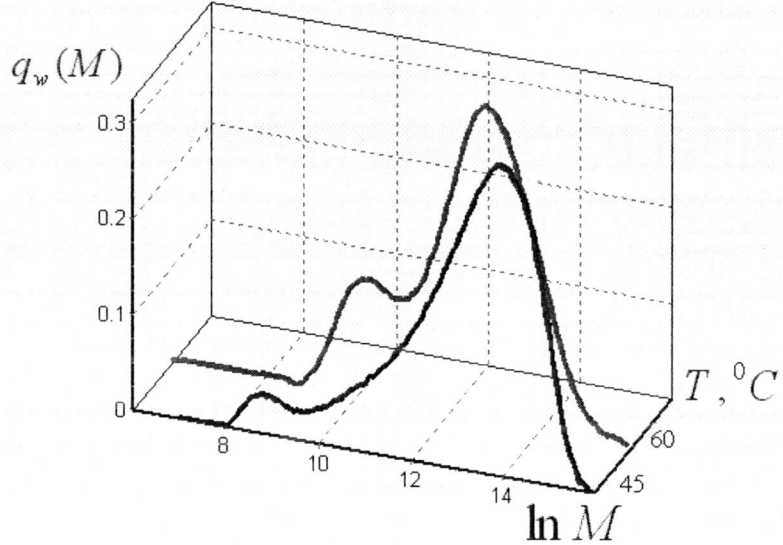

Figure 7. Molecular mass distribution curves of polystyrene synthesized using zirconocene-benzoyl peroxide initiating system at the T_{pm} – 45 (1) и 60 °C (2). The monomer conversion: 5%

Thus, the kinetic nonuniformity of radical polymerization of styrene initiated by metallocene –benzoyl peroxide involving free and complex growth radicals was demonstrated.

ACKNOWLEDGMENT

This work was supported by the Russian Foundation for Basic Research (project nos. 02-03-33315 and 02-01-97903), the program "Leading Scientific Schools" (grant NSh 728.2003.3).

REFERENCES

[1] Yu.I.Puzin., E.M.Prokudina, R.Kh.Yumagulova, R.R. Maslukhov, C.V.Kolesov // *Dokl. RAN*.386. №1, 69 (2002).
[2] Yu. I. Puzin, R. Kh. Yumagulova, V. A. Kraikin, et al., Vysokomol. Soedin., Ser. B 42, 691 (2000) [*Polymer Science*, Ser- B 42, 90 (2000)].
[3] R. Liu, X. Zhou, and S. Wu, *Gaofenzi Xuebao*, No. 3, 374(1994).
[4] N. N. Sigaeva, S. V. Kolesov, E. M. Prokudina, et al., *Dokl. Akad. Nauk* 386, 785 (2002).
[5] D. F. Grishin and L. L. Semenycheva, *Usp. Khim.* 70, 486(2001).
[6] D. F. Grishin and A. A. Moikin, Vysokomol. Soedin., Ser. A 40, 1266 (1998) [*Polymer Science, Ser.* A 40, 775 (1998)]
[7] N. N. Sigaeva, T. S. Usmanov, E. A. Shirokova, et al., *Dokl. Akad. Nauk.* 365, 221 (1999).
[8] N. N. Sigaeva, T. S. Usmanov, V. P. Budtov, et al., Vysokomol. Soedin., Ser. B 42, 112 (2000) [Polymer Science, Ser. B 42, 1 (2000)].
[9] A. N. Tikhonov and V. Ya. Arsenin, *Methods for Solving Ill-Posed Problems* (Nauka, Moscow, 1986) [in Russian].
[10] A. N. Tikhonov, A. V. Goncharskii, V. V. Stepanov, and A. G. Yagola, *Numerical Methods for Solving Ill-Posed Problems* (Nauka, Moscow, 1990) [in Russian].
[11] V. A. Kabanov and V. P. Zubov, *Complex-Radical Polymerization* (Khimiya, Moscow, 1987) [in Russian].

Chapter 9

THE STRUCTURAL TREATMENT OF FLUCTUATION FREE VOLUME IN AMORPHOUS STATE OF POLYMERS

G. V. Kozlov, G. E. Zaikov[1] and Yu. S. Lipatov[2]

[1]Institute of Biochemical Physics, Russian Academy of Sciense,
Moscow – 119991, Kosygin str., 4, Russia
[2]Institute of Macromolecular Chemistry of the
National Academy of Sciences of Ukraine, 48 Kharkov Shaussee,
Kiev – 02160, Ukraine

ABSTRACT

It is shown, that fluctuation free volume in polymers has fractal structure. The microvoids of free volume may be simulated by a D_f-dimensional sphere. The microvoids sizes are controlled by the volume necessary for accumulation of energy of thermal fluctuations, required for their formation. The absolute values of relative fluctuation free volume f_g may be considered as characteristics of the nonequilibrium state of the polymer structure. For quasi-equilibrium structures the value f_g coincides with data, following from the equation Williams-Landel-Ferry. The microvoids of fluctuation free volume form structure, which reflects the polymer structure. These microvoids structure may be represented as a specific pulsating percolation cluster due to thermodynamically nonequilibrium state of a polymer.

The multifractal nature of semi-crystalline polyethylene structure is verified and the estimation of its multifractal characteristics is given on the basis of diffusion properties .. The melt of polyethylene is the Euclidean object with monodisperse distribution of microvoid sizes. These circumstances define the different form of dependencies of a diffusivity on themolecular size of a gas-diffusant. The microvoid itself is a specific multifractal, whose fractal properties are not determined by its structure or by its environment.

Keywords: Fluctuation free volume, structure of polymers, amorphous state, fractal analysis, diffusivity, multifractal object, percolation, microvoid.

INTRODUCTION

The model of free volume originated in classical works by Frenkel and Eiring [1-6] has received the broad applications in physics of liquid and solid states of mutter. Some concepts allowing to improve the concept of fluctuation free volume were proposed [7-12] during last 10 years. However, there is the other aspect of the problem, which practically were not discussed earlier. As a rule, the application of the free volume theory for the description of properties of amorphous bodies is based on a supposition, that the free volume characterizes the structure of these bodies. This postulate is essentially caused by the absence of quantitative model of structure of amorphous state of polymers. This structure is meant as distribution of body elements in space. It is evident, that microvoids of free volume couldn't be considered as structural elements and they only mirror the structural state of the bodies. Introduction into consideration of some structural elements (relaxators, see, for example, [7]) contributes practically nothing into the better structural representation of free volume.

The development of the cluster model of structure of amorphous polymers [13-15] allows to consider the free volume as a mirror of the polymer structure. Let's remind, that within the framework of the cluster model the structure of amorphous polymers may be represented by the local order regions (clusters), consisting of several collinear close-packed segments of different macromolecules and immersed in a loose-packed matrix, in which all fluctuation free volume is concentrated. The clusters have the thermofluctuational nature, i. e., the number of segment in a cluster is dependent on the temperature, decreasing with its growth [13]. In such a treatment the detaching of some segments from cluster means the formation of a fluctuation free volume microvoid, whereas an association of segments brings about the "collapse" of this microvoid, the energies of these two processes being equal [16] Besides, it turned out that the relative fluctuation free volume f_g is a linear function of a clusters relative fraction φ_{cl}, and the increase in φ_{cl} means the decrease of f_g [17]. At limiting values $\varphi_{cl}=0$ and 1 the expected limiting values $f_g=0,140$ [18] and 0 are obtained. The said above allows to assume, that the free volume is not simply a set of microvoids in a condensed state, but represents some structural build-up mirroring of structure of an amorphous state and is connected to it. This problem will be main in the present paper.

Besides there are some problems of interest at the description of polymers free volume. As it was already mentioned, the fluctuation free volume, and the clusters have the thermofluctuational nature. Apparently, that energy of thermal fluctuations can be expressed as kT (where k is Boltzman's constant, T – temperature), while energy of microvoid formation ε_v equal [3]:

$$\varepsilon_v = kT_g \ln(1/f_g), \qquad (1)$$

where T_g is glass transition temperature, f_g is the fluctuation free volume at glass transition temperature.

As the value $f_g \leq 0.159$ [18], and $T<T_g$, it is apparent, that in case of glassy polymers the energy of thermal oscillations about $\sim kT$ is not sufficient for a microvoid formation ($kT<\varepsilon_v$). Second problem requiring explanation and repeatedly discussed [3, 5-7], is the absolute value f_g, which within the framework of the kinetic theory is estimated from the equation [3, 5]:

$$f_g \approx 0.017\left(\frac{1+\nu}{1-2\nu}\right), \quad (2)$$

where ν is Poisson's ratio.

Values $f_g \approx 0.050-0.100$ were obtained [17] for different polymers, that much more than generally accepted value $f_g=0.025\pm0.003$ for the most of polymers within the framework of the concept Williams, Landel and Ferry (WLF) [2-4].

The other problem is connected to the physical nature of the free volume microvoids. The presence of such microvoids in polymers is confirmed by numerous experiments, for example, by a positron spectroscopy [19-22]. Diameter of these microvoids makes the value about several Angstroms and now it becomes apparent, that the modeling of microvoids by three-dimensional sphere with smooth walls [3] is far from reality [21]. It is also apparent enough, that the microvoid walls represent fragments of macromolecules oscillating near their equilibrium positions. It was proposed [23, 26] to simulate a microvoid as D_f-dimensional sphere, where D_f is the dimension of the region of localization of extra energy [27]. In this treatment the value D_f characterizes an energetic exitation degree of loose-packed matrix within the framework of cluster model [13]. Therefore, there exists a problem of the dimension of the microvoid "walls" (for smooth walls this dimension is equal 2) which should be solved to better define the physical sense of a microvoid concept.

Finally, the last problem, intimately connected to the free volume, is the diffusion in polymers. It is known, that the diffusion of gases through polymeric membranes is realized by passing of gases molecules through free volume microvoids and in this case free volume is an analogue of porosity in crystalline solids [28, 29]. However, for passing of diffusant molecules through polymer the formation of through channels is necessary, i.e. the formation of the percolation framework of microvoids. In the case of overlapping spheres the percolation threshold is equal 0.34 ± 0.01 [30]. This value is much higher than the maximum value of $f_g=0.159$ [18]. This fact may be explained in two ways, which will be considered below. In the present paper the attempt to answer the put problems has been done using, as examples, three main classes of polymers: amorphous glassy, semi-crystalline and cross-linked polymers. The fractal analysis is chosen as the main theoretical model for their solution [24, 25, 31].

EXPERIMENTAL

The experimental data for two sets of epoxy polymers (EP), based on diglycidyl ether of bisphenol A (DGEBA) and hardened by 3.3'-dichloro-4.4'-diaminodiphenylmethane (EP-1) and methyl tetrahydrophtalic anhydride (EP-2) are used, which were tested directly after producing, and the data for a set of specimens EP-1, which were exposed to aging on air at 293 K within two years (EP-3). Each of the three indicated sets consisted of 5 specimens with the different hardener: oligomer ratio K_{st} ($K_{st}=0.50-1.50$). Methods of producing and testing of specimens are quoted in [15, 32], and calculation of structural parameters within the framework of cluster model – in [15].

The data of mechanical testing of two amorphous glassy polymers – polycarbonate (PC) and polyarylate (PAr) are also used [33]. These polymers were diluted in methylene chloride and the films of thickness 0.05 mm were obtained from 5 % solutions by a pouring method on the glass substrate. For moisture and solvent removing the films were dried in vacuum at ~ 400 K during 2 days. Then for uniaxial tensile tests with the aid of template the specimens were cut with working width 5 mm and base length 40 mm . These tests were carried out under the strain rate $\dot{\varepsilon}=10^{-3}$ s^{-1} and in the temperature range 293-413 K for PC and 293-473 K for PAr.

As semi-crystalline polymers : commercial polyethylenes of high (HDPE) and low (LDPE) density having a molecular mass $\overline{M}_w \approx 1.5 \times 10^5$ and crystallinity degrees 0.723 and 0.507, accordingly, were used. Specimens for impact testing by sizes of 4×6×50 mm were produced by injection moulding [34].

The impact tests by Sharpy method were carried out using a pendulum impact machine, equipped with a piezoelectric load sensor connected to a recording oscillograph, model S 8-13. This unit allows to receive the oad-time diagram (P-t), from which the elasticity modulus E [35] and yield stress σ_Y [36] could be calculated. The fractal dimension d_f of structure of all studied polymers was calculated from the equation [27]:

$$d_f = (d-1)(1+\nu) \qquad (3)$$

where d is dimension of Euclidean space in which the fractal is considered (equal in our case 3), ν is the value of Poisson's ratio, which can be calculated from the results of mechanical tests according to the equation [3]:

$$\frac{\sigma_Y}{E} = \frac{(1-2\nu)}{6(1+\nu)}. \qquad (4)$$

The Fractal Analysis of the Fluctuation Free Volume

Now it is known [24, 25, 31, 37], that the structure of a polymer is fractal (generally-multifractal) with fractal dimension d_f ($2 \leq d_f < 3$). Therefore it is possible to expect that the fluctuation free volume, being a mirror of polymer structure, also has fractal properties. If this assumption is correct, the general relationship should be valid [38]:

$$N_v \sim r_v^{D_f}, \qquad (5)$$

where N_v is the number of microvoids, r_v is their characteristic size, D_f is fractal dimension of fluctuation free volume.

The value r_v may be approximated by $V_v^{1/3}$, where V_v is the volume of microvoid. The values V_v and N_v are determined as follows [3]:

$$V_v = \frac{3(1-2\nu)kT_g}{f_g E},\quad (6)$$

$$N_v = \frac{f_g}{V_v}.\quad (7)$$

In Fig. 1 displays the dependencies of N_v (r_v) in double log-log coordinates in correspondence with the relationship (5). It is see that they are linear, that is the characteristic sign of the fractal behavior [38] and one may determine the value D_f from their slopes. For the sets EP-1 and EP-2 D_f is equal ~3.75, and for EP-3 D_f is equal ~2.70 [26].

Let's consider the physical sense of the model and determined parameter D_f. As it is well known, both length of a macromolecule segment between the cross-links having a molecular weight M_s [39] and the sizes of free volume microvoids [20] for cross-linked network have a definite distribution. For the set of epoxy polymers under investigation the cross-linking densities ν_s are: for EP-1 and EP-3 ~2.5×10^{26} m^{-3} and for EP-2 ~8.5×10^{26} m^{-3}. These data well agree with the distribution width M_s for one epoxy polymer [39]. Thus, the offered treatment in its essence simulates one epoxy polymer with allowance for constancy of its chemical constitution (DGEBA resin), having some distribution M_s and V_v, by the epoxy polymers sets with varied ν_s and, therefore, M_s and V_v. Let's mark, that the free volume microvoids do not form any structure, for which the condition $d_f<3$ is obligatory in three-dimensional Euclidean space, though such condition is necessary for elements of polymer structure, for example, of statistical segments [37]. His circumstance once again underlines that f_g is not the direct structural characteristic feature [24].

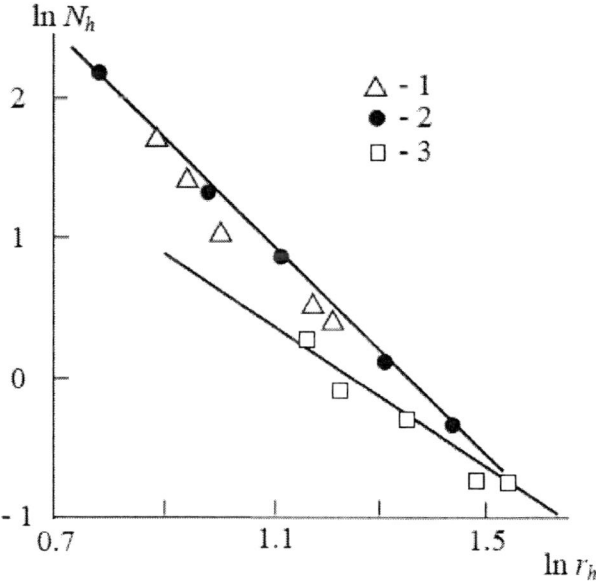

Figure 1. Dependencies of the microvoids number N_v on size r_v in double log-log coordinates for epoxy polymers EP-1 (1), EP-2 (2) and EP-3 (3)

Apparently the values D_f calculated are the average characteristics, but, nevertheless, they allow the important conclusion to be drawn: microvoids of fluctuation free volume are formed D_f-dimensional fractal cluster or the collection of such microvoids has the definite (fractal) structure in general. The power distribution of particles by sizes is widely spread in the nature. Eddy cascades in a developed turbulence [40], caverns on fracture surfaces of ductile metals [41], aggregates of filler particles in polymeric composites [42] etc obey this distribution. However exact fractality definition of this or that object (or their collection) alongside with carrying out a general relationship (5) needs the proof of its self-similarity [43, 44]. This problem will be in detail discussed below.

The absolute values D_f are also not occasional, and have deep enough physical sense. They are close to corresponding values of dimension of the excess energy localization regions D_f', defined as follows [27]:

$$D_f' = \frac{2(1-\nu)}{(1-2\nu)}. \tag{8}$$

Besides the value D_f' is equal to structure automodelity coeffcient Λ_i [45]:

$$D_f' = \Lambda_i = \frac{L_{i+1}^{DS}}{L_i^{DS}}, \tag{9}$$

where L_i^{DS} and L_{i+1}^{DS} are space scales of dissipative structures (DS) of adjacent levels.

For amorphous polymers the values L_i^{DS} and L_{i+1}^{DS} are identified as follows. DS, formed in amorphous state, are clusters [46]. The cluster model [13-15] postulates, that length of segments, included in a cluster, is equal to a statistical segment length l_{st} and consequently $L_i^{DS} = l_{st}$ [24]. One should accept the distance between clusters R_{cl} as the next structural scale and consequently $L_{i+1}^{DS} = R_{cl}$ [24]. In such treatment the fractal dimension D_f is intimately connected to the characteristics of the polymers structure. And at last, earlier the next relationship was obtained [24]:

$$D_f' \approx C_\infty, \tag{10}$$

where C_∞ is characteristic ratio, which is a measure of statistical flexibility of a polymeric chain [47].

Thus, the relationship (10) provides the correlation of the free volume characteristics and polymers molecular characteristics. Similar correlation was obtained in [48] before. The correlation D_f and D_f' will be described below in more detail.

Now we shall consider the energetic criterion of the formation of a microvoid of size V_v. For this purpose the concept of free fracture of rigid bodies is used [49], which postulates, that it is possible to accumulate the restricted amount of potential energy with density A^* in restricted volume V. To form a new surface it is necessary to spend the work equal to the product of the surface area S by a specific surface energy σ_s to form new surface. The limiting case of the free fracture takes place under condition [49]:

$$A^*V = \sigma_s S. \tag{11}$$

With the reference to the creation of a microvoid of size $V=V_v$ included in the relationship (11), the following parameters can be presented by rewriting this relationship in the form [27]:

$$\frac{V}{S} = \frac{\sigma_s}{A^*}. \tag{12}$$

If to present the microvoids of the fluctuation free volume as D_f-dimensional sphere, the relationship [27] is correct is valid:

$$\frac{V}{S} = \frac{r_v}{D_f}. \tag{13}$$

In turn, the value σ_s can be expressed by parameters of "holes" model [3]:

$$\sigma_s \approx \frac{\varepsilon_v}{2V_v^{2/3}}. \tag{14}$$

Defining $A^*=kT/V_v$ from the combination of the equations (1) and (12)-(14) we l receive the final relationship:

$$D_f = \frac{4\pi T}{\ln(1/f_g)T_g}. \tag{15}$$

It follows from the data of Fig. 1 that the aging of epoxy polymers EP-1 results in decrease of D_f and the value D_f, close to 3 is obtained for aged systems EP-3. As the process of physical aging of polymers can be considered as an approaching to the state of thermodynamic equilibrium [50], it is necessary to assume, that the condition $D_f=3$ responds to some quasi-equilibrium state of polymer structure [51] and allows the simulation of microvoids as three-dimensional sphere ($D_f=d=3$). Such treatment allows to make some conclusion. At first, the increase in under other constant conditions means the increase of D_f and, therefore, increase of the thermodynamic nonequilibrium degree of polymer structure.

From the equation (8) follows that the growth in D_f means the increase in ν, that explains the well known increase of the Poisson ratio for polymers due to increasing temperature [52]. Secondly, the increase in ν shows the lowering of the local ordering degree of amorphous polymers [52], that maniests itself in the decrease of the clusters relative fraction φ_{cl}. Thus, as it was expected [32, 51], the increase of φ_{cl} (the order parameter of amorphous polymers [25]) reflects the tendency of polymer to reaching a thermodynamic equilibrium. Thirdly, from the equation (15) under condition $D_f=3$ it is possible to calculate the quasi-equilibrium value $f_g(f_g^{qe})$, which appeared to be equal $f_g^{qe} \approx 0.028 - 0.042$ for the studied polymers. This value is close enough to the value of the free volume value in the WLF theory. Therefore, the estimation of f_g according to equation (2) is correct, and the large absolute values of f_g for the epoxy polymers show a high level of their thermodynamic nonequilibrium.

Though the parameters f_g and D_f also are not the direct characteristics of the structure, the interrelation between d_f and D_f is evident [27]

$$D_f = 1 + \frac{1}{d - d_f}. \tag{16}$$

From equation (8) follows that theoretically the value D_f' can be infinitely large (at $\nu=0.5$). However, the value of Poisson's ratio for real solids is terminated above: $\nu \leq 0.475$ [27]. Therefore, from equation (8) follows that for real solids, including polymers, the limiting value D_f' is equal to 21. Thus, the dimension D_f' characterizes not only the structure, but its energetic exitation degree as well, i. e., the degree of divergence from a thermodynamic equilibrium, being connected to the structural characteristic d_f.

Using the equations (8) and (15), it is possible to demonstrate the postulated above closeness of absolute values D_f' and D_f (Fig. 2). As it is seen, the close correspondence between them is really obtained, and some discrepancy of absolute values D_f' and D_f is due to the mentioned above approximation. This correspondence displays, that f_g can really be the mirror of the polymer structure.

In summary we shall consider the problem of self-similarity of a cluster of free volume microvoids. As is known [43], the linearity of the dependencies corresponding to the relationship (5), in double log-log coordinates and non-whole value of the scale of D_f, obtained from the slope of these dependencies, is not the proof of the indicated cluster fractality. Besides for a strict proof of this property it is necessary to confirm the self-similarity of the object and to consider a scale interval of this self-similarity [43, 44].

As it was shown [43], for the self-similar fractal objects when using method based on the relationship similar to (5) at $r_{v_i}/r_{v_{i+1}} = const$ the following condition should be satisfied:

$$N_{v_i} - N_{v_{i+1}} \sim r_{v_i}^{-D_f}. \tag{17}$$

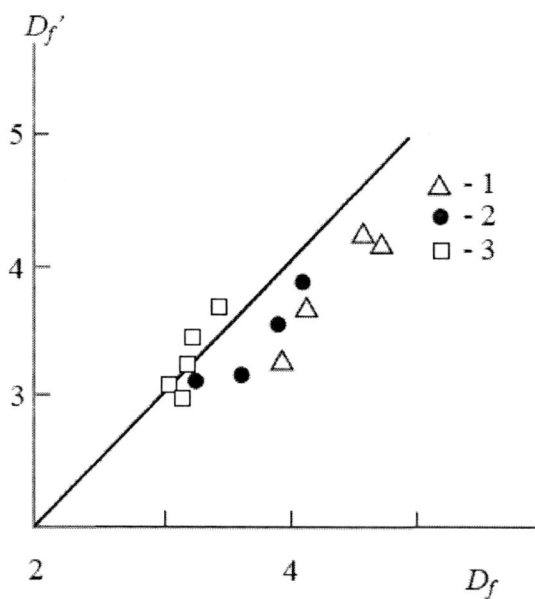

Figure 2. Relationship between the dimensions of fluctuation free volume D_f (equation (15)) and excess energy localization regions D_f' (equation (8)) for epoxy polymers EP-1 (1), EP-2 (2) and EP-3 (3)

In Fig. 3 the dependencies corresponding to the relationship (17), for three sets of studied epoxy polymers are shown. It is seen that these dependencies are linear and pass through the origin of coordinates. According to the relationship (17), it confirms the self-similarity of a microvoid cluster of the fluctuation free volume. As for the scales range of self-similarity, with allowance for the equation (16), giving correlation between D_f and d_f, it is necessary to assume, that this range coincides with a similar range for polymer structure. For the structure of amorphous glassy polymers the indicated range according to the experimental data [53] is spread from several Ångströms up to several tens of Ångströms. From equation (9) follows this range is spread from l_{st} up to R_{cl} [24], that quantitatively corresponds to the experimental limits of the indicated range [54].

It was shown also [55, 56] that the minimum range of values r_{v_i} should contain at least one iteration of self-similarity. In this case the condition [56] should be satisfied:

$$r_{v_{max}} / r_{v_{min}} > 2^{1/D_f} \qquad (18)$$

For epoxy polymers EP-1 and EP-2 the value $2^{1/D_f} \approx 1.21$, for EP-3 this value is equal ~ 1.29. At the same time the ratio $r_{v_{max}} / r_{v_{min}}$ is equal: ~ 1.42 for EP-1, ~ 1.93 for EP-2 and ~ 1.46 for EP-3. Thus, from the comparison of the mentioned estimations the satisfaction of the criterion (18) follows.

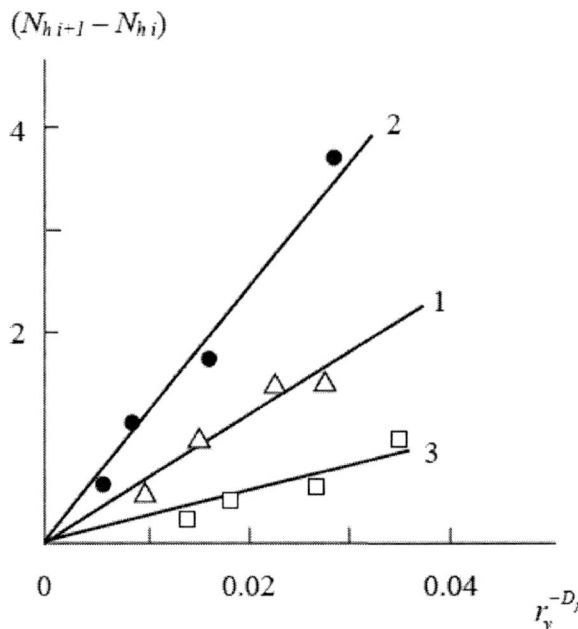

Figure 3. Dependencies of $\left(N_{v_{i+1}} - N_{v_i}\right)$ on $r_v^{-D_f}$, corresponding to relationship (17), for epoxy polymers EP-1 (1), EP-2 (2) and EP-3 (3)

Alternatively the number of self-similarity iterations n can be determined from the equation [56]:

$$\left(r_{v_{max}} / r_{v_{min}}\right)^{D_f} = 2^n. \tag{19}$$

The value n for studied epoxy polymers are equal: $n \approx 1.90$ for EP-1, $n \approx 3.57$ for EP-2 and $n \approx 1.49$ for EP-3, i.e., in our experiment the number of self-similarity iterations is more than unity, that again confirms the correctness of the value D_f estimation [55-57].

Now let's consider the applicability of the D_f-dimensional sphere model to the description of fluctuation free volume microvoids in more detail. The kinetic theory of fluctuation free volume [3] allows the theoretical estimation of its main parameters: a relative fraction f_g and size (radius r_v or volume V_v) of this microvoid (see equations (2) and (6)). Now there is an experimental technique (annihilation of positrons), permitting to make the same estimations [19-22]. Therefore, the comparison of the results of both techniques therefore is of interest. From the theoretical point of view the correspondence of the theory and experiment shows the correctness of the kinetic theory of fluctuation free volume [3]. From the practical point of view such correspondence allows to use the convenient formulas of the kinetic theory for an estimation of values f_g and r_v (or V_v) instead of a complicated and laborious technique, connected with annihilation of positrons. This comparison will be done on the example of typical amorphous glassy polymer – polycarbonate based on bisphenol A (PC).

In [58] the calculation of a relative fraction of the free volume from the results of positron annihilation by two methods was done: in the first of them the thermal expansion coefficient of the PC below the glass transition temperature was used, in the second – the same value above this temperature. It was discovered about linear increase of free volume f_g in accordance with increase of testing temperature T. For the computational method the relative free volume is less systematically (~10%) as compared with the appropriate value obtained in the second method (Fig. 4, straight lines 1 and 2, accordingly). In Fig. 4 the results of calculation f_g by the equation (2) (points) are also shown. As follows from the data given in Fig. 4, the theory and experiment agree very well. Theoretical values of f_g are well corresponded to the straight line 1 in the range $T=293-373$ K and straight line 2 – in a range $T=383-413$ K (the transition is indicated by a dashed line). This observation may be explained within the framework of the cluster model of structure of polymers. [13-15]. As it was mentioned above, all the fluctuation free volume is concentrated in the loose-packed matrix, which is kept in a glassy state by a network of small (unstable) clusters [59]. At the temperature T_g', equal about T_g - 50 K [60], the unstable clusters are decomposed and loose-packed matrix passes in to a rubber state. This effect, realized namely at T_g' for PC (Fig. 4), results in the jumpy increase of f_g, because within the framework of the cluster model the detaching of a segment from a cluster id associated with the formation of a microvoid of fluctuation free volume [16].

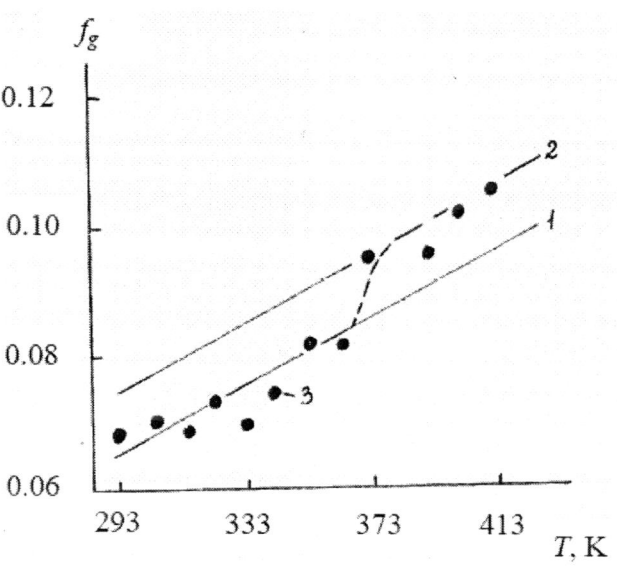

Figure 4. Dependence of relative fluctuation free volume f_g on temperature T for PC. 1, 2 – the data [58] at using the thermal expansion coefficient lower (1) and above (2) T_g, 3 – the calculation according to equation (2)

Besides the transition of the theoretical data from a straight line 1 on to a straight line 2 at $T_g' = 373K$ for the PC is lawful, as at $T<T_g'$ loose-packed matrix is in a glassy state, that corresponds to conditions of obtaining of a straight line 1, and at $T>T_g'$ - in a rubber state, that corresponds to conditions of obtaining of straight line 2. Thus, the estimation of f_g by the equation (2) has appeared much more sensitive to thereal structural changes in the PC, than the method of positrons annihilation [19-22].

The calculation of the volume V_v of a fluctuation free volume microvoid may be done within the framework of the kinetic theory according to the equation (6). Within the framework of the conventional simulation by three-dimensional sphere a characteristic size, namely, the microvoid radius r_v can be determined from the known geometric equation:

$$V_v = \frac{4}{3}\pi r_v^3. \tag{20}$$

In Fig. 5 the comparison of the dependencies of $r_v(T)$ for the PC, obtained experimentally [21, 58] and calculated from the equation (20) is shown. First of all systematic (approximately by 8%) discrepancies of the experimental values r_v, obtained for PC, produced by different corporations pay attention to themselves.. This discrepancies visually demonstrate the importance of the application of identical samples for comparing their properties, obtained with the help of different techniques. The comparison of these dependencies with the corresponding dependence $r_v(T)$, calculated from equations (6) and (20), displays two important disparities. Firstly , the values r_v, estimated by equations (6) and (20), by their absolute value are sufficiently less than experimental ones (at high T the discrepancy reaches 28%) and, secondly, different tendencies of the change of r_v are observed in accordance with the variation in T. On the first sight it assumes an incorrectness of the kinetic theory of fluctuation free volume and, in particular, of equation (6) for calculation of its microvoid size.

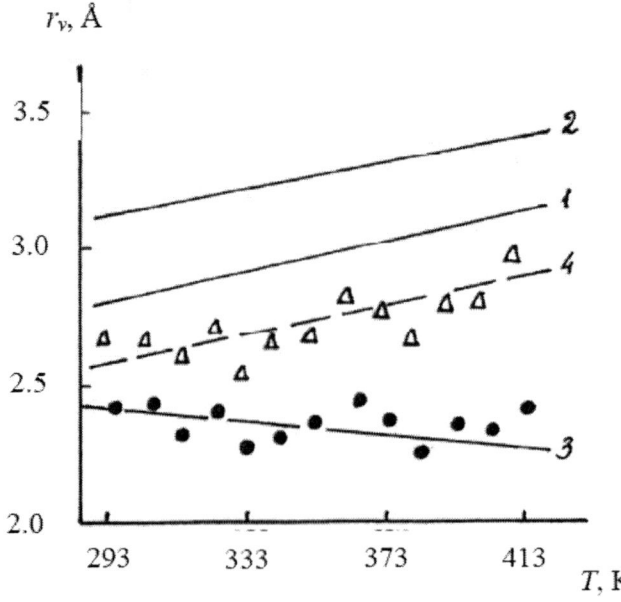

Figure 5. Dependencies of the radius of fluctuation free volume microvoid r_v on temperature T for PC. 1 – the data of [58], 2 – the data of [21], 3- the calculation according to equation (6), 4 – the calculation according to equation (23)

However, the fractal concept of the free volume allows to explain and quantitatively to account for these discrepancies. First of all we shall mark, that even in the theory of positrons annihilation the simulation of microvoid by a three-dimensional sphere is put under doubt [58]. Let's remind, that by derivation of equation (6) the multiplicand 3 (1-2v) in numerator comes up from the known in the mechanics of continuum interrelation between a modulus of dilatation B and Young's modulus E [3]:

$$B = \frac{E}{3(1-2v)}. \tag{21}$$

Within the framework of the fractal analysis the equation (21) is written as [61]:

$$B = \frac{E}{d(d-d_f)}. \tag{22}$$

The comparison of the equations (3), (21) and (22) displays that the factor 3 in a denominator of the equation (21) and, therefore, in numerator of the equation (6), is dimension of space $d=3$. However, it was above exhibited, that the microvoid of fluctuation free volume should be simulated by a D_f-dimensional sphere and the value D_f can be calculated from equation (8). It means that equation (6) is correct for a special case $D_f=3$ or $v=0.25$ and generally should look like the following for the arbitrary value D_f:

$$V_v = \frac{D_f(1-2v)kT_g}{f_g E} = \frac{2(1-v)kT_g}{f_g E}. \tag{23}$$

The calculation of dependence r_v (T) from equations (20) and (23) is also shown in Fig. 5. In this case the discrepancy of experimental and theoretical values r_v does not exceed 8%, and the tendencies of change r_v with a variation T for the indicated dependencies are identical. The discrepancy of absolute values r_v for dependencies 1 and 4 Fig. 5 is explained by the different techniques of samples producing. Samples for the positron spectroscopy were produced by moulding, and (for mechanical tests) by the method of a solution pouring (in our case). The last method assumes obtaining more equilibrium structure and, accordingly, smaller values V_v (or r_v).

Let's mark, that usage of dimension D_f, describing an energetic exitation degree of loose-packed matrix [24], is corresponded with the mechanism of microvoid creation by displacement of atoms (segments) as a result of fluctuations of density [8, 10, 12, 62]. Apparently, the stronger the energetic exitation of the matrix and the more D_f are, the larger are the mean square relative fluctuations of the density $\overline{(\Delta \rho/\rho)}^2$. Let's mark also, that by the temperature rise the simultaneous increase D_f [24], $\overline{(\Delta \rho/\rho)}^2$ [63] and V_v (or r_v) [21, 58] is observed.

Thus, the fluctuation free volume of polymeric glasses is a parameter widely used when making the phenomenological description of their properties [1-3]. As it was demonstrated [4], the fluctuation free volume reveals the fractal properties in the interval of linear scales from several Ångströms up to several tens of Ångströms, which complies with the intervals of fractality of the polymer structure [24] and the sizes of the cluster structure [14]. This correspondence enables to consider the glassy polymer in the specified interval scales as a porous solid, in which microvoids of fluctuation free volume play the role of pores. There are possible two regimes of the behavior: static and dynamic. These two regimes are conditioned by the corresponding behavior regimes of fluctuation density [64, 65] which are closely connected with the notion of fluctuation free volume [8, 62, 63]. As is known [65], the temperature dependence of the fluctuation density disintegrates into two regimes from 0 K up to some temperature T^* and from T^* up to the temperature of glass transition T_g. In the first temperature interval the value of thermal fluctuation of density is constant. In the second one it is a linear function of temperature [64, 65]. Within the frames of fractal analysis T^* is considered as the temperature of obtaining the quasi- equilibrium state of polymer structure[9]. The fractal dimension of structure $d_f \approx 2,5$ (or $D_f \approx 3$, $v \approx 0,25$) [24] corresponds to this state. That's why, the relative fraction of fluctuation free volume f_g is constant (static regime) if $T<T^*$. If $T^*<T<T_g$ then the volume f_g is function T (dynamical regime). Such treatment makes it possible to use the fractal analysis for the estimation of the value f_g [67].

Fractality of polymer structure in the said interval of linear scales assumes that the value f_g can be defined by analogy with microporosity in the following way [68]:

$$f_g = A(l_1 / l_2)^{d-d_f}, \qquad (24)$$

where A - constant, l_1 and l_2 - lower and upper linear scales of a range of polymer structure fractal behavior.

Let's survey the choice of parameters in equation (24). For the case of sandstones it was assumed that the value A is approximately equal 1 [68]. It means that if $d_f=3=d$ then the porosity of a sandstone is equal to 1, which is hypothetically possible, as the value of porosity apparently has no physical limits. However, the magnitude of f_g for glassy polymers is restricted from the top at T_g by some value :either $\sim 0,113$ [69], or $\sim 0,159$ [18]. That's why we have chosen the largest of the specified values in the case under study, in particular, $A \approx 0,159$ [67].

As the intervals of linear scales of fractal behavior and cluster structure of polymers practically coincide [54] and there is no technique for the theoretical estimation of the former, we have used the lower and the upper limit as l_1 and l_2 for the cluster structure. We shall remind that according to the cluster model [6] the structure of amorphous state of polymers is an area of local order (cluster) submerged into a loosely packed matrix. The clusters in their turn are composed from several collinear densely packed segments of different macromolecules and the length of a segment in a cluster is adopted as equal to the length of a statistical segment of a polymer l_{st}. So in this treatment value $l_1=l_{st}$ and l_2 is adopted as being equal to the distance between clusters R_{cl}. The values l_{st} and R_{cl} are defined in the following way [5]:

$$l_{st} = l_0 C_\infty, \qquad (25)$$

$$R_{cl} = 18(\frac{v_{cl}}{2F})^{-\frac{1}{3}}, \quad \overset{0}{A} \qquad (26)$$

where l_0 is the length of the skeleton linkage of the main chain, C_∞ is the characteristic ratio, v_{cl} is the density of cluster network of macromolecular entanglements, F is the cluster functionality. The values v_{cl}, F and μ for the polymers, studied in this paper, (polycarbonate (PC)) and polyarylate (PAr)), are adopted according to the data [5, 13, 15] and the values l_0 and C_∞ are taken from [5]. As it was found [54] the magnitude R_{cl} for PC and PAr practically doesn't depend on the temperature.

In figures 6 and 7 the comparison of temperature dependencies of f_g is given. They were calculated according to the equations (2) and (24) for PC and PAr respectively. We will not discuss the reasons of quantitative discrepancies of calculations according to these two equations. There are many of them (for example, an arbitrary choice of A in the interval 0,113÷0,159). For both polymers this variance doesn't exceed 20%. Fig. 6 gives also the results of calculation for PC made according to the equation (24) with A=0,113 which [69] have been appeared lower than values of f_g, calculated by equation (2). However, it is remarkable that the results of the kinetic theory give the values which are in the interval of two fractal estimations at A=0,113 and 0,159 [67].

Besides, the value f_g has been calculated within the framework of the lattice theory [70]:

$$f_g = 0,0985 + 0,00013(T - 416,5), 300K < T < T_g \qquad (27)$$

where T and T_g are the temperatures of tests and glass transition respectively.

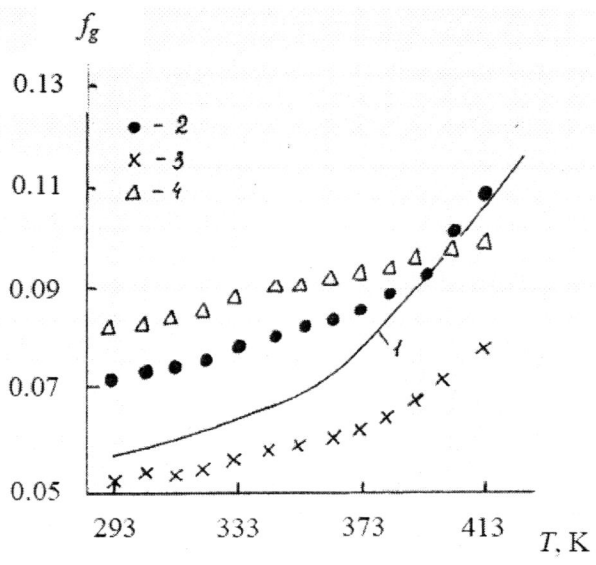

Figure 6. Temperature dependencies of the relative fluctuation free volume f_g for PC. Calculation is made according to equations: 1 – (5); 2 – (1) when A=0.159; 3 – (1) when A=0.113; 4 – (6)

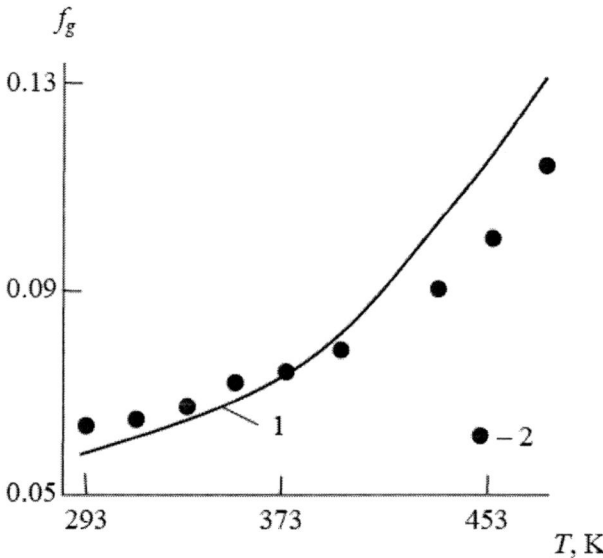

Figure 7. Temperature dependencies of the relative fluctuation free volume f_g for PAr. Calculation is made according to equations: 1 – (5); 2 – (1) when A=0.159

The comparison for PC, given in Fig. 6, has shown a good enough correspondence of calculations (~ 15%) performed according to equations (24) and (27), though the data of the fractal analysis show a stronger dependence f_g on T, than the lattice model does [68].

Thus, the fractal concept of the structure of glassy polymers in the interval of linear scales ~ 3÷50 Å as a sort of dynamic porous system is justified. The values f_g, calculated according to equation (24), satisfactorily correspond to the calculations done within the frames of other models in relation to the temperature dependence and absolute values. There is another fundamental conclusion which follows from equation (24). As the choice of f_g at T_g as A is apparent is evident that denotes an opportunity of obtaining this value only if d_f=d =3, that is for Euclidean space. Any deviations of the value f_g in the direction of decreasing denote the transition to the condition d_f< 3, that is the transition to the fractal behavior.. This statement, in its turn, means that the glassy state of polymers is always fractal.

Percolation Model of the Fluctuation Free Volume

It was earlier shown [71, 72], that cluster network of entanglements forms a percolation framework at the glass transition temperature T_g (or melting temperature T_m for semi-crystalline polymers), where the relative fraction of clusters φ_{cl} is an order parameter for polymers structure, and the classification may be considered as the second-order transition. On the basis of this model, it is necessary to expect to carry out the percolation theory relationship for holes in a percolation cluster structure [73]:

$$\xi \sim |\tau|^{-\nu}, \qquad (28)$$

where ξ is the length of percolation, which above the percolation threshold characterizes the hole sizes in percolation cluster, $\tau=(x-x_c)/x_c$ (where x is the joints' concentration, x_c is percolation threshold), ν is critical index.

The value ν is connected to fractal dimension d_f of percolation cluster by relationship [74]:

$$\nu = \frac{2}{d_f}, \qquad (29)$$

where the coefficient 2 in the equation (29) defines that the structural component of a percolation cluster, which is responsible for effects connected to critical index ν. This component is the second subset of a percolation fractal cluster surrounding its chain backbone [72]. For models [13-15] the chain framework may be identified as a cluster network of entanglements, and the second subset (mesh structure) – as a loose-packed matrix. Then it is possible to consider a microvoids of fluctuation free volume concentrated in loose-packed matrix [17] as holes with a characteristic size ξ.

The density of a cluster network decreases with temperature and at T_m for amorphous and semi-crystalline polymers the complete decomposition of clusters is realized [75]. It allows to consider cluster structure of semi-crystalline polymers noncristalline areas as a thermal cluster (i.e., cluster, whose equilibrium configuration is defined by both geometrical, and thermal interactions), for which [76]:

$$|\tau| = |\varepsilon| = \left|\frac{T_{cr} - T}{T_{cr}}\right|, \qquad (30)$$

where for semi-crystalline polymers $T_{cr}=T_m$, and ε is order parameter ($\varepsilon=\varphi_{cl}$).
The relationship (28) with allowance for equation (30) can be written as follows:

$$V_v^{1/3} \sim \varepsilon^{-\nu} = \left(\frac{T_m - T}{T_m}\right)^{-\nu}, \qquad (31)$$

where the value $V_v^{1/3}$, describing a linear size of a microvoid (equation (20)) is accepted as ξ as the first approximation.

In Fig. 8 the dependencies $V_v^{1/3}\left(\varepsilon^{-1}\right)$, corresponding to the relationship (31), in double log-log coordinates for LDPE and HDPE are given. As follows from the given dependencies, they are well approximated by straight lines, that allows to estimate the value of a critical index ν from their slopes, which are equal to 0.77 and 0.85 for HDPE and LDPE, accordingly (table 1). These values are close to the classical value ν, which for $d=3$ is accepted equal to 0.88 [73] or 0.8-0.9 [74]. From the known values of ν according to the equation (29) it is possible to estimate the values d_f for polyethylenes, which are also given in table 1. Good correspondence (discrepancy is less ~8%) of the values d_f, obtained according to the equations (3) and (29) is observed..

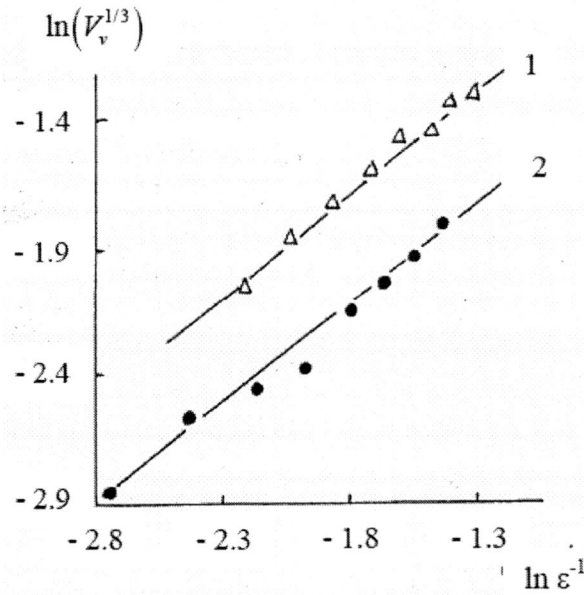

Figure 8. Dependencies of specific size of microvoids $V_v^{1/3}$ on parameter ε^{-1} in double log-log coordinates, corresponding to equation (31), for HDPE (1) and LDPE (2)

As was shown above, the microvoid of fluctuation free volume generally should be simulated by a D_f-dimensional sphere. The value D_f can be calculated by two methods: by using the equations (8) and (16) or equation (15). The comparison of the average values D_f, is also given in table 1, from which follows, that the equations (8) and (15) yield close values of parameter D_f. Besides the absolute values D_f are close to 3; that fact allows to approximate microvoids in studied polyethylenes by three-dimensional spheres.

As follows from the data of Fig. 8, the dependencies $V_v^{1/3}(\varepsilon^{-1})$ in double log-log coordinates are not approximated by a unique straight line. It means, that the constant coefficients in the relationship (31) at the replacement of a sign of proportionality by a sign of equality for LDPE and HDPE will be different. The estimations have shown, that these coefficients are equal to ~1.93 and 1.41, accordingly. It is easy to see, that these values are well corresponded with the reciprocal of crystallinity degree K_ρ, determined from the density of polyethylenes ($K_\rho^{-1} = 1.97$ and 1.38 for LDPE and HDPE, accordingly). Thus, the combination of the equations (29) and (31) with allowance for the considered above method of determination of constant coefficient in the relationship (31) allows to write the following equation for the estimation of the value V_v:

$$V_v = K_\rho^{-3} \varepsilon^{-6/d_f}, \tag{32}$$

where the parameters K_p and d_f characterize the structure of semi-crystalline polymers, and value ε – its position on a temperature scale in relation to critical temperature (temperature of phase second-order transition) T_m.

In Fig. 9 displays the comparison of values V_v, calculated according to the equations (6) and (32) is given. The good correspondence is seen between the values of fluctuation free volume microvoids sizes, calculated by various indicated ways. The equation (32) directly includes the structural characteristics (K_p and d_f), whereas the equation (6) estimates the value V_v in main outgoing from properties of polymers (ν, E, T_g and f_g).

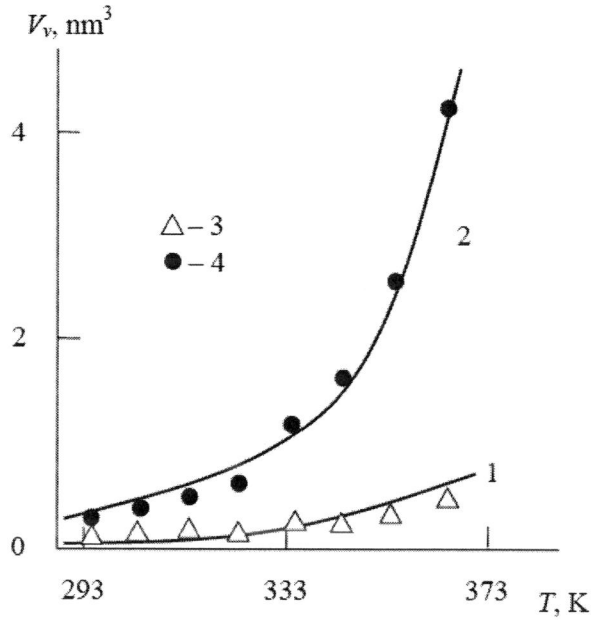

Figure 9. Comparison of the dependencies of volume of fluctuation free volume microvoids V_v on temperature T, calculated according to equations (32) (1, 2) and (6) (3, 4) for HDPE (1, 3) and LDPE (2, 4)

Multifractal Model of Fluctuation Free Volume

Now it is well known [77] that the diffusivity D for polyethylene varies suddenly at its melting point Tm. This effect has received various explanations in the different theoretical concepts. So, [77] the jump may be explained from the structural positions, namely, by the disappereance of crystallites at Tm. From the positions of a cluster model the indicated jump is the result of decomposition of the "frozen" local order in amorphous phase of polyethylene at Tm [78] may be responsible. And at last, from the positions of the fractal analysis this jump is explained by the transition of polyethylene structure from fractal (solid state) to Euclidean (melt) [79].

However, there is one more fundamental structural aspect of the problem, which up to now now was not given the proper attention. This aspect originates from two well-known experimental observations: the dependencies of the value D for the same polymer on the size (diameter) d_m of gas-diffusant molecules [28, 29] and the various degrees of this dependence for solid semi-crystalline polyethylene and its melt [77]. In the last case the dependence $D(d_m)$ is expressed much more weakly than in the first one. These observations allow the following conclusion to be drawn. In semi-crystalline polyethylene there is the distribution, typical of polymers, of microvoids size d_v of free volume (where d_v is a diameter of microvoids) by their dimensions.. Such distribution (or multifractality) of free volume microvoids is a consequence of the multifractality of the very structure of a polymer. [80]. If diffusion in semi-crystalline polymers proceeds at condition $d_m \leq d_v$, then with increase of d_m less microvoids with size d_v, , corresponding to the diffusion condition, remain. This circumstance leads to the diminishing value D. The PE melt is an Euclidean object with monodisperse distribution of d_v. This leads to the leveling off the differences in diffusion of molecules of gas-diffusant with different d_m. The verification of this assumption correctness will be presented below.

To determine the diffusivity D the following equation is used [77]:

$$D = D_0 \exp\left(-\frac{E_a}{RT}\right), \qquad (33)$$

where D_0 is a constant for each gas-diffusant, E_a is the activation energy of diffusion process, R is a universal gas constant, T is the temperature.

The values D_0 and E_a for semi-crystalline polyethylene and its melt were taken from [77]. It was supposed that for polyethylene is $T=293$ K, and for its melt is $T=383$ K [77]. The values d_m were selected from [29, 81].

The fractal concept of gas diffusion through polymeric membranes assumes that the diffusivity D can be calculated from the following equation [79]:

$$D = D_0' f_g \left(\frac{l_1}{l_2}\right)^{2(D_f - d_s)/d_s}, \qquad (34)$$

where D_0' is a constant, f_g is a fractional free volume, l_1 and l_2 are lower and upper scales of fractal behavior of the system, correspondingly, D_f is a dimension of the localization regions of extra energy, controling processes of gas transport, d_s – a spectral (fracton) dimension.

With the reference to gas diffusion the scales l_1 and l_2 should be defined as follows. The upper scale l_2 is the size (diameter) of a free volume microvoid d_v, and the lower scale l_1 – molecular diameter of a gas-diffusant d_m [79]. From equation (34) it follows, that under the conditions of D_0'=const, f_g=const and $l_2=d_v$=const the dependence $D(1/d_m)$ in double log-log coordinates should be linear and its slope allows to determine an exponent $2(D_f/d_s)/d_s$. The value d_s owing to its small variation can be accepted as a constant value and equal ~1.45 [79].

Fig. 10 presents the dependence $D(1/d_m)$ for diffusion of 9 gases [77] in semi-crystalline polyethylene and its melt in double log-log coordinates, value D being determined according to equation (33). This dependence sharply differs for two states of polymer: if for semi-crystalline polyethylene the strong decrease D is observed in accordance with increasing d_m, for a melt any dependence is practically absent. The dependence D $(1/d_m)$ for semi-crystalline polymer is curvilinear. From this fact follows the change of the index $2(D_f - d_s) / d_s$ with variation of the measurement scale d_m or, with allowance for the condition d_s=const, the change of dimension D_f. Value D_f is connected to a fractal dimension of polymer structure d_f via relationship (16). From the equation (16) follows that the change in D_f with a scale means the similar change of d_f. Approximating the curvilinear dependence $D(1/d_m)$ by the rectilinear segments, from their slope it is possible to calculate the value of the exponent $2(D_f-d_s)/d_s$, n the dimension D_f and from equation (16) – value d_f for each such part. Fig. 11 demonstrates the dependence $d_f(d_m)$, from which the increase of d_f follows in accordance with the increase of d_m or the increase of a measurement scale. It is a typical sign of the multifractality of semi-crystalline polyethylene structure [37]. Let's mark that from Fig. 10 and equation (16), the similar diagram may be constructed for D_f which means that multifractality of a free volume in semi-crystalline polyethylene h is a consequence of the multifractality of the structure of this polymer (Fig. 11). Fig. 11 shows the decrease in d_f in accordance with the decrease of a scale d_m, i.e., increase of the density of structure with the decrease of a scale [24]. It, in essence, is the definition of a fractal by relationship [82]:

$$\rho \sim R_g^{d_f - d}, \qquad (35)$$

where ρ is the density of a fractal object, R_g is its radius of gyration. As it is known [24], the number of microvoids N_v with the size d_v can be estimated from the relationship:

$$N_v \sim d_v^{-d_f}. \qquad (36)$$

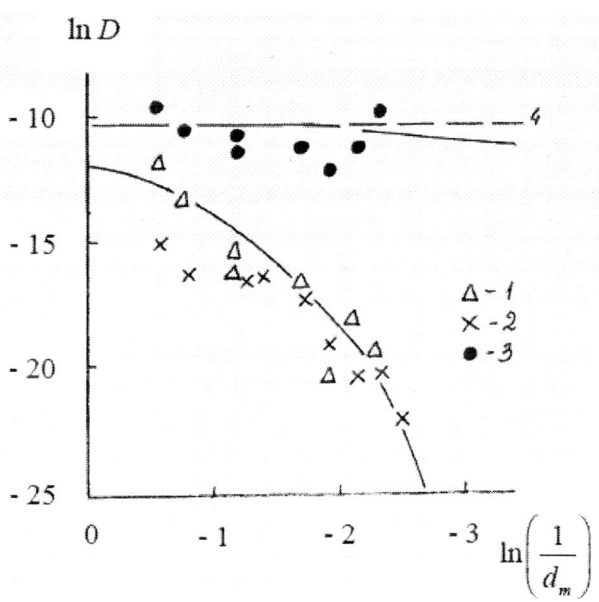

Figure 10. Dependencies of the diffusivity D on reciprocal molecular diameter of a gas-diffusent d_m in double log-log coordinates for semi-crystalline polyethylene (1,2) and its melt (3,4). Calculation: by equations (33) (1,3), (34) (2) and (40) (4)

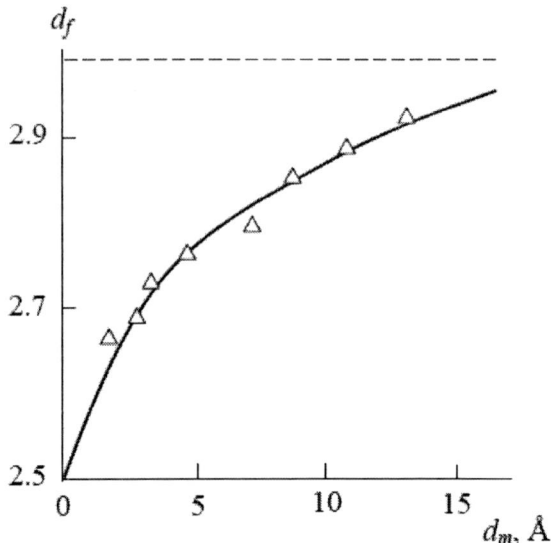

Figure 11. Dependence of the fractal dimension d_f on the measurement scale d_m for semi-crystalline polyethylene. The dashed line indicates the maximum value $d_f=d=3$

If to assume that N_v corresponds to a relative fraction of microvoids of the size d_v, it is possible (after the appropriate normalization of the value N_v) to construct the distribution d_v, shown in Fig. 12. As follows from this Figure, the probability P_v (equal to a relative fraction of microvoids with the size $\geq d_v$) of the microvoid detection with the size more than d_v is fast decreased with the increase d_v. So, the detection probability of a microvoid with $d_v \approx 12$ Å is equal to 0.001.

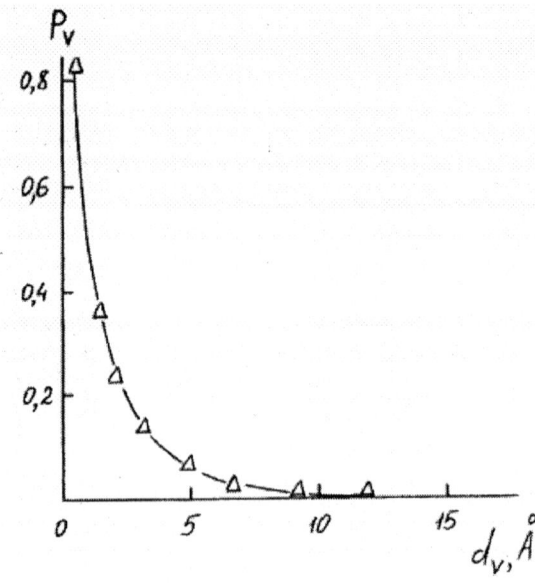

Figure 12. Dependence of the probability P_v of detection of a microvoid of free volume by the size d_v on the value d_v for semi-crystalline polyethylene

Now we consider the dependence $D(1/d_m)$ for a polyethylene melt, also shown in Fig. 10. At the first approximation its slope is accepted to be zero and from it the condition follows:

$$2(D_f - d_s)/d_s = 0. \tag{37}$$

or $D_f = d_s$. Fractal objects have $d_s \leq 2$ [83], and minimum value $d_f = 2$ [24]. From the equation (16) follows, that the minimum value D_f is also equal to 2. It is possible to determine the magnitude d_s from the following equation [83]:

$$d_s = \frac{2d_f}{1+d_f}. \tag{38}$$

The equation (38) assumes the variation $d_s = 1.33-1.50$ at the greatest possible variation d_f ($2 \leq d_f < 3$ [27]). Thus, it is impossible to receive the condition (37) in the assumption of the fractality of a polyethylene melt. Therefore, it is necessary to assume that the melt is an Euclidean object, in which the dimension D_f loses its sense. From the formal point of view from equation (16) follows: at $d_f = d$ the value $D_f \to \infty$. From the structural positions it means the following: the dimension D_f characterizes the energetic exitation degree of the loose-packed regions of polymer. In a melt both crystallites, and "frozen" local order are absent, owing to which there is no need in the separate characterization of the different structural components and the whole melt is characterized by the dimension $d=3$. For Euclidean objects the identity is correct [84]:

$$d_s = d_f = d = 3. \tag{39}$$

Therefore the condition (37) for Euclidean objects is trivially fulfilled at the replacement D_f on $d_f = d$.

Therefore, from the above mentioned, the equation (34) in case of a melt becomes:

$$D = D_0' f_g, \tag{40}$$

where the value f_g can be accepted according to Boyer [67] equal ~0.113. Then, for a melt of polyethylene, D up to the constant is equal to D_0' and constant. The value D_0' in this case equal to 2.45×10^{-4} cm^2/s, that is close to the theoretical value D_0^T [77]:

$$D_0^T = \frac{\nu \lambda^2}{6} \approx 10^{-3} \div 10^{-4} \; sm^2/s, \tag{41}$$

calculated from the assumption about a jump of gas-diffusant molecule on the distance λ by some Angstroms with the frequency $\nu = 10^{12}-10^{13}$ Hz. Calculated on the equation (40)

Values D calculated from equation (40) at the mentioned-above values of constants D_0' and f_g are shown in Fig. 10 by the dashed line. The theoretical dependence agrees with experiment which show the absence of the multifractality (distribution) of the microvoid sizes of free volume in a melt. The latter effect is the consequence of losses by the polymer structure of fractal properties.

The dependence of parameter D_0' for semi-crystalline polyethylene calculated by equation (34) is shown in Fig. 13. The large values D_0' (about 10^{-5} cm^2/s) are observed only for two inert gases with the least molecule size (He and Ne). For remaining gases-diffusants the value D_0' is approximately constant and equal about 10^{-9} cm^2/s. It gives the possibility for the theoretical estimation of diffusivity D for semi-crystalline polyethylene by equation (34) under the following conditions: $D_0'=2\times10^{-9}$ cm^2/s, $f_g \approx 0.07$, $d_s=1.45$, $d_v=12$ Å and the values D_f are accepted with in the interval 4.06-9.78 according to their estimation from slopes of curvilinear dependence D ($1/d_m$), Fig. 10. The comparison of the theory and the experiment shown in Fig. 10, demonstrates their good correspondence, except for the data for He and Ne for the mentioned-above reasons. Therefore, the correspondence of the theory and the experiment for semi-crystalline polyethylene is reached when to account for the dependence d_f (or D_f) on the measurement scale d_m and by the inclusion in the calculation scheme of these scales (d_m and d_v). This fact implies the multifractality of the structure (and, therefore, of free volume) of this polymer. To obtain the correspondence between the theory and experiment in the case of a polyethylene melt, which is an Euclidean object, the application of the mentioned-above conditions is not required. Let's mark also that the dimension d_f =2.67 corresponds to the probability $P_v \approx 0.84$. It means that the value d_f in the monofractal description of semi-crystalline polyethylene structure should be close to the indicated value d_f. The estimation of the value d_f for polyethylenes by the independent methods confirms this assumption [85].

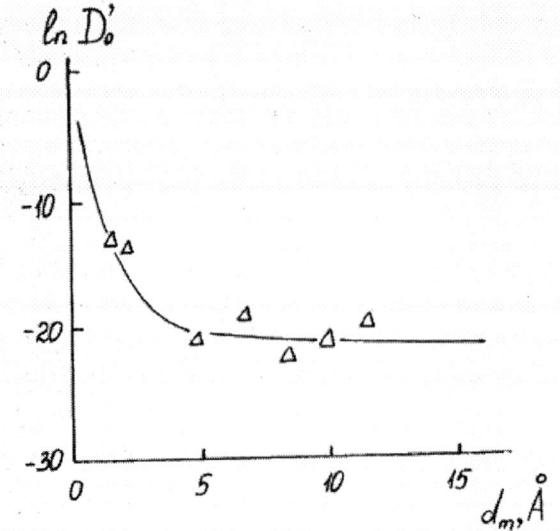

Figure 13. Logarithmic dependence of a constant D_0' on molecule diameter d_m of gas-diffusant for semi-crystalline polyethylene

Further it is possible to construct the multifractal diagram for semi-crystalline polyethylene structure in terms α-f, where α is a scaling index describing the concentration of singularities, f is a dimension of singularities α [86]. For this purpose the elementary variant is used [87]:

$$P_i \approx l_i^{\alpha_i}, \qquad (42)$$

where P_i is the probability of detection of a microvoid with the relative size l_i.

The value P_i is accepted equal to P_v (Fig. 12) and value l_i is determined from the construction of Cantor's set ("dust") as follows. From the data of Fig.12 it follows that the interval of the variation d_v can be accepted equal 0-12 Å. Then the value d_m should be normalized by the value of an interval of the variation d_v, i.e., 12 Å. Further in the construction of Cantor's set from an initial segment of length of 1 its median part (one third) is taken away then the length of two rest parts will be equal ~0.667 [86, 88]. This value should be divided proportionally to the size between the surfaces of adjacent microvoids and the size of a microvoid of free volume under the condition $d_v=d_m$ and the latter of the indicated parts corresponds to l_i [88, 89]. The value f is calculated in such a way [88]:

$$f = d_f - 2. \qquad (43)$$

The constructed in such a manner diagram α-f for semi-crystalline polyethylene is shown in Fig. 14, where the diagram α-f for a polyethylene melt (trivial case) is also shown: the latter is a straight line with ordinate f=1 ($d_f=d=3$) parallel to axis α.

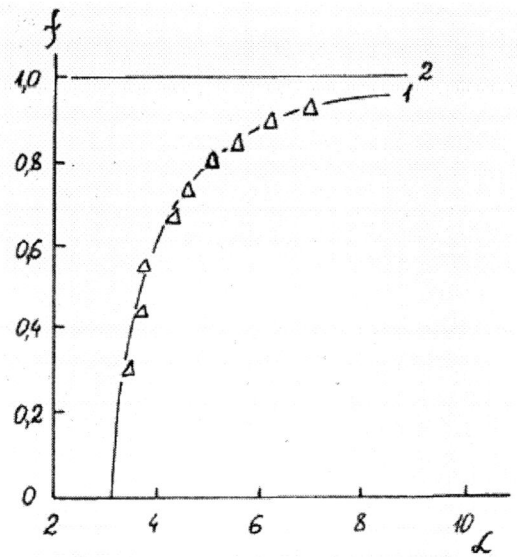

Figure 14. The multifractal diagrams of the structure of semi-crystalline polyethylene (1) and its melt (2) (trivial case) in coordinates α-f

The diagram α-f for semi-crystalline polyethylene has the bell-shaped form typical for such diagrams [37, 80, 86-89]. Its physical sense consists of the following: the curve α-f shows the probability of detection in polymer of microvoids with the size higher than an arbitrary size d_v at the arbitrary d_f.

Now let's consider the physical nature of a separate microvoid of fluctuation free volume [90]. To estimate the dimension of the walls of free volume microvoid D_w (condition $D_w=2$ means a smooth wall, $D_w>2$ – rough), the following technique is used. It is supposed that the solubility of simple gases which are not interacting with a polymer, is realized by deposition of their molecules on the walls of free volume microvoids. That is, we consider the solubility as "true sorption" [91]. The comparison of a diameter d_v of free volume microvoids in HDPE (d_v=5.83-7.32 Å in an interval T=293-373 K [22]) and diameters d_g of gases molecules (d_g=1.77-3.15 Å [29]) show that these molecules practically can be deposited on the walls of microvoids only as a monlayer. This circumstance allows to use the following equation for calculation D_w [92]:

$$\sigma \sim S_g^{-D_w/2}, \tag{44}$$

where σ is solubility, S_g is cross section area of a of gas-penetrant molecule, calculated from known values d_g.

The temperature dependence of solubility σ of three gases (helium, neon and nitrogen) in HDPE is determined according to the equation [77]:

$$\sigma = \sigma_0 \exp(-\Delta H / RT), \tag{45}$$

where σ_0 is a constant, ΔH is apparent heat of dissolution, R is universal gas constant, T is temperature.

The values S_0 and ΔH for three pointed gases are accepted according to the data [77]. The diameters of molecules d_g of these gases were taken from [29].

In Fig. 15 the dependence σ (S_g) in the double logarithmic coordinates appropriates to a relationship (44) for HDPE and three above mentioned gases at five temperatures of tests are shown. It is seen that the dependencies are linear that allows to determine values D_w from their slopes. The estimated values D_w lay in an interval 0.96-1.53 for an interval T=393-373 K. Such values $D_w<2$ hardly have any physical sense [37]. Therefore it is necessary to notice that till now we have simulated microvoids by the three-dimensional spheres, whereas they should be simulated by D_f-dimensional spheres. It means that the estimated values D_w (denote them as D_w3) should be recalculate for the space with dimension D_f. The recalculation of values D_w3 (for d=3) in $D_w^{D_f}$ for space with dimension D_f can be carried out under the formula [93]:

$$D_w^{D_f} = \frac{\left[D_f + D_w 3 \pm \left((D_f - D_w 3)^2 - 2\right)^{1/2}\right]}{2} \tag{46}$$

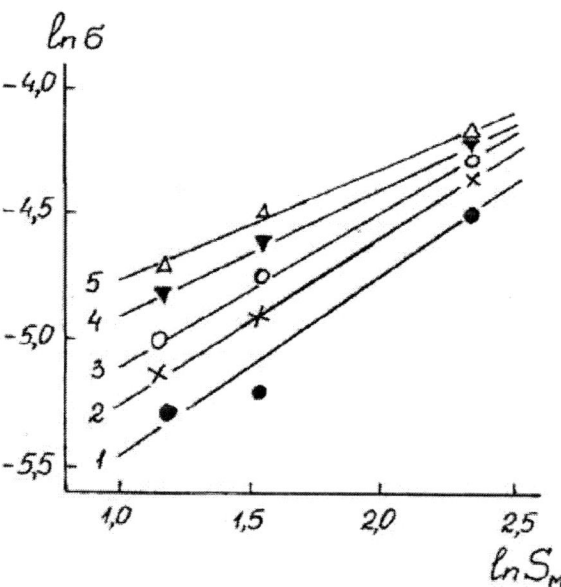

Figure 15. Dependencies of the solubility σ on the area of cross section S_m of gas-penetrants molecules in double logarithmic coordinates for HDPE at temperatures: 293 (1), 313 (2), 333 (3), 353 (4) and 373 K (5)

The comparison of the temperature dependencies of D_f and $D_w^{D_f}$ for HDPE is given in Fig. 16. It is seen the value $D_w^{D_f}$, changing similar to D_f, has absolute values in an interval 3.28-4.85, which are lower than the absolute magnitudes of D_f. If we consider a microvoid as a certain material object, it completely corresponds to the fractal theory where the dimension of a surface of object d_{surf} is connected with the dimension of the object d_f as [37]:

$$d_f - 1 \leq d_{surf} < d_f. \tag{47}$$

It the relationship (47) the sign of equality is carried out for regular fractals.

Here it is necessary to note the interesting fact. Was shown [94], for multifractals the following relationship is correct:

$$D_n (n=1) = D_n (n=0) - 9\varepsilon n / 64, \tag{48}$$

where ε is defined from the formula [94]:

$$\varepsilon = 4 - d, \tag{49}$$

where d is the dimension of Euclidean space in which multifractal is considered, n – index of dimension D_n ($n=0,1,2,...$).

Following Williford [88], it is possible to assume that the dimension D_f corresponds to the Hausdorff dimension (the greatest) for multifractal ($n=0$) and dimension of a surface, that is, $D_w^{D_f}$ is the information dimension ($n=1$). Then the calculation by equations (48) and (49) gives the following result [95]:

$$D_w^{D_f} = D_f - 0.14. \qquad (50)$$

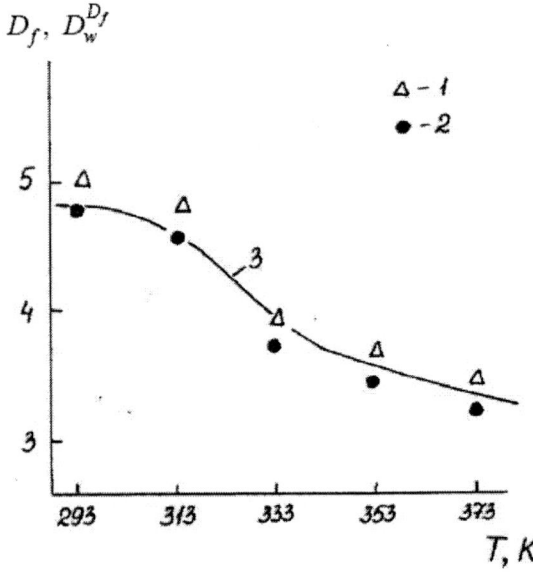

Figure 16. Temperature dependencies of dimension D_f (1) and calculated from equations (46) (2) and (48) (3) $D_w^{D_f}$ for HDPE

Let's note that in this case the multifractal (the microvoid and its walls) is considered in Euclidean space with dimension $d=3$. At such a method of calculation (the dependence calculated on the equation (50) $D_w^{D_f}$ on temperature is shown in Fig. 16 by continuous line) values $D_w^{D_f}$ calculated on the equations (46) and (50), demonstrate excellent correspondence (maximal disagreement 2.4 %, average – 1.3 %). It means that the microvoid of fluctuational free volume can be considered (at least formally) as the multifractal with Hausdorff dimension D_f and information dimension $D_w^{D_f}$. Such multifractal is very specific and its basic features may be presented as follows: [90, 95]:.

(1) Specified multifractal does not consist of any elements, constructing it, and is non-material (phantom) multifractal. The single condition therefore is correct:

$$D_f > d = 3. \qquad (51)$$

(2) A microvoids acquires the fractal properties due to itsenvironment, i.e., the oscillating atoms (segments) of macromolecules. At the same time the surface with dimension $D_w^{D_f}$ cannot be attributed to any other object but to microvoid.

(3) The microvoid is "dynamic" multifractal and the high roughness of its walls is caused by both the atomic scale of the sizes and by thermal fluctuations of its environment. It defines the absolute value $D_w^{D_f} > 3$.

(4) Despite some specificity of microvoid modeling as multifractal with Hausdorff dimension D_f, such method has obvious practical advantages. It is easy to see that the procedure, opposite to offered, allows to predict the temperature dependence S on known polymer structural parameter v for simple gases with known d_g (or S_g) according to the equation (44).

And in summary we shall consider the problem of gas transport through polymeric membranes. As it was mentioned above, the microvoids of fluctuation free volume can not form a percolation backbone because the value f_g is lower than a percolation threshold, and consequently there exist two opportunities of this process realization.

Within the framework of the percolation theory the calculation of a percolation threshold x_c has been done for passing through the overlapped statistically disposed spheres. The following result has been gained[74]:

$$x_c = \frac{N_c \pi d_{sp}^3}{6} = 0.34 \pm 0.01, \qquad (52)$$

where N_c is the critical number of spheres of a diameter d_{sp} in the bulk.

It is evident that in case under consideratione as value x_c it is necessary to accept the value of relative fluctuation free volume f_g, as N_c – a number of its microvoids per unit volume N_v and as d_{sp} – a microvoid diameter d_v. It is possible to estimate these parameters from the equations (2), (6), (7) and (20).

From the equation (52) follows, that the exponent 3 for a sphere with diameter d_{sp} assumes its simulation as three-dimensional object. The replacement of this exponent on D_f allows the simulation of microvoid as D_f- dimensional sphere According to equation (52) in this case it is possible to receive the fractal value of a percolation threshold f_g^{fr}. In Fig. 17 the dependence $f_g^{fr}(T)$ for HDPE in a range T=173-293 K is given. It is seen that at 53-273 K the value f_g^{fr} becomes more then a percolation threshold x_c=0.34±0.01 and it means that in such treatment the microvoids will form through channels facilitating gas diffusion through polymer. In Fig. 17 the calculated from equation (33) dependence $D(T)$ for a diffusion CO_2 in HDPE (D_0=7.4×10^{-2} sm^2/s, E_a=7.9±0.3 kcal/mole [77]) is also shown. As it follows from the comparison of the dependencies of $f_g^{fr}(T)$ and $D(T)$ for diffusion of CO_2 in HDPE there is observed their similarity and the sharp increase D takes place in the same temperature range where the condition $f_g^{fr} = x_c = 0.34$ is fulfilled. Thus, there are two modes of a diffusion satisfying the conditions $f_g^{fr} < x_c$ and $f_g^{fr} > x_c$. In the first mode the gas molecule "expects" the formation of a fluctuation free volume microvoid next to the microvoid, where it is situated at given moment is, then its transfer on the distance in some Ångströms, approximately equal d_v, is realized [77]. It is possible to call this mode "series". In the second case the diffusion of a gas molecule through percolation channels is realized, which form D_f-dimensional spheres. This effect is essentially (by some orders of magnitude) increases D. It is possible to call the second mode as "parallel".

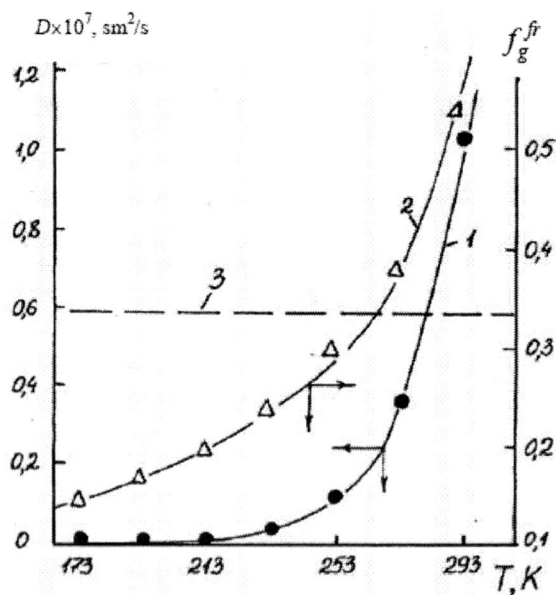

Figure 17. Dependencies of the diffusivity on CO_2 D (1) and the fractal percolation threshold f_g^{fr} (2) on the temperature T for HDPE. The dashed line (3) point out the value of percolation threshold x_c

The problem of a gas diffusion in polymers is closely connected to molecular mobility problem [96]. The increasing molecular mobility enhances diffusion. Quantitatively the degree of molecular mobility is characterized by fractal dimension D_{ch} of a chain segment between joints or cross-links ($1 \leq D_{ch} \leq 2$).[97] The value D_{ch} can be found as [97]:

$$\frac{2}{\varphi_{cl}} = C_\infty^{D_{ch}}, \qquad (53)$$

where φ_{cl} is the relative fraction of the local order regions (clusters) estimated as [71]:

$$\varphi_{cl} = 0.03(T_m - T)^{0.55}. \qquad (54)$$

Fig. 18 displays the dependence $D_{ch}(t)$ for HDPE from which follows thag the increase D_{ch} occurs in accordance with the increase T for HDPE. Special interest for the present description have the temperature limits, at which the limiting values $D_{ch}=1$ and $D_{ch}=2$ are reached. In the case of the temperature limit T_1, which may be found from the extrapolation of the dependence $D_{ch}(T)$ to $D_{ch}=1.0$, T_1 is equal ~138K, that corresponds to temperature of freezing of β-mobility T^* in polymers (for HDPE T^*=140-150 K [65]). Within the framework of the fractal analysis this temperature corresponds to temperature T_{qs} of realization of a quasi-equilibrium state of structure, for which $D_{ch}=1.0$, i.e., the chains are frozen (for HDPE $T_{qs}\approx 160$ K [66]). The calculation by equation (33) displays, that at $T=140$ K $D=4.13\times 10^{-8}$ cm^2/s, and at $T=273$ K $D=3.9\times 10^{-8}$ cm^2/s. In such a way, the freezing of molecular mobility decreases D approximately by six orders of magnitude, i.e., practically up to zero. At $D_{ch}=2.0$ the extrapolation of dependence $D_{ch}(T)$ yields $T_2\approx 485$ K, that is about equal to temperature of so-called "liquid 1 – liquid 2" transition, T_{ll}, defined as [65]:

$$T_{ll} \approx (1.20 \pm 0.05)T_m . \qquad (55)$$

The transition to an unstructured liquid takes place at such temperature, i.e., to true rubber, for which $D_{ch}=2.0$ [24]. At $T=485$ K the value D, calculated according to [77], is equal for a diffusion CO_2 in HDPE ~ 3.6×10^{-5} cm^2/s, i.e., by three orders higher than D at $T=273$ K . These results allow to receive a simple empirical relationship between D and D_{ch}:

$$D \approx 7.66 \times 10^{-14} D_{ch}^{28.1}, \qquad (56)$$

which demonstrates strong dependence of a gas diffusion through polymers on the level of molecular mobility. Apparently the approximate equation (56) is correct only for a diffusion CO_2 in HDPE.

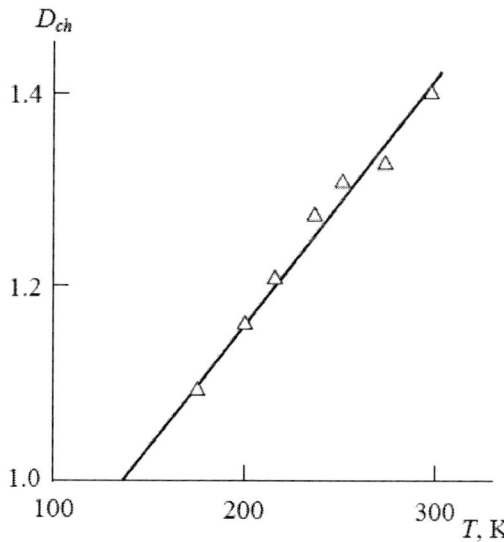

Figure 18. Dependence of the fractal dimension D_{ch} of the chain portion between entanglements on the temperature T for HDPE

At last we shall consider the simple physical interpretation of D_f-dimensional sphere. The walls of a fluctuation free volume microvoid represent the surfaces of segments of macromolecules oscillating around the equilibrium positions. The higher is the testing temperature, the higher fractal dimensions D_f and D_{ch} are These dimensions characterize energetic exitation level of structure and molecular mobility degree. Accordingly the larger sizes can be reached by the microvoids, before at particular temperature they will form n oscillating (pulsating) percolation backbone, the higher are D_f and D_y. These effect essentially increase the diffusivity. Therefore, the increase of D_f and D_{ch} means an actual increase of microvoid size. Let's mark, that the process of a diffusion in polymers is described by the dimension D_f, instead of fractal dimension of structure d_f (the equation (34)) [79, 98].

CONCLUSIONS

The results of the present investigation have shown, that the fluctuation free volume in polymers has fractal structure, owing to which the microvoids, forming it, should be simulated by a D_f-dimensional sphere. The microvoids sizes are controlled by volume necessary for accumulation of thermal fluctuations energy, required for their formation. The absolute values of relative fluctuation free volume f_g may serve as the characteristic of the thermodynamic nonequilibrium state of polymer structure . For quasi-equilibrium structures the value f_g coincides with the data, following from the equation Williams-Landel-Ferry equation.

The microvoids of fluctuation free volume form structure, which is a mirror of the polymer structure. This microvoids structure represents a specific pulsating percolation cluster due to thermodynamically nonequilibrium structure of polymer.

The multifractal nature of structure (and, as consequence, free volume) of semi-crystalline polyethylene is verificated and the estimation of its multifractal characteristics is given on the basis of diffusion properties in this polymer. The polyethylene melt is the Euclidean object with monodisperse distribution of microvoid sizes. The indicated circumstance defines the different forms of dependencies of a diffusivity on the molecules size of a gas-diffusant. The microvoid itself is specific multifractal, whose fractal properties are defined not by its structure, but by its environment.

Table 1. Fractal characteristics of polyethylenes

Polymer	v, equation (28)	d_f equation (3)	d_f equation (29)	D_f equation (8)	D_f equation (15)
LDPE	0.85	2.543	2.350	3.19	3.46
HDPE	0.77	2.657	2.597	3.92	3.52

REFERENCES

[1] Ya.I. Frenkel. *The kinetic theory of liquids* (in Russian). Izd. AN SSSR. Moscow-Leningrad, 1945.

[2] J.D. Ferry. *Viscoelastic properties of polymers*. Wiley, New York – London, 1970.

[3] D.S. Sanditov, G.M Bartenev. The physical properties of nonordered structures (in Russian). *Nauka*, Novosibirsk, 1982.
[4] Yu.S. Lipatov. *Uspekhi Khimii*, 42, 332 (1978).
[5] G.V. Kozlov, D.S. Sanditov. Anharmonic effects and physical-mechanical properties of polymers (in Russian). *Nauka*, Novosibirsk, 1994.
[6] D.S. Sanditov, G.V. Kozlov. *Fizika i khimiya stekla*, 21, 547 (1995).
[7] V.S. Nechitailo. *Zhurn. fizich. Khimii*, 65, 1979 (1991).
[8] D.S. Sanditov, G.V. Kozlov. *Fizika i khimiya stekla*, 22, 97 (1996).
[9] D.S. Sanditov, S.Sh. Sangadiev. *Vysokomolek. Soed.* A, 40, 12, 1 (1998).
[10] D.S. Sanditov, S.Sh. Sangadiev. *Fizika i khimiya stekla*, 24, 417 (1998).
[11] D.S. Sanditov, S.Sh. Sangadiev. *Fizika i khimiya stekla*, 24, 741 (1998).
[12] D.S. Sanditov, S.Sh. Sangadiev, G.V. Kozlov. *Fizika i khimiya stekla*, 24, 758 (1998).
[13] V.N. Belousov, G.V. Kozlov, A.K. Mikitaev, Yu.S. Lipatov. *Dokl. Akad. Nauk SSSR*, 313, 630 (1990).
[14] D.S. Sanditov, G.V. Kozlov, V.N. Belousov, Yu.S. Lipatov. Ukrain. *Polymer J.*, 1, 241 (1992).
[15] G.V. Kozlov, V.A. Beloshenko, V.N. Varyukhin, Yu.S. Lipatov. *Polymer*, 40, 1045 (1999).
[16] D.S. Sanditov, G.V. Kozlov, V.N. Belousov, Yu.S. Lipatov. *Fizika i khimiya stekla*, 20, 3 (1994).
[17] V.N. Belousov, V.A. Beloshenko, G.V. Kozlov, Yu.S. Lipatov. *Ukrain. Khim. Zhurn.*, 62, 62 (1996).
[18] I.C. Sanchez. *J. Appl. Phys.*, 45, 4204 (1974).
[19] B.D. Malhotra, R.A. Pethrick. *Eur. Polymer J.*, 19, 457 (1983).
[20] Q. Deng, F. Zandiehnadem, Y.C. Jean. *Macromolecules*, 25, 1090 (1992).
[21] J.-E. Kluin, Z. Yu, S. Vleeshouwers, J.D. McGervey, A.M. Jamieson, R. Simha. *Macromolecules*, 25, 5089 (1992).
[22] D. Lin, S.J. Wang. *J. Phys.: Condens. Matter.*, 4, 3331 (1992).
[23] G.V. Kozlov, V.N. Novikov, D.S. Sanditov, M.A. Gazaev, A.K. Mikitaev. *Article deposited at the VINITI*, No. 3073-B95 (Moscow, 1995).
[24] G.V. Kozlov, V.N. Novikov. Synergetics and fractal analysis of cross-linked polymers (in Russian). *Klassika,* Moscow, 1998.
[25] V.N. Novikov, G.V. Kozlov. *Uspekhi Khimii,* 69, 572 (2000).
[26] B.D. Sanditov, G.V. Kozlov, D.S. Sanditov, Yu.S. Lipatov. In *Baikal reading on mathematical modelling of processes in synergetic systems* (Mater. Rep. All-Russian conf.), Ulan-Ude, 1999, p. 330.
[27] A.S. Balankin. *Synergetics of deformable body* (in Russian). MO SSSR, Moscow, 1991.
[28] S.A. Reitlinger. Permeability of polymeric materials (in Russian). *Khimiya*, Moscow, 1974.
[29] N.I. Nikolaev. Diffusion in membranes (in Russian). *Khimiya*, Moscow, 1980.
[30] B.I. Shklovski, A.L. Efros. *Usp. Fiz. Nauk*, 117, 401 (1975).
[31] V.U. Novikov, G.V. Kozlov. *Uspekhi Khimii*, 69, 378 (2000).
[32] G.V. Kozlov, V.A. Beloshenko, M.A. Gazaev, Yu.S. Lipatov. *Vysokomolek. Soed.* B, 38, 1423 (1996).
[33] G.V. Kozlov, V.D. Serdyuk, I.V. Dolbin. *Materialovedenie*, (12), 2 (2000).
[34] N.I. Mashukov, G.P. Gladyshev, G.V. Kozlov. *Vysokomolek. Soed.* A, 33, 2538 (1991).

[35] G.V. Kozlov, R.A. Shetov, A.K. Mikitaev. *Vysokomolek. Soed.* A, 29, 1109 (1987).
[36] G.V. Kozlov, R.A. Shetov, A.K. Mikitaev. *Vysokomolek. Soed.* A, 29, 2012 (1987).
[37] J. Feder. *Fractals.* Plenum, New York, 1988.
[38] A.B. Mosolov, O.Yu. Dinariev. *Probl. Prochn.*, (1), 3 (1988).
[39] G.I. Sandakov, L.P. Smirnov, A.I. Sosikov, K.T. Summanen, N.N. Volkova. *J. Polymer Sci.: Part B: Polymer Phys.*, 32, 1585 (1994).
[40] B.B. Mandelbrot. *The fractal geometry of nature.* W.H. Freeman and Company, San-Francisco, 1982.
[41] K. Ishikawa. *J. Mater. Sci. Lett.*, 9, 400 (1990).
[42] G.V. Kozlov, A.K. Mikitaev. Mekhan. Kompos. *Mater. I Konstr.*, 2, 144 (1996).
[43] D. Farin, S. Peleg, D. Yavin, D. *Avnir. Landmuir*, 1, 399 (1985).
[44] E. Hornbogen. *Intern. Mater. Rev.*, 34, 277 (1989).
[45] A.S. Balankin. Pis'ma v Zh. *Techn. Fiz.*, 16, (7), 14 (1990).
[46] G.V. Kozlov, G.B. Shustov, K.B. Temiraev. Vestnik KBGU, ser. khim. nauki, (2), 50 (1997).
[47] V.P. Budtov. Physical chemistry of polymer solutions (in Russian). *Khimiya,* St. Peterburg, 1992.
[48] Yu.S. Lipatov, V.P. Privalko. *Vysokomolek. Soed.* A, 15, 1517 (1973).
[49] E.I. Shemyakin. *Dokl. Akad. Nauk SSSR*, 300, 1090 (1988).
[50] S.E.B. Petric. *J. Macromol. Sci. – Phys.*, B12, 225 (1976).
[51] G.V. Kozlov, V.A. Beloshenko, Yu.S. Lipatov. *Ukrain. Khim. Zhurn.*, 64, (3), 56 (1998).
[52] D.S. Sanditov, G.V. Kozlov. *Fizika i khimia stekla*, 19, 593 (1993).
[53] M.G. Zemlyanov, V.K. Malinovski, V.N. Novikov, P.P. Parshin, A.P. Sokolov. *Zhurn. Eksp. I Teor. Fiz.*, 101, 284 (1992).
[54] G.V. Kozlov, V.N. Belousov, A.K. Mikitaev. *Fiz. Tekhn. Vys. Davl.*, 8, (1), 101 (1998).
[55] P. Pfeifer. *Appl. Surf. Sci.*, 18, 146 (1984).
[56] D. Avnir, D. Farin, P. Pfeifer. *J. Colloid. Interface Sci.*, 103, 112 (1985).
[57] G.V. Kozlov, Yu.G. Yanovski, A.K. Mikitaev. *Mekhan. Kompozit. Mater.*, 34, 539 (1998).
[58] L. Xie, D.W. Gidley, H.A. Hristov, A.F. Yee. *J. Polymer Sci.: Part B: Polymer Phys.*, 33, 77 (1995).
[59] G.V. Kozlov, V.A. Beloshenko, M.A. Gazaev, V.U. Novikov. *Mekhan. Kompozit. Mater.*, 32, 270 (1996).
[60] V.N. Belousov, B.H. Kotsev, A.K. Mikitaev. *Dokl. Akad. Nauk SSSR*, 280, 1140 (1985).
[61] A.S. Balankin. *Dokl. Akad. Nauk SSSR*, 319, 1098 (1991).
[62] D.S. Sanditov, G.V. Kozlov. *Vysokomolek. Soed.* A, 38, 1389 (1996).
[63] V.D. Serdyuk, D.S. Sanditov, N.I. Mashukov, G.V. Kozlov, V.U. Novikov, G.E. Zaikov. *Intern. J. Polymer Mater.*, 42, 65 (1998).
[64] E.W. Fisher, M. Dettenmaier. *J. Non-cryst. Solids*, 31, 181 (1978).
[65] V.A. Bernstein, V.M. Egorov. *Differential scanning calorimetry of polymers*: physics, chemistry, analysis, technology. Ellis Horwood, New York, 1994.
[66] G.V. Kozlov, G.E. Zaikov. In *Fractals and local order in polymeric materials* (G.V. Kozlov, G.E. Zaikov, ed.), Nova Science Publishers, Inc., New York, 2001.

[67] G.V. Kozlov, S. Ozden, V.N. Shogenov. In *Fractals and local order in polymeric materials* (G.V. Kozlov, G.E. Zaikov, ed.), Nova Science Publishers, Inc., New York, 2001.
[68] A.J. Katz, A.H. Thompson. *Phys. Rev. Lett.*, 54, 1325 (1985).
[69] R.F. Boyer. *J. Macromol. Sci. – Phys.*, B7, 487 (1973).
[70] S Vleeshouwers, J.-E. Kluin, J.D. McGervey, A.M. Jamieson, R. Simha. *J. Polymer Sci.: Part B: Polymer Phys.*, 30, 1429 (1992).
[71] G.V. Kozlov, M.A. Gazaev, V.U. Novikov, A.K. Mikitaev. *Pis'ma v Zh. Techn. Fiz.*, 22, (16), 31 (1996).
[72] G.V. Kozlov, V.U. Novikov, M.A. Gazaev, A.K. Mikitaev. *Inzh.-Fiz. Zh.*, 71, 241 (1998).
[73] I.M. Sokolov. *Usp. Fiz. Nauk*, 150, 221 (1986).
[74] A.N. Bobryshev, V.N. Kozomazov, L.O. Babin, V.I. Solomatov. *The synergetics of composite materials* (in Russian). ORIUS, Lipetsk, 1994.
[75] V.N. Belousov, G.V. Kozlov, N.I. Mashukov. *Dokl. Adyg. Akad. Nauk*, 2, (1), 74 (1996).
[76] F. Family. *J. Stat. Phys.*, 36, 881 (1984).
[77] V.A. Tochin, R.A. Shlyakhov, D.N. Sapozhnikov. *Vysokomolek. Soed.* A, 22, 752 (1980).
[78] V.V. Afaunov, N.I. Mashukov, G.V. Kozlov, D.S. Sanditov. Izv. Vyssh. Uchebn. Zaved., Sev.-Kav. Region, *Estestv. Nauki*, (4), 69 (1999).
[79] G.V. Kozlov, V.V. Afaunov, N.I. Mashukov, Yu.S. Lipatov. *Dokl. Nat. Akad. Nauk Ukrainy*, (10), 140 (2000).
[80] Y. Hayakawa, S. Sato, M. Matsushita. *Phys. Rev.* A, 36, 1963 (1987).
[81] V.V. Teplyakov, S.G. Durgar'yan. *Vysokomolek. Soed.* A, 26, 1498 (1984).
[82] R.M. Brady, R.S. Ball. *Nature*, 309, 225 (1984).
[83] P. Meakin, H.E. Stanley. *Phys. Rev. Lett.*, 51, 1457 (1983).
[84] R. Rammal, G. Toulouse. *J. Physiq. Lettr.*, 44, L13 (1983).
[85] G.V. Kozlov, V.U. Novikov. *Materialovedenie,* (7), 22 (1999).
[86] T.C. Halscy, M.H. Jensen, L.P. Kadanoff, I. Procaccia, B.I. Shraiman. *Phys. Rev.* A, 33, 1141 (1986).
[87] J.L. McCauley. *J. Modern. Phys.* B, 3, 821 (1989).
[88] R.E. Williford. *Scr. Metal.*, 22, 1749 (1988).
[89] V.U. Novikov, G.V. Kozlov, A.V. Bilibin. *Materialovedenie*, (10), 14 (1998).
[90] G.V. Kozlov, I.V. Dolbin, G.E. Zaikov. In *Fractals and local order in polymeric materials* (G.V. Kozlov, G.E. Zaikov, ed.), Nova Science Publishers, Inc., New York, 2001.
[91] C.E. Rogers. In *Engineering design for plastics* (E. Baer, ed.), Reihold Ltd., New York – London, 1965.
[92] D. Farin, A. Volpert, D. Avnir. *J. Amer. Chem. Soc.*, 107, 3368 (1985).
[93] G.V. Vstovsky, L.G. Kolmakov, V.F. Terent'ev. *Metally*, (4), 164 (1993).
[94] M.E. Cates, T.A. Witten. *Phys. Rev.* A, 35, 1809 (1987).
[95] G.V. Kozlov, I.V. Dolbin, G.E. Zaikov. *Russian Polymer News*, 6, (4), 35 (2001).
[96] N.M. Emanuel, A.L. Buchachenko. *The chemical physics of aging and stabilization of polymers* (in Russian). *Nauka*, Moscow, 1982.

[97] G.V. Kozlov, K.B. Temiraev, R.A. Setov, A.K. Mikitaev. *Materialovedenie*, (2), 34 (1999).
[98] G.V. Kozlov, V.V. Afaunov, N.I. Mashukov, Yu.S. Lipatov. In *Fractals and local order in polymeric materials* (G.V. Kozlov, G.E. Zaikov, ed.), Nova Science Publishers, Inc., New York, 2001.

Chapter 10

CONCENTRATION DEPENDENCE OF MELT DENSITY OF ETHYLENE-VINYLACETATE CO-POLYMERS BINARY BLENDS

O. V. Stoyanov, A. E. Zaikin, R. M. Khyzakhanov, R. Ya. Deberdeev, G. E. Zaikov, Ya. V. Kapitskaya, S. Yu. Sofina

Kazan State Technological University,
420015, Russia, Kazan, K.Marx str., 68

ABSTRACT

The melt density of binary blends of copolymers ethylene and vinyl-acetate has been study.

INTRODUCTION

Change of polymer properties and structure under their mixture is very complex [1, 2]. Density can give very important information about the changes in the structure of substances and materials, including polymer mixtures [1, 2]. One can judge the molecules packing change by density at least. The interpretation of polymers and their blends in a solid state density change is, as a rule, connected with some difficulties, depending on their over-molecular structure change under cooling and on non-balanced receiving systems. The melt density value of polymers mixtures gives more objective information. Approximately, we can consider that density's change under the mixing of two liquids reflects the efficiency of inter-molecular interaction in the system [3]. We took an interest in studying the application of the last assertion for polymers mixtures.

EXPERIMENTAL PART

Binary mixtures of ethylene and vinyl-acetate co-polymers with different content of vinyl-acetate links were chosen as the objects of study. High pressure polyethylene (HPPE) marked as 15313-003 (TC 605-186-79); co-polymers of ethylene with vinyl-acetate (CEVA) marked as 11104-030, 11306-075, 11507-070, 11708-210, containing different amount of complex-ether groups, produced by OAO "NEFTEKHIM-SEVILEN" (TC 6-05-1636-97), were used. The main characteristics of CEVA and HPPE are shown in the table.

Table. Characteristics of the initial HPPE and CEVA

Characteristics	HPPE 15313-003	CEVA 11104-030	CEVA 11306-075	CEVA 11507-070	CEVA 11708-210
Conditional sign	HPPE	CEVA-7	CEVA-14	CEVA-22	CEVA-29
Content of vinyl-acetate, %	0	7	14	22	29
Index of the melt fluidity (2,16 kg), g/10 min	0,29 (190°C)	2,4 (190°C)	6,5 (190°C)	6,1 (125°C)	18,7 (125°C)
Density, kg/m^3	921	924	934	936	945
Destructive tension under the stretch, MPa	20,4	22,4	16,9	14,6	5,5
Relative extension under the rupture, %	660	710	760	740	824
Module of the elasticity, MPa	80,5	58	51	31	11

The researched binary mixtures of CEVA with different content of vinyl-acetate groups and molecular mass were received by blending on the rollers for 4-5 minutes after feeding the initial CEVA mechanically blended beforehand. The temperature of forge rolling was identical for all compositions (110-115°C). The speed of the rolls rotation was 12,5 m/min, the friction was 1:1,2. The initial CEVA were flared under the same conditions.

The melt density was defined by the viscosity method (GOST 11645-73) under the temperatures of 120-160°C by. The essence of this method consists in estimate polymers volume and mass squeezed out through a capillary under the given constant temperature. A cargo is chosen so as 0,5-1 g of a polymer can be squeezed out per minute. An indicator fixes a shift of a piston. Polymer's melt density is calculated according to the formula:

$d = G/(t \cdot F)$, kg/m^3;

G is the weight of out squeezed polymer, kg; t is a piston's shift during a squeezing out of one specimen (the reading of an indicator), m; F is the piston's square, m^2.

RESULTS AND DISCUSSION

Firstly, the melt density of CEVA and HPPE was measured. It turned out that (fig. 1) under the temperatures of 120-160°C the density is linearly depends on vinyl-acetate content by a parallelism of the respective straights.

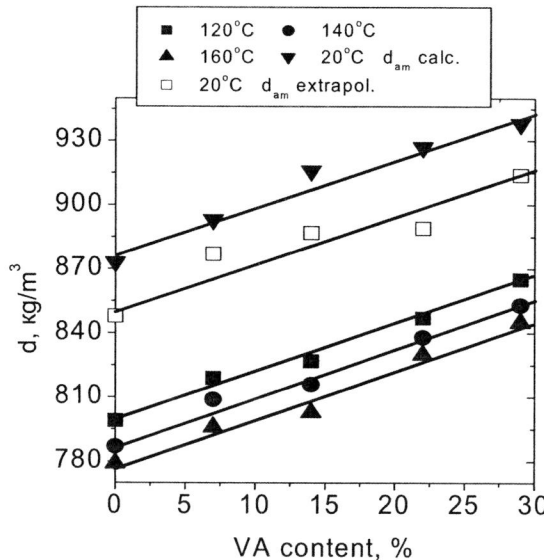

Figure 1. Melt density's dependence upon the vinyl-acetate content in CEVA

It means, that a density's growth in the case of going from a polyethylene to a CEVA realizes only on an account of the increase of a mole part of "heavy" and polar vinyl-acetate links from the law of additivity.

We evaluated a hypothetic density of an amorphous phase or, in the other words, "a density of an overcooled melt" at 20°C by an extrapolation (fig. 2) of the density's temperature dependencies. The received dependence (fig. 1) is linear and parallel to the straights based on the experienced values. It testifies that this data according to the dependence character correlates with the experiment.

Using experimental values of a crystallinity degree (DSC method) and density (hydrostatic weighing method) [4] we calculated the density of an amorphous phase of CEVA and HPPE according to additivity relationship:

$$d = d_{am}(1-k) + d_{cr}k;$$

d – is a specimen's density defined experimentally; k – is a crystallinity degree; d_{am} – is an amorphous phase density; d_{cr} – is a crystal phase density taken as 1000 kg/m^3 [5, 6]. The calculated values of d_{am} received by this method are represented on fig. 1. One can see, that they have higher value in comparison with extrapolated values, but they also have the linear character and the similar angle to an ordinate axis, that is they differ only in the constant odds in absolute values. Obviously, it is caused by a difference in the approach to its estimation for different techniques and also by an imperfection of the methods of such values receiving, consisting in the necessity of assumptions. Nevertheless, the received data are qualitatively coordinate with each other.

Thus, the received data's analysis for the initial CEVA and HPPE allows to make a conclusion: the results, received by the used method of a melt density measuring, are balanced and are not run counter to the logic. This received data analysis gives an opportunity to realize the analysis of the mixtures CEVA-29+HPPE, CEVA-29+CEVA-7, CEVA-

22+CEVA-14. The dependencies of a melt density for the mixture CEVA-29+HPPE are shown on fig. 3. The curves have slight negative diversion from additivity with a tendency to the diversion reduction with a temperature's growth.

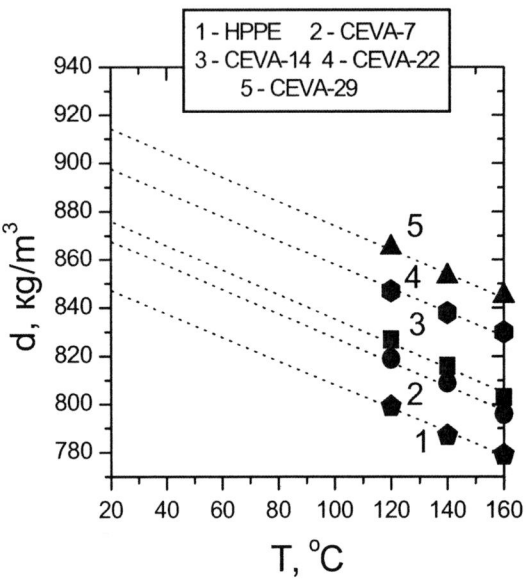

Figure 2. Temperature's dependence of HPPE and CEVA melt density

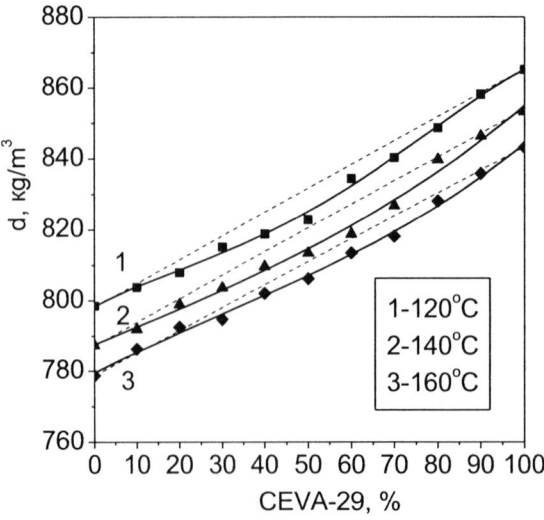

Figure 3. Concentration dependence of the melt density for the system HPPE+CEVA-29

The similar dependence for the system CEVA-7+CEVA-29 are shown on fig. 4. The curves are S-figurative and with 30% content of CEVA-7 the absolute density's minimum, degenerated at 160°C, are observed. Under CEVA-29 concentration over 70% slight excess of an additive values, which is also degenerate at 160°C, is observed.

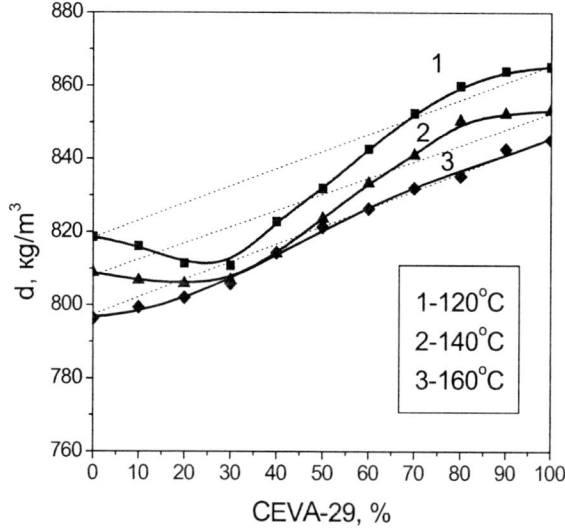

Figure 4. Concentration dependence of the melt density for the system CEVA-7+CEVA-29

And in this case the over-all tendency is the curve's flattering with a temperature of measuring growth, and at 160°C the dependence is close to the additive with a small diversion in the fields of prevailed CEVA-7 concentrations.

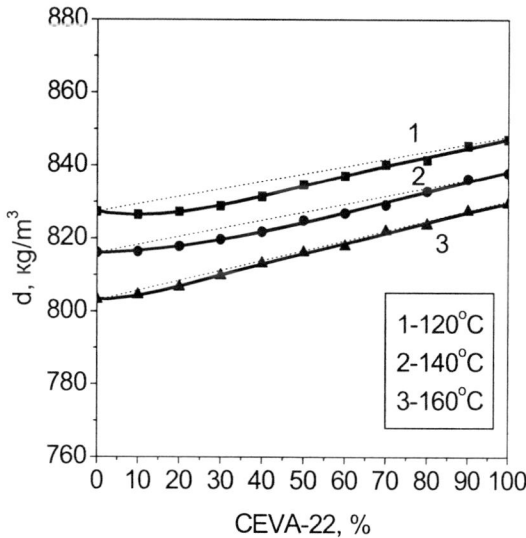

Figure 5. Concentration dependence of the melt density for the system CEVA-14+CEVA-22

Concerning the system CEVA-22+CEVA-14, it has the minimal diversion of dependence from an additive in comparison with the other systems and the similar approaching to that with the temperature's growth. In this case the S-figurative excess of the additive values under the prevalence of a most dense component as it was observed for the mixture with CEVA-29 is absent. Slightly expressed minimum at 120°C in the field of 30% CEVA content degenerates with a temperature growth and at 169°C a diversion from the additivity is comparable to an experiment error.

All above mentioned is illustrated on fig. 6 where the dependencies of a diversion of a density's value from the additive values are shown. The maximum diversion is observed for the system CEVA-22+CEVA-7, the minimum – for the system CEVA-22+CEVA-14.

We shall analyze the received data, taking into account that the relative solubility and the value of inter-molecular interaction in the row of the systems CEVA-29+HPPE, CEVA-29+CEVA-7, CEVA-22+CEVA-14 increases as their chemical structure becomes similar.

It is known, that an additivity of density is characteristic for heterogeneous systems with phases, which do not influence each other [1]. The additive dependence is also possible for homogeneous systems. But in the majority of cases for homogeneous mixtures [1] of polymers an excess of density over the additive values was observed. It is explained by the increased inter-molecular interaction of the diverse macromolecules, promoted the better packing of macromolecules. The similar phenomenon is characteristic for polymer solutions in low-molecular liquids [7].

In our case the growth of a value of component's relative solubility with temperature's growth (because this systems are the systems with top critical temperature of solution [8]), and with concentration of vinyl-acetate links in co-polymers a convergence takes place. For all investigated systems the relative influence of the components on their density, both in the solid phase [9] and in the melt (fig. 6), is observed. Moreover, in majority of cases the negative diversion of melt density from the additive values takes place.

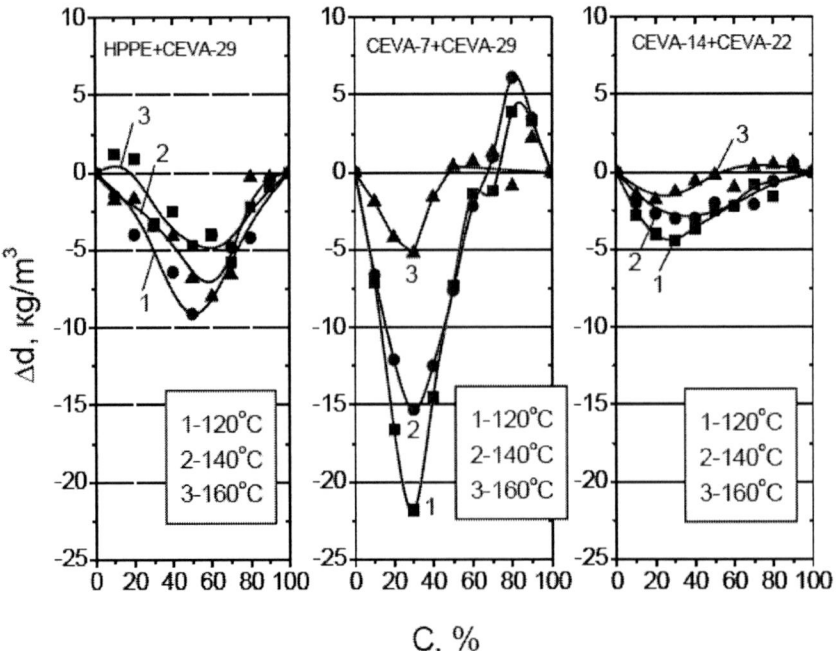

Figure 6. Concentration dependence diversion of melts density from the additivity

Let's observe the features of the investigated systems. The mixture HPPE+CEVA-29 is the least compatible of all considered mixtures. The value of relative solubility of its components at the studied temperatures does not exceed several percent [8]. Hence, it is heterogeneous in the majority fields of compounds. It is of no doubt that density of a heterogeneous mixture consists of the phases density and of s density of inter-phase and frontier areas. We have the information [10, 12] that density of frontier areas can differ from phase density in volume. It is connected with revolted influence of an adjacent phase. The maximal negative diversion of density from the additive values realizes in the field of the medium contents. An area of an inter-phase contact has the maximum in this case, and therefore, the mutual influence of the components is maximal. The mutual influence of the components in this case consists in a breaking up of molecules packing.

When changing HPPE on CEVA-7 the mutual solubility and intermolecular interaction increase. From this we should expect an increase of packing density of macromolecules in melt when mixing. Practically, the density's growth observes only in the field of dominant CEVA-29 concentrations (fig. 4, 6), with the maximum at 20% CEVA-7 content. In the majority field of the compounds the negative diversion of a density from the additivity takes place (fig. 6), moreover, at 30% CEVA-29 content at 120 and 140°C obviously expressed the absolute minimum of melt density observes (fig. 4). So complex dependence of the mixture's melt density from the component's ratio can be explained only if one supposes, that a diversion from the additivity both co-existing phase density (saturated solution of the one polymer component in the other) and relative indignative phase influence on the frontier areas density take place. It seems that in the field of dominant CEVA-29 concentrations a solution of polymers, which has an increased density because of the high intermolecular interaction, forms. In the field of the medium polymer's ratio the system is heterogeneous. An inter-phase tension in this system is considerably lower than in the system HPPE with CEVA-7 because of the proximity of the polymer's compounds and co-existing phases. The difference in component's viscosity's value is also lower. As a result, the sizes of the polymer's phases in the mixture CEVA-7 with CEVA-29 should be [2] considerably lower than in the mixture HPPE+CEVA-29. More developed phase's surface causes more intense their mutual revolted influence. Hence, the looseness of the mixture CEVA-7 with CEVA-29 is more significant from the all studied systems.

As to the system CEVA-14+CEVA-22, as it was already mentioned, the components of the mixture have the minimum distinctions and are maximally compatible with each other. Extrapolating the data [8] about the temperature's and vinyl-acetate content in CEVA influence on its solubility in HPPE one can expect that the system CEVA-14+CEVA-22 under the researched conditions is homogeneous in the whole field of the compounds. At the increased compatibility could be supposed that more high level of intermolecular interaction and the increased packing density take place. Really, for this system, the negative diversion of melt density from the additive values in the fields of prevailing concentrations of CEVA-14 at 120 and 140°C (fig. 5, 6) is characteristic. But this diversion is minimal in comparison with the other studied systems and at 160°C the diversion correlates with the experiment error. For all that the packing density of this system is higher in comparison with the system CEVA-7+CEVA-29 that is evidently connected with the higher level of macromolecules interaction in the system CEVA-14+CEVA-22.

Conclusion

All given above data show, that a connection between the melt density for the mixture of two polymers with the value of intermolecular interaction of its components and with the value of their mutual solubility, is not simple. The change of packing density of the mixture could not be explained only basing on the energy of intermolecular interaction and phase conversions. This is especially clearly seen on the mixture CEVA-7+CEVA-29, when only a change of the component's ratio causes a change of the positive diversion of the melt density from the additive values into the negative.

The available information shows [13-15] that for the binary mixtures of the polymers the negative diversion of their melt density from the additive values in the field that is so called "the critical concentration" takes place. The critical concentration is the concentration of the polymer that is slightly lower than its limit of solubility in the second component of the mixture [13-15]. It is supposed that because of the system's proximity to the unmixed it's loosening takes place and exuberant free volume comes up [13-15]. In this case the maximum of loosening corresponds to transferring of the mixture from the homogeneous into the heterogeneous state [13-15]. After the unmixed the melt density of the mixture comes nearer to the additive values again. As the binary mixture has two critical concentrations on the curve of its melt density dependence from the component's ratio two extremums (minimum) presents [13-15].

Another view point [2] takes place: the extremums or transfers on the dependence of the mixture's melt density from its content are not connected with phase transferring in the polymer's mixtures. Experimental data given in this paper confirms the last point of view.

References

[1] Maccait V., Karash F., Frid Dg. Phase and relaxation transferrings in the solid polymers. In the book: *Polymer mixtures* / Edition of D.Pole and S.Numen. M.: Mir, 1981, v.1, ch.5, p.266-267. (in Russia)

[2] Kuleznev V.N. Polymer blends. M.: *Chemistry,* 1980, p.184-185. (in Russia)

[3] Frenkel Ya.I. Kinetic theory of liquids. L.: *Science*, 1975, 592 p. (in Russia)

[4] Kapitskaya Ya.V. *Adhesion materials on the base of the CEVA's mixtures.* Auto-abstract of the dissertation of candidate of technical sciences. 2004. Kazan. KSTU. 18 p. (in Russia)

[5] Godovsky Yu.K. The thermal methods of the polymer researching. M.: *Chemistry*, 1976. 216 p. (in Russia)

[6] Privalko V.P. The basis of thermo-physics and rheo-physics of the polymer materials / V.P.Privalko, V.V.Novikov, Yu.V.Yanovsky. – Kiev: *Naykova dumka*, 1991. 232 p. (in Russia)

[7] Tager A.A. Physics-chemistry of polymers. M.: *Chemistry*, 1978, 544 p. (in Russia)

[8] Druz N.I., Chalykh A.E., Aliev A.D. The influence of molecular mass of co-polymers on phase balance in the system polyethylene-co-polymer of ethylene with vinylacetate. *High molecular compounds*. 1987. Ser. B. v.29. №2. p.101-104. (in Russia)

[9] R.M.Khyzakhanov, O.V.Stoyanov, E.R.Mykhamedzyanova, S.N.Rusanova, A.E.Zaikin The properties of the industrial CEVA's mixtures. VUZ news. *Chemistry and chemical technology.* 2002. Vol.45. №.5. P.103-105. (in Russia)

[10] Lipatov Yu.S., Moisa E.G., Semenovitch G.N. The research of the macromolecules packing density in the frontier areas of the polymers // *Macromolecular compounds.* 1977. Ser. A. V.19. №1. P.125-128. (in Russia)

[11] Kuleznev V.N., Dogadkin B.A., Klykova V.D. // *Colloid journal.* 1968. V.30. №1. P.255-257. (in Russia)

[12] Letz J. // *J.Polym.Sci.* pt.A-2. 1970. V.8. №9. P.1415-1424.

[13] Lipatov Yu.S. Colloid chemistry of the polymers. Kiev: *Naykova dumka.* 1984. P.226-262. (in Russia)

[14] Lebedev E.V. *Polymer's modificators of the composite polymeric materials and foretransitive state of the polymer-polymeric systems.* Auto-abstract of the dissertation of doctor of chemical sciences. Kiev. IChVS, 1982. 34 p. (in Russia)

[15] Lipatov Yu.S., Lebedev E.V., Mamunya E.P., Gladureva N.A. The compound's influence on the change of the specific volume and the coefficient of the thermal extension in the compositions of polyethylene with polypropylene // *Composite polymeric materials.* 1982, №15, P. 3-5. (in Russia)

Chapter 11

PHYSICAL CHEMISTRY OF TOPOLOGICAL DISORDER IN POLYMERS. SOME REMARKS ON THE POLYMER MEMORY

Yu. A. Shlyapnikov and N. N. Kolesnikova
Institute of Biochemical Physics, 119991, Moscow, Russia

ABSTRACT

Long-term structural memory of a polymeric substance is determined by concentration, size and space distribution of the elements of topological disorder present in it. Short-term structural memory is the result of reversible reconstruction of existing elements.

It is well known that the properties of the polymer may depend on pre-history of a polymer sample [1-3]. We will show that there are two types of the structural memory: the long-term, which preserves at many influences on it, and short-term which disappears with time or slight heating.

According to the theory discussed in [4, 5], polymer may be considered as the ideally ordered substance composed from chain-like molecules in which various elements of disorder ξ_i are distributed. The content of these elements may be expressed in the units of concentration [4]:

$$[\xi_i] = (\text{the number of } \xi_i)(m_s N_A)^{-1} \text{ (mol/kg)} \qquad (1)$$

where m_s is the mass of the sample, and $N_A = 6.02 \times 10^{23}$ is the Avogadro number. The total concentration of the elements of disorder is the sum of $[\xi_i]$:

$$[\xi] = \sum_i [\xi_i] \qquad (2)$$

There are several ways to change these elements and their distribution, using which we may try to increase or decrease the total concentration of these elements and its distribution in size or shape. But first we must be able to evaluate both.

To investigate these elements of disorder the method of sorption of low molecular mass additives may be used. The long chain like macromolecules are virtually immobile and to form the solvent shell the additive molecule A must find the element containing the free volume sufficient for A molecule, the shape of which may be easily changed to fit for it. Such experiments were performed in several modifications, by saturation the polymer with A from gas or liquid phase. Using dual sortption model, we assume that the molecules of compound A dissolved in polymer divide into two groups: those truly dissolved in the ordered part of the polymer and present as the complexes with the certain centers Z_i :

$$A + Z_i \leftrightarrow AZ_i \tag{3}$$

Neglecting the concentration of A outside centers Z_i we may write:

$$[AZ_i] = \frac{K_{ai}[Zi][A]}{1+K_{ai}[A]} \tag{4}$$

and total concentration of A in the polymer $[A]_p$ will be:

$$[A]_p = [A] + \sum_i [AZ_i] = [A] + \sum_i \frac{K_{ai}[Zi][A]}{1+K_{ai}[A]} \tag{5}$$

Assuming that the real solubility of A in the ordered part of polymer is negligibly small compared to that in the sorption centers we may neglect [A] compared with the second term of the sum in (5). It was shown in [5, 6] that this second term at various assumption of $[Z_i]$ distribution may be approximated with Langmuir-type formula:

$$[A]_p = \frac{K_a[Z_a][A]}{1+K_a[A]}, \tag{6}$$

and, assuming that the concentration of truly dissolved A in the polymer is directly proportional to that in surrounding medium $[A]_m$:

$$[A] = \gamma_a[A]_m \tag{7}$$

we get the formula similar to (6) in which K_a must be substituted with $\gamma_a K_a$ and [A] with $[A]_m$. This formula written in inverse coordinates $1/[A]_p$ vs $1/[A]_m$ will describe the straight line intersecting the ordinate axis in the point equal to $1/[Z_a]$ – to the reciprocate concentration of centers sorbing compound A:

$$\frac{1}{[A_p]} = \frac{1}{\gamma_a K_a [Z_a]} \times \frac{1}{[A_m]} + \frac{1}{[Z_a]} \qquad (8)$$

To check this dependence the sorption of two compounds absorbing UV light from its vapors, namely of phenyl benzoate and phenyl-β-naphthylamine has been studied. Concentrations of both compounds in the gas phase were measured from their UV spectra.

As seen from Fig. 1, the sorption isotherms, i.e. the equilibrium dependencies of $[A]_p$ on $[A]_m$ at sorption of phenyl benzoate by polyethylene in coordinates of formula (8) are straight lines. It must be emphasized that the lines corresponding to various temperatures from 80° to 180° C intersect the ordinate axis in the same point, that is the limiting value of $1/[Z_a]$ and correspondingly the meaning of sorption centers concentration does not change with temperature, both below and above the melting temperature of the polymer (near 130° C). Such invariably structures can be only topological ones, i.e. based on various knots and interlacements of the polymer chains [4, 5].

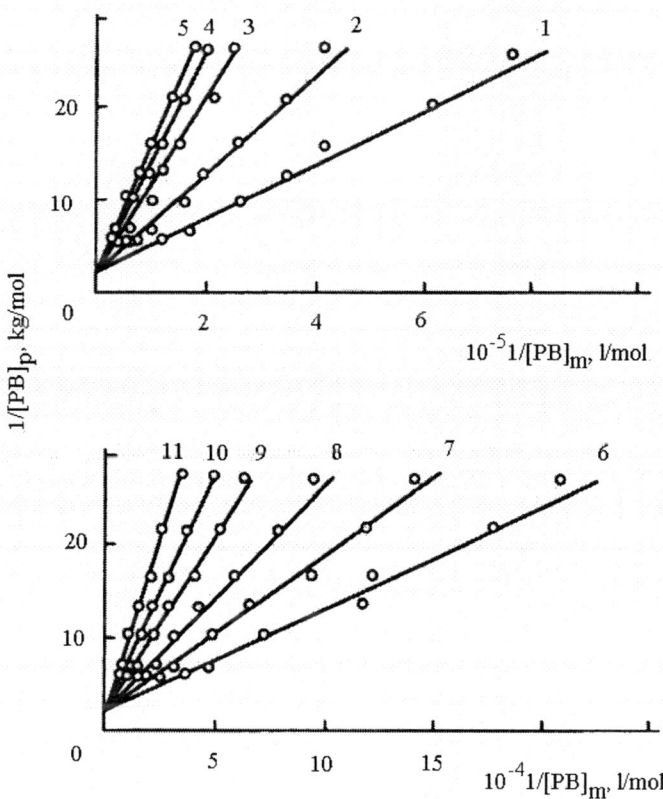

Figure 1. Sorption isotherms of phenyl benzoate by polyethylene from gas phase in the coordinates $1/[PB]_p$ vs $1/[PB]_m$ at the temperatures 80° (1), 90° (2), 100° (3), 110° (4), 120° (5), 130° (6), 140° (7), 150° (8), 160° (9), 170° (10), and 180° (11)

Some experiments were performed on sorption of additives by polyethylene and isotactic polypropylene from solutions of these additives in various solvents. In some studies the substitution of one additive with another that is the process:

$$AZ_i + B \rightarrow BZ_i + A \quad (9)$$

was observed [6]. In some experiments instead of sorption isotherms the limiting additive concentration, corresponding to the additive solubility $[A]_s = \gamma_{as}[A]_m$ has been measured. Similarly the solvent molecules (S) can substitute the additive in the sorption centers:

$$AZ_i + S \rightarrow SZ_i + A \quad (11)$$

In the absence of additive the molecules of solvent, including water, penetrate into the sorption centers. This is why the data of polymer density obtained by hydrostatic weighing are not reliable. The regularities of sorption from solution are the same as from the gas phase, only the values of $[Z_a]$ are considerably smaller because some centers occupied with the solvent molecules are inaccessible for the molecules of the additive. In these experiments it was shown polymer "remembers" from what solvent it was precipitated. Thus the meanings of $[Z_a]$ for isotactic polypropylene precipitated from n-decane was 9 times greater than that for the same polypropylene precipitated from chlorobenzene [7] (Fig. 2). Similarly the polypropylene samples precipitated from the same solvent – p-xylene by rapid cooling and by slow cooling markedly differ in their sorption capacity [8].

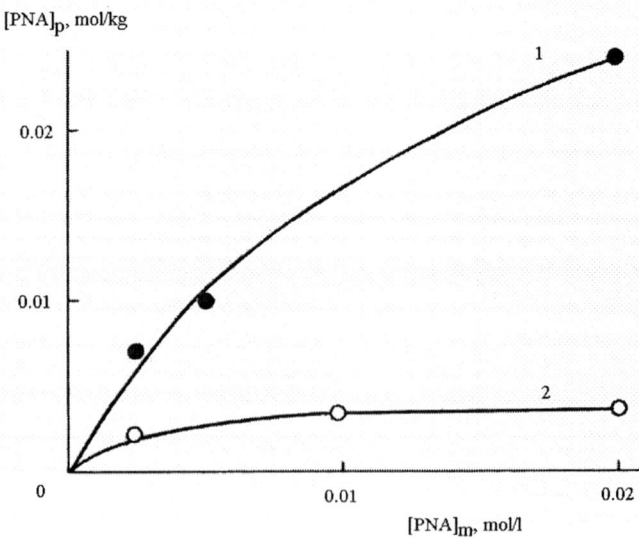

Figure 2. Sorption isotherms of phenyl-β-naphthylamine from heptane solutions by polypropylene precipitated from n-decane (1), and chlorobenzene (2). 40°C

Repeated melting with subsequent crystallization at 150°C leads to decrease of solubility $[A]_s$, the greater the less is molecular mass of the additive [9]. To explain we must assume that the greater is molecular mass, which in compounds consisting of light atoms H, C, N, and O means that the greater is the volume of the molecule, the more complicated must be the element of disorder sorbing it, but in the same time the more complicated is the element the more stable it is in the course of polymer crystallization. In this process the leading facets of the growing crystallites disentangle interlacements of the polymer chains, the easier the simpler this interlacement is. In another experiments polypropylene has been crystallized 12 times in vacuum (heating to 220° C, keeping at this temperature 1.5 hours., slow cooling to 150, keeping 1.5 hours, and cooling to room temperature). In Fig. 3 the sorption isotherms of phenyl-β-naphthylamine from alcohol by non-treated and treated samples are presented. Calculated from data of Fig. 3 the meanings of [Z_a] decreased from 2.5×10^{-2} to 1.3×10^{-3} mol/kg, i.e. 19 times [10]. When slow crystallization can change the sorption centers concentration, the kneading of the polymer melt by repeating pressing and folding does not change it. In our experiments [11] the polyethylene film was folded and pressed at 220° until the calculated formally thickness of initial film became 8×10^{-10} cm, i.e. initial polymer structure was completely destroyed. As seen from Fig. 4, the sorption isotherms of phenyl-β-naphthylamine for initial and treated so polyethylene films virtually coincided: the elements of disorder responsible for sorption centers formation moved during polymer pressing without disentanglement.

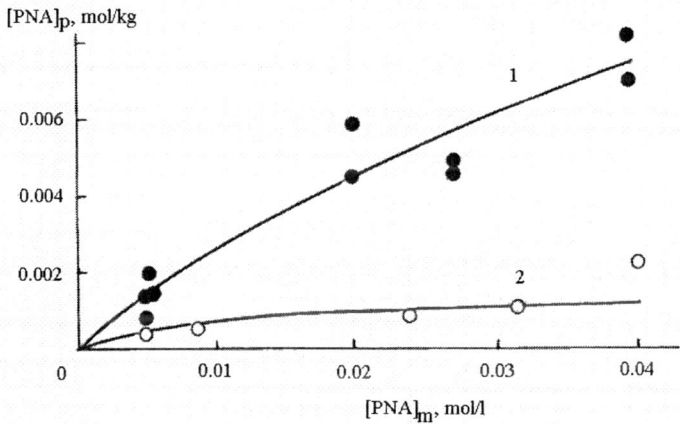

Figure 3. Sorption isotherms of phenyl-β-naphthylamine by isotactic polypropylene from alcohol solutions. 50° C, 1 – usual polymer, 2 – polymer aftere 12 cycles of melting-crystallization [9]

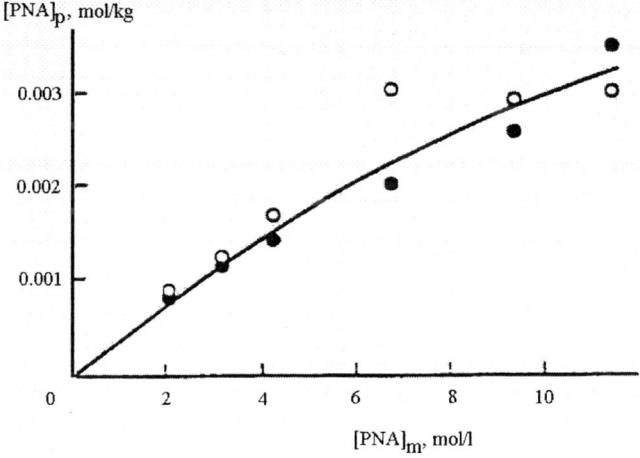

Figure 4. Sorption isotherm of phenyl-β-naphthylamine by polyethylene from heptane solutions. 40° C, light circles – non treated polymer, dark circles – polyethylene after multiple kneading at 220° C

Heating the polymer mixed with the large amount of additive above the melting temperature can temporally change the sorption centers. As seen from Fig. 5, after heating the isotactic polypropylene powder till 220° C with various initial concentrations of phenyl-β-naphthylamine with subsequent cooling to 40° C in the saturated vapors of the same additive the concentration of phenyl-β-naphthylamine in the sample rapidly decreases to a limit depending on initial additive concentration. Analogously changed the concentration of phenyl benzoate (Fig.6).

This limit rises approximately linearly after the certain meaning of initial additive concentration (Fig. 7)., and also depends on the method of polymer precipitation [12, 13]. These additive concentrations in the samples heated in the saturated additive vapors correspond to the additive solubility which, as we may see, depends on the sample pre-history.

The similar dependencies were observed using another additive - phenyl benzoate. To explain this we supposed that at high additive concentration the elements of disorder are transformed into the unstable sorption centers:

$$\xi + A \rightarrow Z^*_{ai} A \tag{10}$$

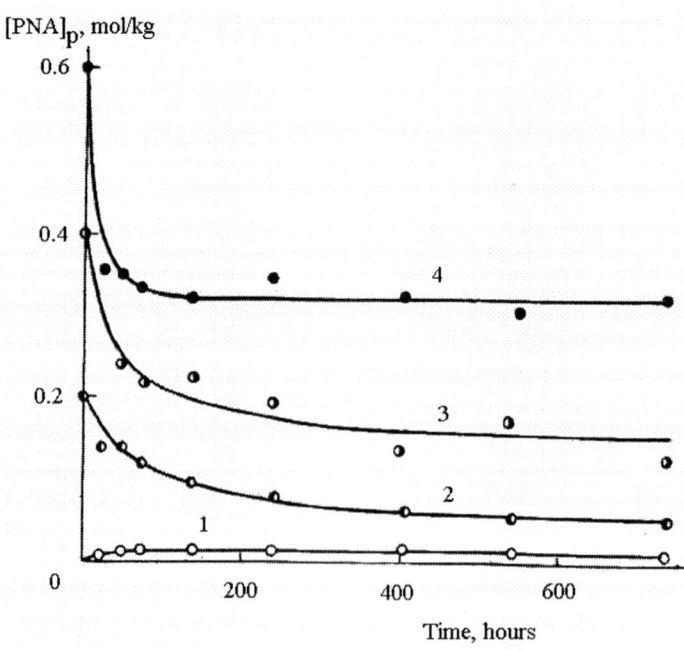

Figure 5. Variation of concentration of phenyl-β-naphthylamine in isotactic polypropylene samples during heating in saturated vapors of this compounds. 40°C, initial additive concentrations are: 0 (1), 0.2 (2), 0.4 (3), and 0.6 mol/kg (4)

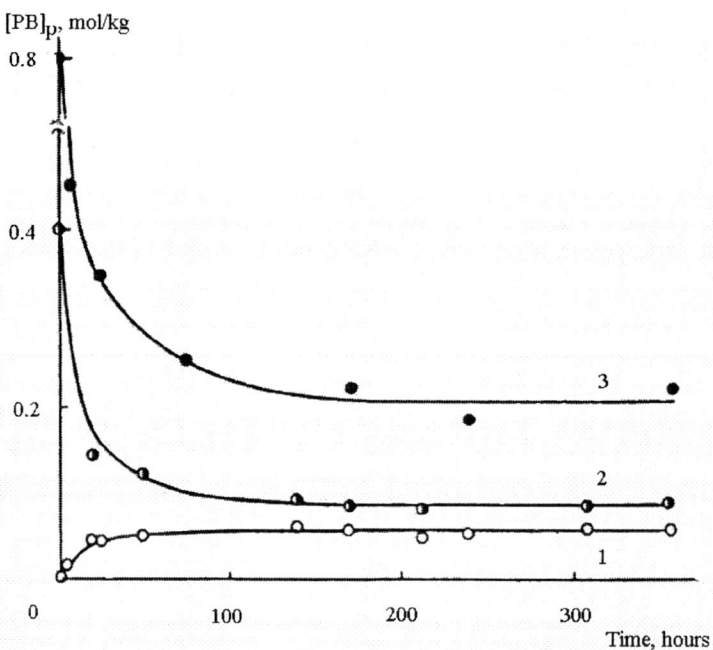

Figure 6. Variation of phenyl benzoate concentration in isotactic polypropylene samples during heating in saturated vapors of this compound. $40°C$, initial additive concentrations are 0 (1), 0.4 (2), and 0.6 mol/kg (3)

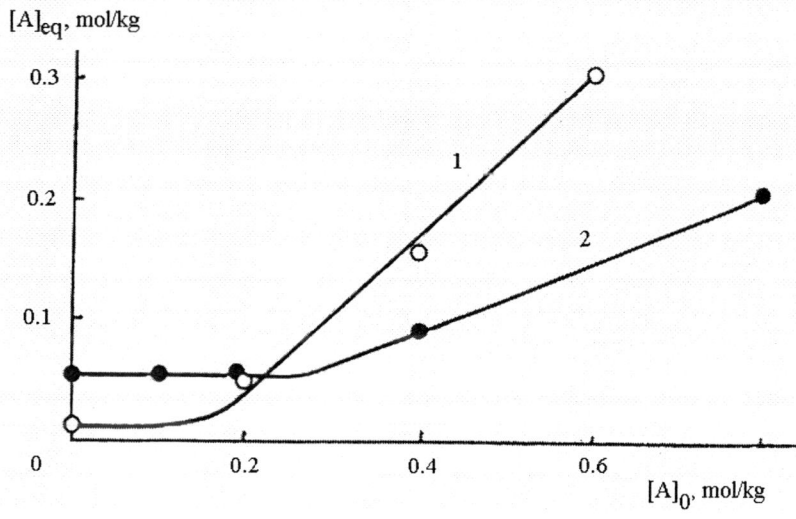

Figure 7. Limiting concentrations of phenyl-β-naphthylamine (1), and phenyl benzoate (2) in isotactic polypropylene as functions of initial additive concentrations in the polymer. $40°C$

The new formed centers possess different stability: a part of them destroy already at low temperatures ($40°$ C), to destroy other elements we must heat the polymer at least to $100°$ C (Fig. 8 [13]). These dependencies demonstrate the examples of short time memory.

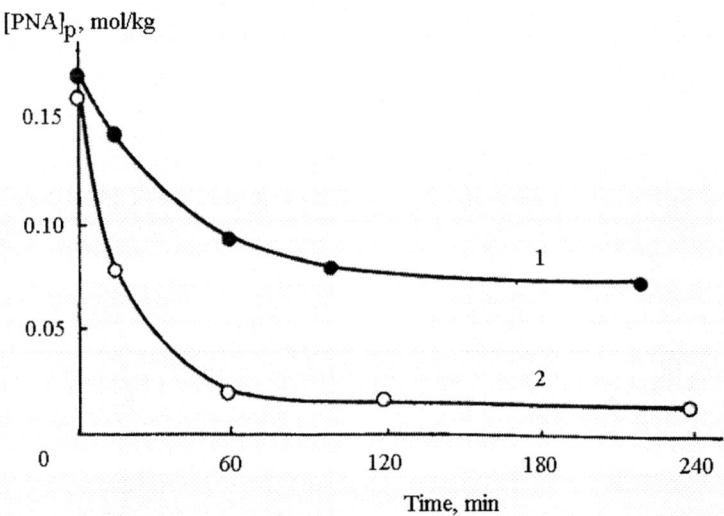

Figure 8. Variation of solubility of phenyl-β-naphthylamine in isotactic polypropylene samples prepared from the melt containing 0.6 mol/kg of this additive during heating at 80° C (1) and 100° C (2) in saturated additive vapors

Thus the total concentration and size distribution of the elements of topological disorder is formed at polymer precipitation and can be decreased by another sedimentation or by special treatment of the polymer – by repeated crystallization, but the properties of these elements may be reversibly changed by treating the polymer with low molecular additives. The processes of relaxation of these elements can differ in their rates.

Here we do not consider the oriented state of polymers.

REFERENCES

[1] Bastiaansen, C.W.M., Meyer, H.E.H., Lemstra, P.G., *Polymer*, 1990, 31(8), p.1435.
[2] Rokudai M.G., *J. Appl. Polym. Sci.*, 1979, 26, 32898.
[3] Lemstra, P.G., Kirshbaum, R, *Polymer*, 1985, 26, 1372.
[4] Shlyapnikov Yu.A., Doklady Akad. *Nauk* SSSR, 1972, 202, 1377.
[5] Shlyapnikov Yu.A., *Russian Chemical Reviews*, 1997, 66 (11), 963,
[6] Shlyapnikov, Yu.A., Giedraityte, G.B., *Polym. Degrade. Stab.*, 1997, 56, 281.
[7] Monakhova, T.V., Bogaevskaya, T.A., Shlyapnikov, Yu.A., Vysokomolek. Soedin., 1995, 37(1), 160.
[8] Shlyapnikov Yu.A., Kolesnikova N.N., *J. Appl. Polym. Sci.*, 1998, 69, 1847.
[9] Tyuleneva, N.K., Shlyapnikov, Yu.A., *Khimicheskaya Fizika* (Rus. Chem.Phys.) 2001, 20(3), 94.
[10] Shlyapnikov, Yu.A., Kolesnikova, N.N., *J. Appl. Polym. Sci.*, 1998, 69, 1847.
[11] Shlyapnikov, Yu.A., Kolesnikova, N.N., *Chem. Phys. Reports*, 1999, 18(2), 313.
[12] Shlyapnikov, Yu.A., Kolesnikova, N.N., *Intertn. J. Polymeric Mater.* 1996, 32, 75.
[13] Shlyapnikov, Yu.A., Kolesnikova, N.N., *Polym. Plast. Technol. Eng.*, 1999, 38(2), 371.

In: Chemistry as Art
Editors: L. Liu and G. E. Zaikov, pp. 267-276
ISBN 1-59454-585-5
© 2006 Nova Science Publishers, Inc.

Chapter 12

ON THE STRUCTURAL MEMORY IN POLYMERS

N. N. Kolesnikova and Yu. A. Shlyapnikov
Institute of Biochemical Physics, 119991, Moscow, Russia

ABSTRACT

Long-term structural memory of a polymeric substance is determined by concentration, size and space distribution of the elements of topological disorder present in it. Short-term structural memory is the result of reversible reconstruction of existing elements.

It is well known that the properties of the polymer may depend on pre-history of a polymer sample [1-3]. We will consider here two types of the structural memory of a polymer: the long-term, which preserves at many influences on it, and short-term which disappears with time or slight heating. We will not discuss the oriented state of polymers.

According to the theory discussed in [4, 5], polymer may be considered as the ideally ordered substance composed from chain-like molecules in which various elements of disorder ξ_i are distributed. The content of these elements may be expressed in the units of concentration [4]:

$$[\xi_i] = (\text{the number of } \xi_i)(m_s N_A)^{-1} \text{ (mol/kg)} \tag{1}$$

where m_s is the mass of the sample, and $N_A = 6.02 \times 10^{23}$ is the Avogadro number. The total concentration of the elements of disorder is the sum of $[\xi_i]$:

$$[\xi] = \sum_i [\xi_i] \tag{2}$$

There are several ways to change these elements and their distribution, using which we may try to increase or decrease the total concentration of these elements and its distribution in size or shape. But first we must be able to evaluate both.

To investigate these elements of disorder the method of sorption of low molecular mass additives may be used. The long chain like macromolecules are virtually immobile and to form the solvent shell the additive molecule A must find the element containing the free volume sufficient for A molecule, the shape of which may be easily changed to fit for it. Such experiments were performed in several modifications, by saturation the polymer with A from gas or liquid phase. Using dual sortption model, we assume that the molecules of compound A dissolved in polymer divide into two groups: those truly dissolved in the ordered part of the polymer and present as the complexes with the certain centers Z_i. It is necessary to discuss the basic principles of the method.

The sorption of additive A by a certain center Z_i may be written as:

$$A + Z_i \leftrightarrow AZ_i \tag{3}$$

To sorb the molecule A the center Z_i must possess enough volume accessible for the sorbed molecule. If the monomeric unit of the polymer is comparable in its volume with the A molecule additional problem of compatibility of shapes of these elements may greatly complicate the sorption, but in the simplest case of sorption by polyethylene there has been found the relatively simple dependence.

Let the element ζ_i contains the free (excess) volume v_i. Assume that the energy of formation of this empty volume is directly proportional to this volume and, if the process of polymeric substance formation proceeds in of solvent S is $\varepsilon_i = v_i(q_p - q_s)$, where q_p and q_s are the energies of monomeric units interaction with the same units and the solvent molecules correspondingly divided by volume.. According to the Boltzmann Law, the ratio of the elements formation of which needs the energy $\varepsilon_i \geq \varepsilon_A$ (otherwise $v_i \geq v_A$) will be:

$$P_A = C\exp(\frac{-\varepsilon_A}{kT}) = C\exp(\frac{-v_A(q_p - q_s)}{RT}), \tag{4}$$

For future consideration we make the natural assumption that the volume of the center Z_A sorbing compound A is proportional to the volume of the molecule of A in the pure state ($v_z = \beta v_A$). Organic compounds used in our investigations have the density in pure state in the range $\rho_A \approx 1.1 \pm 0.1$ which corresponds to the volume the separate molecule with molecular mass M_A $v_A = M_A(\rho_A N_A)^{-1}$. Combining (XX) and (XX), we get the simple formula:

$$[Z_A] = [Z]_o\exp(-\kappa M_A) \tag{5}$$

where

$$\kappa = \frac{\beta(q_p - q_s)}{\rho N_A kT_z} \tag{6}$$

Here T_z is the temperature of the sorption center formation, and $[Z]_o$ is the total concentrations of the elements ζ. It follows from (6) that the sorption centers concentration must depend on the energy of interaction of polymer with the solvent q_s and on the temperature of the polymer sedimentation T_z.

Then we need the method of evaluation of $[Z_i]$. Neglecting the concentration of A outside centers Z_i, it follows from (3):

$$[AZ_i] = \frac{K_{ai}[Zi][A]}{1+K_{ai}[A]}, \qquad (7)$$

and total concentration of A in the polymer $[A]_p$ will be:

$$[A]_p = [A] + \sum_i [AZ_i] = [A] + \sum_i \frac{K_{ai}[Zi][A]}{1+K_{ai}[A]} \qquad (5)$$

Assuming that the real solubility of A in the ordered part of polymer is negligibly small compared to that in the sorption centers we may neglect [A] compared with the second term of the sum in (5). It was shown in [5, 6] that this second term at various assumption of $[Z_i]$ distribution may be approximated with Langmuir-type formula:

$$[A]_p = \frac{K_a[Z_a][A]}{1+K_a[A]}, \qquad (6)$$

and, supposing that the concentration of truly dissolved A in the polymer is directly proportional to that in surrounding medium $[A]_m$:

$$[A] = \gamma_a [A]_m \qquad (7)$$

we get the formula similar to (6) in which K_a must be substituted with $\gamma_a K_a$ and [A] with $[A]_m$. This formula written in inverse coordinates $1/[A]_p$ vs $1/[A]_m$ will describe the straight line intersecting the ordinate axis in the point equal to $1/[Z_a]$ – to the reciprocate concentration of centers sorbing compound A:

$$\frac{1}{[A_p]} = \frac{1}{\gamma_a K_a [Z_a]} \times \frac{1}{[A_m]} + \frac{1}{[Z_a]} \qquad (8)$$

To check this dependence the sorption of two compounds absorbing UV light from its vapors, namely of phenyl benzoate and phenyl-β-naphthylamine has been studied. Concentrations of both compounds in the gas phase was calculated from their UV spectra.

As seen from Fig. 1, the sorption isotherms, i.e. the equilibrium dependencies of $[A]_p$ on $[A]_m$ at sorption of phenyl benzoate by polyethylene in coordinates of formula (8) are straight lines. It must be emphasized that the lines corresponding to various temperatures from 80° to

180° C intersect the ordinate axis in the same point, that is the limiting value of $1/[Z_a]$ and correspondingly the meaning of sorption centers concentration does not change with temperature, both below and above the melting temperature of the polymer (near 130° C). Such invariably structures can be only topological ones, i.e. based on various knots and interlacements of the polymer chains [4, 5].

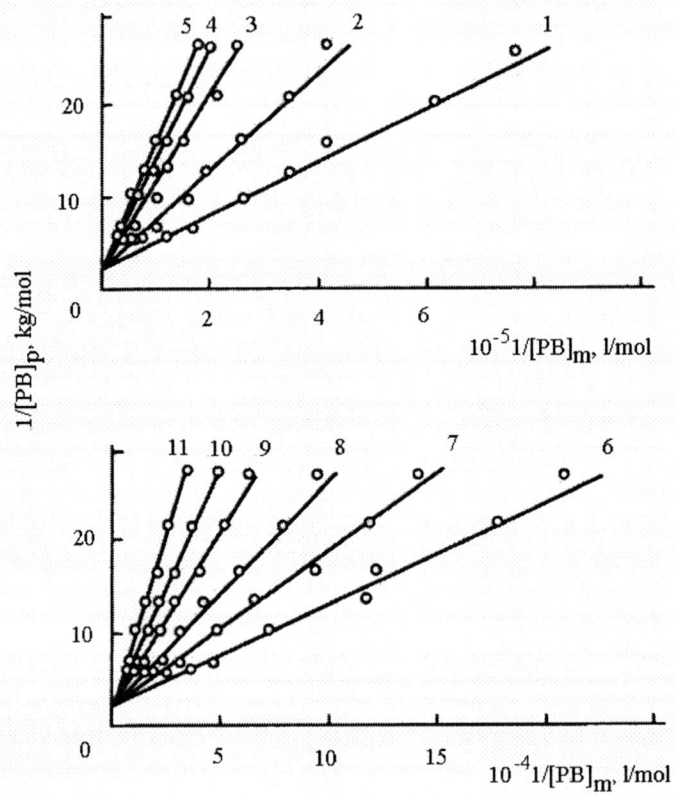

Figure 1. Sorption isotherms of phenyl benzoate by polyethylene from gas phase in the coordinates $1/[PB]_p$ vs $1/[PB]_m$ at the temperatures 80° (1), 90° (2), 100° (3), 110° (4), 120° (5), 130° (6), 140° (7), 150° (8), 160° (9), 170° (10), and 180° (11)

Some experiments were performed on sorption of additives by polyethylene and isotactic polypropylene from solutions of these additives in various solvents. In some studies the substitution of one additive with another that is the process:

$$AZ_i + B \rightarrow BZ_i + A \qquad (9)$$

was observed (Fig. 2) [6]. In some experiments instead of sorption isotherms the limiting additive concentration, corresponding to the additive solubility $[A]_s = \gamma_{as}[A]_m$ has been measured. Similarly the solvent molecules (S) can substitute the additive in the sorption centers:

$$AZ_i + S \rightarrow SZ_i + A \qquad (11)$$

In the absence of additive the molecules of solvent, including water, penetrate into the sorption centers. This is why the data of polymer density obtained by hydrostatic weighing are not reliable.

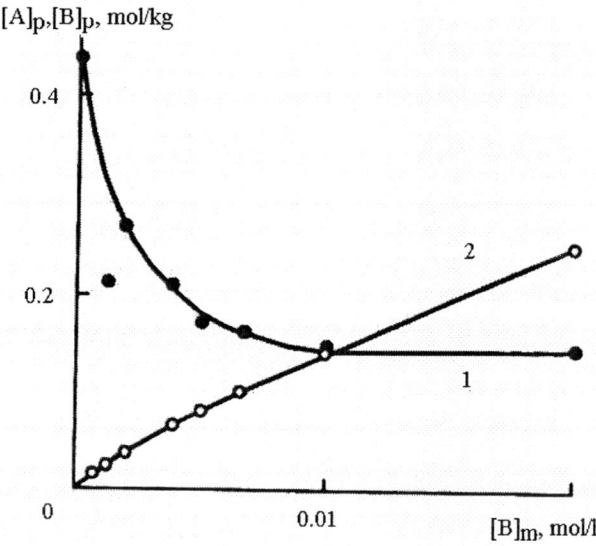

Figure 2. Dependencies of concentration of α-naphthol $[A]_p$ (1) and phenyl benzoate $[B]_p$ (2) in high density polyethylene on the concentration of phenyl benzoate in hexane solution $[B]_m$ at $50°C$ at α-constant naphthol concentration $[A]_m = 0.2$ mol/l

The regularities of sorption from solution are the same as from the gas phase, only the values of $[Z_a]$ are considerably smaller because some centers occupied with the solvent molecules are inaccessible for the molecules of the additive. In these experiments it was shown polymer "remembers" from what solvent it was precipitated. Thus, according to (6), the meanings of $[Z_a]$ for high moleculasr hydrocarbon, isotactic polypropylene, precipitated from low molecular hydrocarbon, n-decane was 9 times greater than that for the same polypropylene precipitated from foreign compound, chlorobenzene [7] (Fig.3). Similarly the polypropylene samples precipitated from the same solvent – p-xylene by rapid cooling and by slow cooling, i.e. when polymer formation proceeds at different T_z, markedly differ in their sorption capacity [8].

Repeated melting with subsequent crystallization at $150°C$ leads to decrease of solubility $[A]_s$, the greater the less is molecular mass of the additive [9]. To explain we must assume that the greater is molecular mass, which in compounds consisting of light atoms H, C, N, and O means that the greater is the volume of the molecule, the more complicated must be the element of disorder sorbing it, but in the same time the more complicated is the element the more stable it is in the course of polymer crystallization. In this process the leading facets of the growing crystallites disentangle interlacements of the polymer chains, the easier the simpler this interlacement is. In another experiments polypropylene has been crystallized 12 times in vacuum (heating to $220°C$, keeping at this temperature 1.5 hours., slow cooling to 150, keeping 1.5 hours, and cooling to room temperature). In Fig. 4 the sorption isotherms of phenyl-β-naphthylamine from alcohol by non-treated and treated samples are presented. Calculated from data of Fig. 4 the meanings of $[Z_a]$ decreased from 2.5×10^{-2} to 1.3×10^{-3}

mol/kg, i.e. 19 times [9]. On the other hand, the kneading of the polymer melt by repeating pressing and folding does not change it. In our experiments [11] the polyethylene film was folded and pressed at 220° until the calculated formally thickness of initial film became 8×10^{-10} cm, i.e. initial polymer structure was completely destroyed.

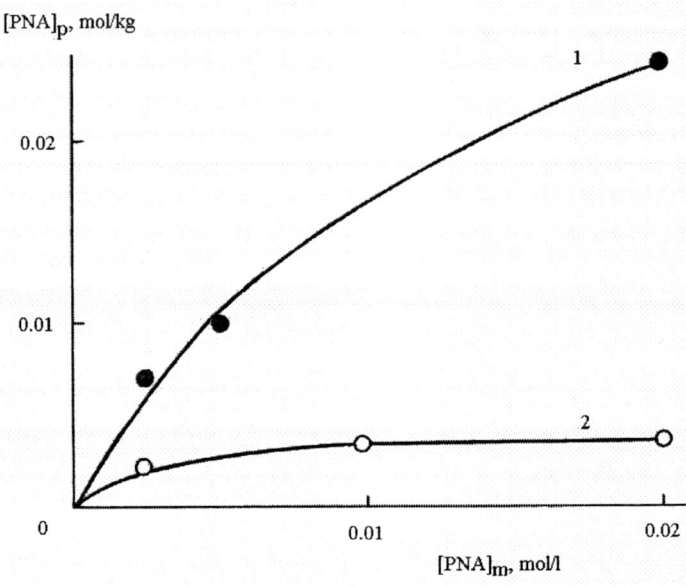

Figure 3. Sorption isotherms of phenyl-β-naphthylamine from heptane solutions by polypropylene precipitated from n-decane (1), and chlorobenzene (2). 40°C

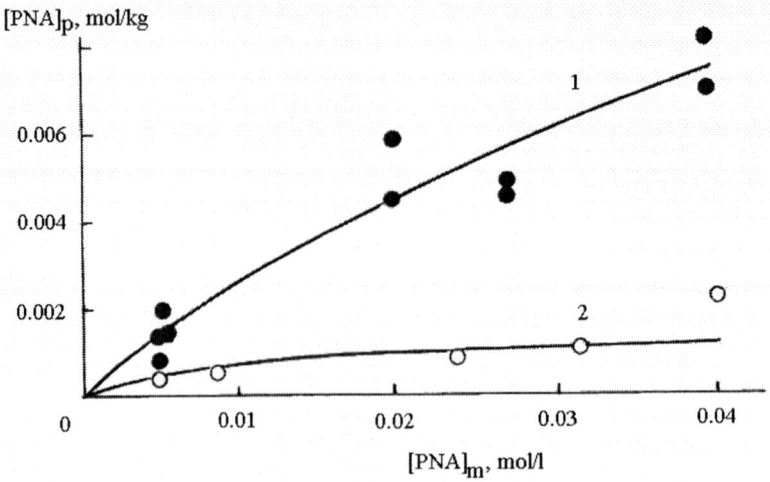

Figure 4. Sortption of phenyl-β-naphthylamine by polypropylene from alcohol solutions. 50°C, 1 – usual polymer, 2 – polymer aftere 12 cycles of melting-crystallization [9]

As seen from Fig. 5, the sorption isotherms of phenyl-β-naphthylamine for initial and treated so polyethylene films virtually coincided: the elements of disorder responsible for sorption centers formation moved during polymer pressing without disentanglement.

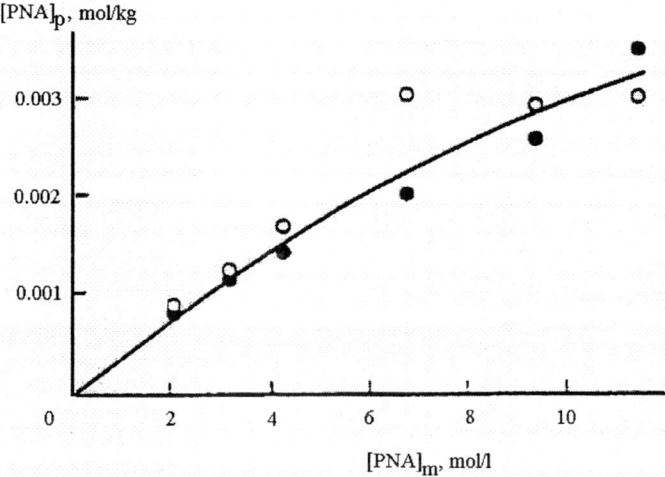

Figure 5. Sorption isotherm of phenyl-β-naphthylamine by polyethylene from heptane solutions. 40° C, light circles – non treated polymer, dark circles – polyethylene after multiple kneading at 220°C [9]

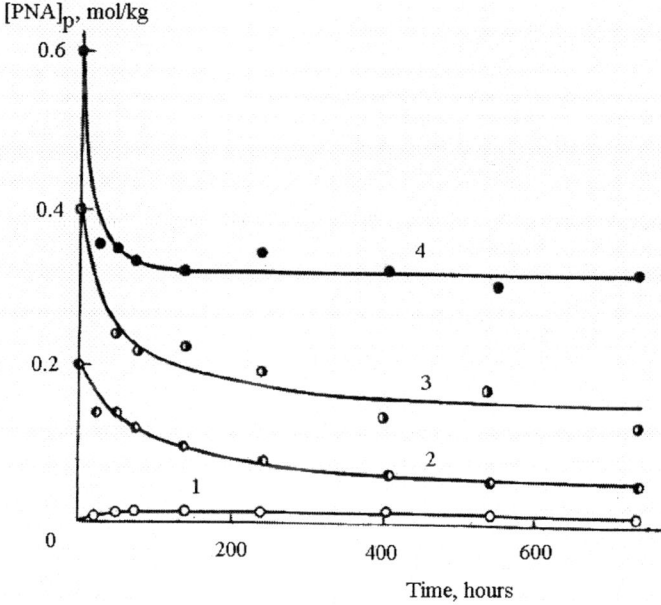

Figure 6. Variation of concentration of phenyl-β-naphthylamine in isotactic polypropylene samples during heating in saturated vapors of this compounds. 40°C, initial additive concentrations are: 0 (1), 0.2 (2), 0.4 (3), and 0.6 mol/kg (4)

Heating the polymer mixed with the large amount of additive above the melting temperature can temporally change the sorption centers. As seen from Fig.6, after heating the isotactic polypropylene powder till 220° C with various initial concentrations of phenyl-β-naphthylamine with subsequent cooling to 40° C in the saturated vapors of the same additive the concentration of phenyl-β-naphthylamine in the sample rapidly decreases to a limit depending on initial additive concentration. Analogously changed the concentration of phenyl benzoate (Fig 7).

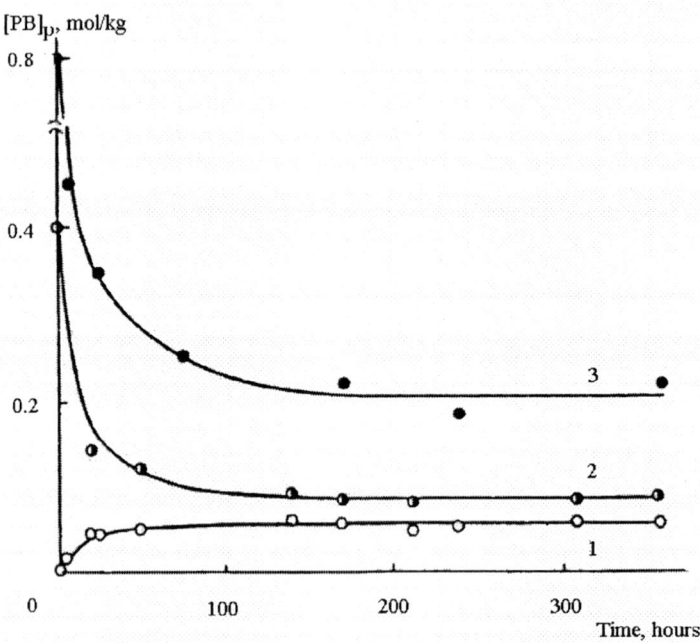

Figure 7. Variation of phenyl benzoate concentration in isotactic polypropylene samples during heating in saturated vapors of this compound. 40°C, initial additive concentrations are 0 (1), 0.4 (2), and 0.6 mol/kg (3)

This limit rises approximately linearly after the certain meaning of initial additive concentration (Fig. 8)., and also depends on the method of polymer precipitation [12, 13]. These additive concentrations in the samples heated in the saturated additive vapors correspond to the additive solubility which, as we may see, depends on the sample prehistory. The similar dependencies were observed using another additive - phenyl benzoate. To explain this we supposed that at high additive concentration the elements of disorder are transformed into the unstable sorption centers:

$$\xi + A \rightarrow Z^*_{ai} A \qquad (10)$$

The new formed centers possess different stability: a part of them destroy already at low temperatures (40° C), to destroy other elements we must heat the polymer at least to 100° C (Fig. 9) [13]. These dependencies demonstrate the examples of short time memory.

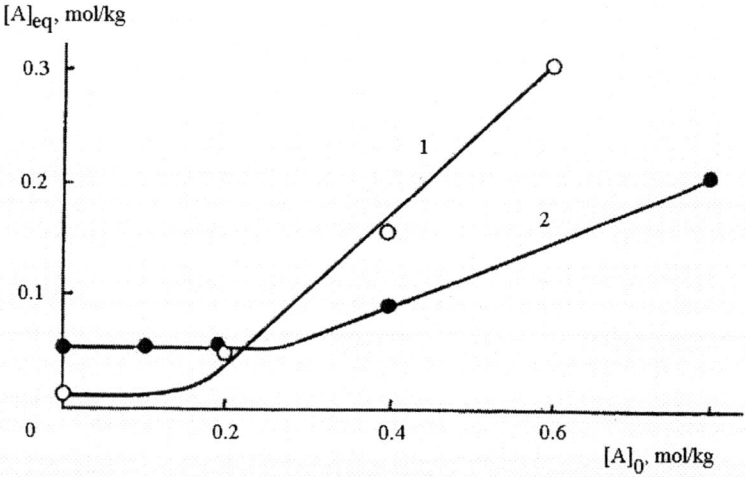

Figure 8. Limiting concentrations of phenyl-β-naphthylamine (1), and phenyl benzoate (2) in isotactic polypropylene as functions of initial additive concentrations in the polymer. 40°C

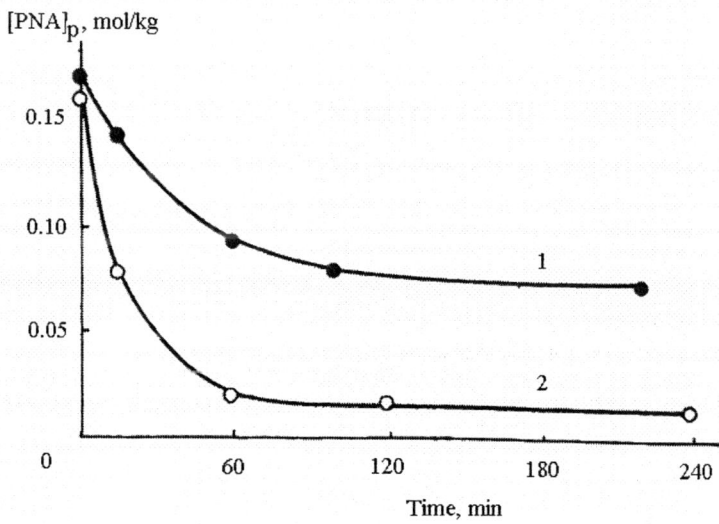

Figure 9. Variation of solubility of phenyl-β-naphthylamine in isotactic polypropylene samples prepared from the melt containing 0.6 mol/kg of this additive during heating at 80°C (1) and 100°C (2) in saturated additive vapors

Thus, the total concentration and size distribution of the elements of topological disorder is formed at polymer precipitation and can be decreased by sedimentation at another condition or by special treatment of the polymer – by repeated crystallization, but the properties of these elements may be reversibly changed by treating the polymer with low molecular additives. The processes of relaxation of these elements can differ in their rates.

REFERENCES

[1] Bastiaansen, C.W.M., Meyer, H.E.H., Lemstra, P.G., *Polymer*, 1990, 31(8), p.1435.
[2] Rokudai M.G., *J. Appl. Polym. Sci.*, 1979, 26, 32898.
[3] Lemstra, P.G., Kirshbaum, R, *Polymer*, 1985, 26, 1372.
[4] Shlyapnikov Yu.A., Doklady Akad. *Nauk* SSSR, 1972, 202, 1377.
[5] Shlyapnikov Yu.A., *Russian Chemical Reviews*, 1997, 66(11), 963.
[6] Shlyapnikov, Yu.A., Giedraityte, G.B., *Polym. Degrade. Stab.*, 1997, 56, 281.
[7] Monakhova, T.V., Bogaevskaya, T.A., Shlyapnikov, Yu.A., *Vysokomolek. Soedin.*, 1995, 37(1), 160.
[8] Shlyapnikov Yu.A., Kolesnikova N.N., *J. Appl. Polym. Sci.,* 1998, 69, 1847.
[9] Tyuleneva, N.K., Shlyapnikov, Yu.A., *Khimicheskaya Fizika* (Rus. Chem.Phys.) 2001, 20(3), 94.
[10] Shlyapnikov, Yu.A., Kolesnikova, N.N., *J. Appl. Polym. Sci.*, 1998, 69, 1847.
[11] Shlyapnikov, Yu.A., Kolesnikova, N.N., *Chem. Phys. Reports*, 1999, 18(2), 313.
[12] Shlyapnikov, Yu.A., Kolesnikova, N.N., *Intertn. J. Polymeric Mater.* 1996, 32, 75.
[13] Shlyapnikov, Yu.A., Kolesnikova, N.N., *Polym. Plast. Technol. Eng.*, 1999, 38(2), 371.

Chapter 13

DESCRIPTION OF CARBON PLASTICS THE HEAT CONDUCTIVITY WITHIN THE FRAMEWORK OF FRACTAL ANALYSIS

G. V. Kozlov., A. I. Burya[1], I. V. Dolbin[2] and G. E. Zaikov[3]

[1]State Agrarian University, Dnepropetrovsk – 49027, Voroshilov str., 25, Ukraine
[2]Researching Institute of Applied Mathematics and Automatization of Kabardino-Balkarian Scientific Center of Russian Academy of Science, Nal'chik – 360000, Shortanov str., 89 "a", Russian Federation
[3]Institute of Biochemical Physical of Russian Academy of Science Moscow – 119991, Kosygin str., 4, Russian Federation

ABSTRACT

It is shown that the heat conductivity of carbon plastics on the basis of phenylone can be described within the framework of a fractal model. Depending on the dimension of filler fibres network (system) such description can be obtained by the application of two limiting cases: random network of resistors (RNR) or random superconducting network (RSN).

Keywords: Composites, carbon plastics, short fibers, heat conductivity, fractal model.

INTRODUCTION

At the theoretical analysis of the conductivity phenomena in composite solid medium the general and the inevitable is the assumption of the full geometric order in the distribution of phases. It is assumed that the fibres are distributed in a matrix evenly, at the same distance and parallel with each another. However, the real composite materials producing as a result of the performation of the whole complex of technological operations have the structure considerably differing from an ideal model [1]. The microscopic studies of the real composite materials show convincingly enough the noneven distribution of fibres, the deviation from the mutual parallelism and the existence of porosity. Besides, the insufficient knowledge of the filler and polymeric matrix properties imposes in its turn the additional limitations on the possibility of the application of theoretical equations for the prediction of the thermodynamic properties of composite materials [1]. That's why the following simple way is often used. In the equation for the calculation of the conductivity coefficients instead of the real physical values characterizing the separate components of the composite material and instead of the variables co-ordinating the experimental and the calculated data the volume conductivity coefficient of the fibre system can be introduced, which takes into account not only the physical properties but the geometrical features of the composite material. Such an approach is not new: fibrous and porous insulants are usually characterized by their volumetric properties [1].

Within the framework of the fractal analysis for the description of particles system (particles aggregate) of a filler the fractal dimension of the network of filler particles D_N is used, which characterizes the density of filling of the polymeric matrix space by particles or fibres of a filler [2, 3]. Then the fractal model considers a random mixture of components A and B, in which there are well and badly conducting area [4]. Such a model entirely corresponds to the polymer composites whose heat conductivities of carbon fibres and polymeric matrix can differ by three orders of value [1]. Two limiting cases of this task [4] claim the special attention:

(1) the random network of resistors RNR. In this case it is assumed that the areas occupied by bad conductor B have the zero conductivity;
(2) the random superconducting network (RSN). In this case the conductivity of good conductor A is infinite.

The purpose of the present paper is the study of the heat conductivity of carbon plastics on the basis of phenylone within the framework of the considered above fractal model [4].

EXPERIMENTAL

As the polymer matrix the aromatic polyamide-phenylone [5] is used and as the filler-carbon fibres (CF) with the diameter 7-9 mcm and the length 3mcm. The mass content of CF makes up 15 % that corresponds to the nominal volume content $\varphi_f \approx 0.115$. The composites are produced by "dry" method including the blending of components in a rotating electromagnetic field. For this in the reactor powdery polymer, CF and nonequiaxial ferromagnetic particles with length of 40 mm were placed. Then the reactor was placed in the end window of the generator of the electromagnetic apparatus. Under the action of rotating electromagnetic field the ferromagnetic particles begin to rotate, colliding between themselves, that results to the equipartition (chaotic) distribution of CF in a polymer matrix. In the result of the collisions the particles are worn down and the products of the wear fall into the composition. For taking away ferromagnetic particles after blending two methods are used: magnetic and mechanical separation [6].

The measurement of the heat conductivity λ were made according to the state standard for plastics on the apparatus UT-λ-400 in the interval of temperatures 323-523 K with interval 25 K. The measurement error is equal to ± 3%.

The thermal properties were determined on the differential scanning colorimeter of model UT-S-400 at heating rate 10 K/min.

RESULTS AND DISCUSSION

As it was shown in paper [4] the heat conductivity λ for two considered above limiting cases is given by the following relationships:

$$\lambda \sim L^{d_u} \tag{1}$$

for RNR and

$$\lambda \sim L^{d_w - D_N} \tag{2}$$

for RSN.

In the relationships (1) and (2) L is a cluster size, d_u is a fractal dimension of its unscreened perimeter, d_w is a walk dimension of fractal.

The necessary for further calculations dimensions can be calculated according to the following method. The value of d_u is determined according to the equation [4]:

$$d_u = (D_N - 1) + (d - D_N)/d_w, \tag{3}$$

where d is a dimension of Euclidean space, in which the fractal is considered (it is obvious, that in our case $d=3$) and the dimension d_w can be estimated according to Alexander-Orbach rule [7]:

$$d_w = \frac{3}{2} D_N. \tag{4}$$

And finally, the dimension D_N for the studied carbon plastics can be calculated according to the equation [3]:

$$D_N = 2 + \frac{\varphi_{int} d_{surf}}{1.20}, \tag{5}$$

were φ_{int} is a relative fraction of the interfacial regions, d_{surf} is a fractal of the surface of filler fibre. As for the considered carbon plastics in virtue of the production technology the aggregation process of fibres is expressed weakly [8] then it was accepted d_{surf}=const=2.13 [9]. In its turn, the value φ_{int} is determined according to the equation [10]:

$$\varphi_{int} = 1 - \frac{\Delta C_p^c}{\Delta C_p^p}, \qquad (6)$$

where ΔC_p^c and ΔC_p^p are the values of jump of the heat capacity at the constant pressure at the glass transition temperature for composite and matrix polymer, accordingly.

In Fig. 1 the dependence λ on parameter L^{d_u} where value L is accepted arbitrary equal to 5 relative units is shown. As it can be seen, in case of the composite modeling by RNR the half of data points lies down on the straight line passing through the origin (in Figure it is not shown), i.e., it agrees with the mentioned model but the half of points has the large scatter, i.e., it does not agree with the model of RNR. This circumstance allows to assume that second group of points can be described by RSN model. Actually, the adduced on Fig. 2 dependence $\lambda(L^\xi)$, where ξ is a conductivity exponent, which is equal to d_u in RNR case and $(d_w - D_N)$ in RSN case, confirms this assumption. The obtained data are approximated by one straight line for both mentioned models, which passing through the origin. The definite scatter of the data for the dependence $\lambda(L^\xi)$ can be due, as minimum, to two factors: the arbitrary choice of L and the condition L=const and also approximate estimation d_w according to the Alexander-Orbach rule. Another cause of the mentioned scatter can be the variations of the heat conductivity of a polymer matrix.

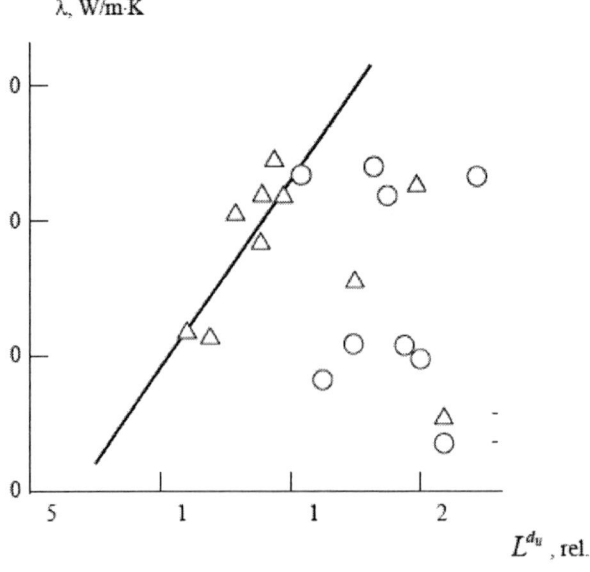

Figure 1. The dependence of heat conductivity λ on parameter L^{d_u} for carbon plastics on the basis of phenilone produced with using of magnetic (1) and mechanical (2) separation

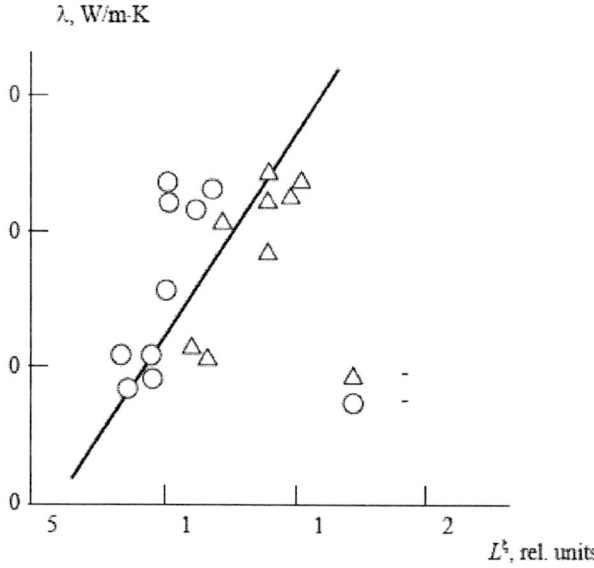

Figure 2. The dependence of heat conductivity λ on parameter L^ξ at $\xi=d_u$ (1) and $\xi=(d_w - D_N)$ (2) for carbon plastics on the basis of phenilone

The transition from one model of the heat conductivity of composites to another occurs at $D_N \approx 2.62$: at $D_N < 2.62$ the RNR model is correct, and at $D_N > 2.62$ the RSN model is correct. It is necessary to mark that the value D_N is connected to the operating parameter of the synergetic structure of carbon plastics – the factor of fibres orientation η by the following simple relationship [11]:

$$\eta = 0.506(D_N - 2). \qquad (7)$$

Then, from the equation (7) it follows that the RNR model is correct for $\eta < 0.313$ and the RSN model – for $\eta > 0.313$.

Let's consider the physical premises of the observed transition from the RNR model to the RSN. In the RNR limit the heat transport does not have the geometric limitations and can be realized both in a polymer matrix and in a network (system) of fibres. Therefore, the value λ is controlled by the dimension d_u or a number of the accessible for this process sites of the filler fibres network [4]. In the RSN case the heat transport in the areas with the zero conductivity, i.e., in a polymer matrix, is impossible and the value λ is controlled by the dimension D_N, i.e., the fibres network dimension. In Fig. 3 the dependence $\lambda(D_N)$ is adduced, which is broken up on two linear secters and whose boundary is the dimension $D_N \approx 2.62$ (the vertical shaded line in Fig. 3). For $D_N < 2.62$ (RNR limit) the dependence $\lambda(D_N)$ is approximated as follows:

$$\lambda = 0.90(D_N - 2), \text{W/m·K}, \qquad (8)$$

and for $D_N > 2.62$ (RSN limit) approximation of the dependence $\lambda(D_N)$ has the following form:

$$\lambda = 0.51(D_N - 2), \text{W/m·K.} \tag{9}$$

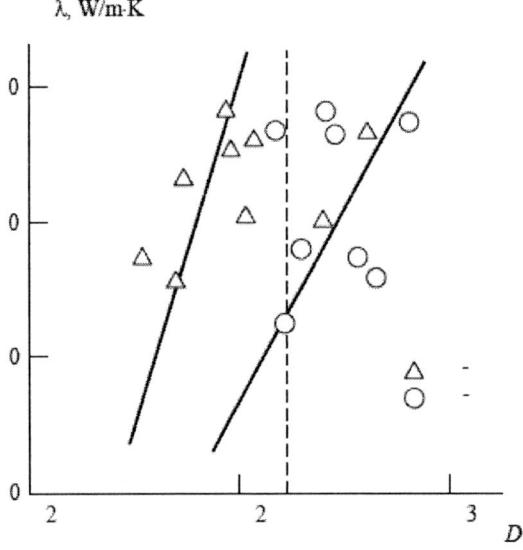

Figure 3. The dependence of heat conductivity λ on fractal dimension of filler fibres network (system) D_N for carbon plastics on the basis of phenylone produced with using of magnetic (1) and mechanical (2) separation. The vertical shaded line points the value D_N, which is boundary for limiting cases RNR and RSN

Hence, in the RNR limit one can observe more fast growth of λ at D_N increase than in the RSN limit. Such conclusion follows directly from the comparison of the relationship (1) and (2), as for the considered carbon plastics $d_u > (d_w - D_N)$.

CONCLUSIONS

Therefore, the results of the present paper showed that the heat conductivity of carbon plastics on the basis of phenylone can be described within the framework of a fractal model. Depending on the dimension of filler fibres network (system) such description can be obtained by the application of two limiting cases: random network of resistors (RNR) or random superconducting network (RSN).

REFERENCES

[1] Zibland H. In book: Polymer Engineering Composites. Ed. Richardson M. London, *Applied Science Publishers* LTD, 1978, p. 284-319.

[2] Kozlov G.V., Mikitaev A.K. *Mekhanika kompoziczionnykh materialov i konstrukczii*, 1996, v. 2, № 3-4, p. 144-157.
[3] Novikov V.U., Kozlov G.V. *Mekhanika kompozitnykh materialov*, 1999, v. 35, № 3, p. 269-290.
[4] Stanley H.E. In book: *Fractals in Physics*. Ed. Pietronero L., Tosatti E. Amsterdam, Oxford, New York, Tokyo, North-Holland, 1986, p. 463-477.
[5] Sokolov L.B., Kuznetsov G.A., Gerasimov V.D. *Plast. Massy*, 1967, № 9, p. 21-23.
[6] Bueya A.I., Kozlov G.V. *Trenie i iznos*, 2003, v. 24, № 3, p. 279-283.
[7] Alexander S., Orbach R. *J. Phys. Lett.* (Paris), 1982, v. 43, № 14, p. L625-L631.
[8] Burya A.I., Chigvintseva O.P., Suchilina-Sokolenko S.P. Polyarylates. Synthesis, Properties and Composite Materials (in Russian). Dnepropetrovsk, *Nauka i Obrasovanie*, 2001, 152 p.
[9] Avnir D., Farin D., Pfeifer P. *Nature*, 1984, v. 308, № 5959, p. 261-263.
[10] Lipatov Yu.S., Physical Chemistry of Filled Polymers (in Russian). Moscow, *Khimiya*, 1977, 304 p.
[11] Burya A.I., Kozlov G.V., Sverdlikovskaya O.S. *Voprosy khimii i khimicheskoi tekhnologii*, 2004, № 4, p. 109-112.

Chapter 14

ANISOTROPY OF A FIBER STRUCTURE AND THE FRICTIONAL WEAR OF COMPOSITES ON THE BASIS OF POLYARYLATE

G. V. Kozlov, A. I. Burya and G. E. Zaikov

Agrarian State University
Voroshilov st., 25, Dniepropetrovsk, 49 027, Ukraine
Institute of Biochemical Physics Russian Academy of Sciences
Kosygin str. 4, 119991, Moscow, Russia

ABSTRACT

An influence of the anisotropy of surface structure of short fibers on an interfacial layer structure in polymer composites is shown. In its turn, mentioned changes of structure cause an essential variation of frictional wear for these materials. In this aspect the important role plays the existence of a hydrogen bonds polymer-filler.

Keywords: Polymer composite, fiber, friction, wear, structure of surface, cluster model, fractal dimension.

INTRODUCTION

A reinforcement of polymers of polymers by oriented fibers has several specific features in comparison with a reinforcement by mineral fillers [1]. In paper [2] a supposition is made, that the fibrillar structure of a fiber surface results in the more frictional stability in comparison with an isotropic structure of analogous fibers. Besides, the authors [2] by the methods of IR-spectroscope demonstrated the existence of hydrogen bonds on the interface of fibers uglen and vniivlon with polyarylate (PAr), which is used as matrix polymer. The existence of these bonds assumes an anisotropy of PAr macromolecules on the interface, that must also influence on the frictional stability of composite. Such "copy" of the main features of fiber surface by interfacial layers of polymer is well illustrated on the example of system

PAr-terlon. This fiber has crystalline structure, that causes the PAr crystallization up to cristallinity degree ~ 0.20 [2], whereas amorphous uglen and vniivlon do not give that effect. Therefore a purpose of this paper is to study the influence of surface anisotropy of the fibers iglen and vniivlon on the frictional wear of composites on the basis of PAr filled by mentionad fibers.

EXPERIMENTAL

The composites on the basis of PAr of trade mark DV-102 with content of uglen and vniivlon 5, 15, 25 and 35 mass. % are used. The density of uglen is equal to equal 1240 kg/m^3, of vniivlon – 1100 kg/m^3 [2]. Composites are produced by "dry" method, which includes a blending of components in a rotating electromagnetic field with the help of nonequiaxial ferromagnetic particles [3]. Tribological properties of composites are studied on the disk-shaped apparatus of friction according to the method of the paper [4]. As a contrbody material is used the steel 45, quenched up to HRC 45-48 with surface roughness R_a=0.32 μm. The experiments were carried out after the final attrition of samples to a constant friction coefficient. The intensity of linear wear is calculated according to the paper [5].

The X-ray structural analysis is made on the apparatus DRON-3 according to the method of Bragg-Brentano in angles interval $2Q$=10-70° (Cu-radiation intensity is made on the points with introduction in to computer memory [6].

RESULTS AND DISCUSSION

In Fig. 1 a diffraction X-ray curves for PAr, uglen and carboplastics on the basis of PAr with various volume content of uglen φ_f is shown. As one can see at the increase of φ_f the decrease of intensity I_r of carboplastics amorphous halo and its displacement in the direction of more scattering angles $2Q$ is observed. That change of diffraction curves assumes an increase of a relative part of local order regions (clusters) φ_{cl} in polymer matrix structure and the decrease of interchain distance at the increase of φ_f [7]. It is necessary to mention, that in the considered case the generalized definition of clusters is adopted: these are structure parts, make from segments with "freezing" molecular mobility [7]. Besides, on the densification of polymer matrix structure at increase φ_f the measurements of composites density ρ point out: the experimental value ρ is more than additive value this difference being increased at the increase of φ_f [2]. Between the parameters I_r and φ_{cl} the following relationship [7] is received:

$$\varphi_{cl}^{-1} = CI_r, \qquad (1)$$

where C is constant.

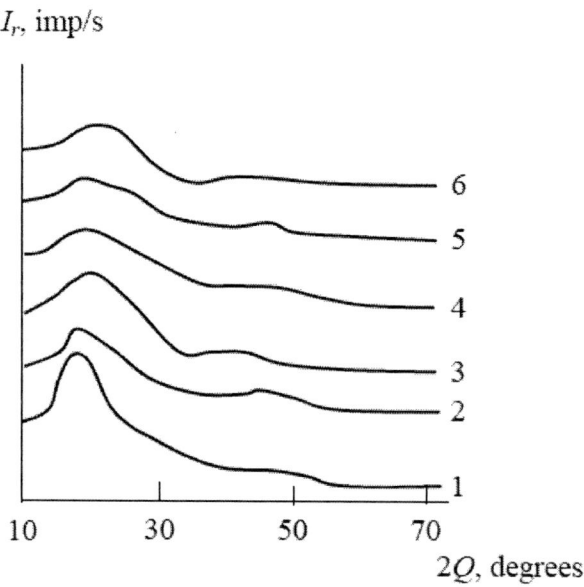

Figure 1. A diffraction X-ray curves for polyarylate (1), composites on the basis of polyarylate with volume content of uglen 0.038 (2), 0.114 (3), 0.195 (4), 0.282 (5) and uglen (6)

The estimation of the value C for studied composites can be made by the following method. For initial PAr the value φ_{cl} can be determined from the following percolation relationship [7]:

$$\varphi_{cl} = 0.03(T_g - T)^{0.55} \qquad (2)$$

where T_g and T are glass transition temperature and testing temperature, accordingly. For PAr T_g=468 K [2], T=293 K. Then the value C is defined as a reciprocal of product of calculated by that method value φ_{cl} and experimental magnitude I_r for PAr.

On the basis of above mentioned it can be assumed, that the polymer matrix structure densitification (the estimations according to the equation (1) gave an increase φ_{cl} from 0.514 for initial PAr up to 0.887 for composite PAr-uglen with volume content of uglen equal 0.282) is determined by the anisotropy of PAr macromolecules in the interfacial layer polymer-filler, which is due to the existence of a structure anisotropy of fiber surface and hydrogen bonds between uglen and PAr [2]. Differently speaking, the drawing of PAr macromolecular on the surface of a filler fibers is assumed the degree of this drawing can be estimated with the help of fractal dimension D of a chain part between its fixation points (cross-links, clusters, sites of hydrogen bonds, ets.) [8]. The value D is varied in the limits $1.0 < D \leq 2.0$, where D=1.0 assumes full drawing of a chain part and molecular mobility loss and D=2.0 assumes maximum possible mobility of this part, which is typical for devitrificated polymers [9]. The value D can be determined according to the following equation [9]:

$$\frac{2}{\varphi_{cl}} = C_\infty^D, \tag{3}$$

where C_∞ is characteristic ratio, which is an index of a statical flexibility of a polymer chain [10]. For PAr C_∞ is equal to 2.3 [11].

In Fig. 2 the dependence of calculated according to the equation (3) dimension D on volume content of uglen and vniivlon in studied composites is shown. As it should be expected, a decrease D at an increase φ_f is observed, that assumes a process of chain PAr drawing on fibers surface (fibllization). At $\varphi_f \approx 0.25$ the asymptotic value $D \approx 1.05$ is reached, that assumes practically fullloss of molecular mobility for chains of PAr. It is necessary to point out, that for the studied composites a minimal value of linear wear intensity I is reached approximately at $\varphi_f \approx 0.20$-0.22 [2].

Some more factor, which is influenced on the polymer fibrillization on the fibers surface, is the area of their summary surface, which is proportional to value φ_f – the more this area is, the more article sites concentration is, which are able to formation of hydrogen bonds polymer-filler. On the basis of above mentioned it, should be expected, that the decrease D and increase φ_f bring to a reduction I. In Fig. 3 the dependence I on the value of ratio D/φ_f is shown, which is approximated well enougt by linear correlation passing through the origin of coordinates. Such course of dependence $I(D/\varphi_f)$ shows, that the frictional wear intensity decreases at the increase of hydrogen bonds number and drawing degree (fibrillization) of polymeric chain on the surface of anisotropic fibres in polymer composites.

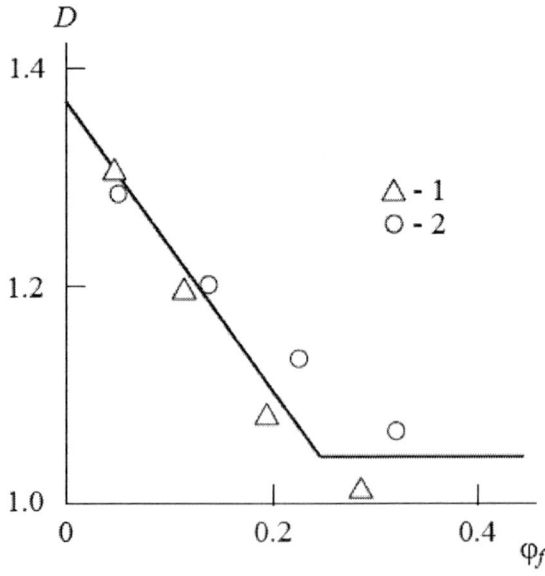

Figure 2. The dependence of fractal dimension D of chain part on volume content of fibers φ_f for composites PAr-uglen (1) and PAr-vniivlon (2).

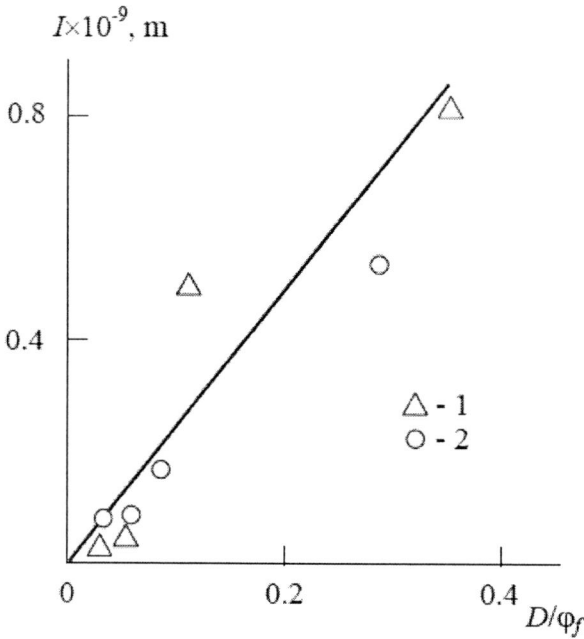

Figure 3. The dependence of linear wear intensity I on the value of ratio (D/φ_f) for composites PAr-uglen (1) and PAr-vniivlon (2)

CONCLUSIONS

Thus, the results of the present paper have shown the influence of the anisotropy of the surface structure of short fibers on an interfacial layer structure in polymer composites. In its turn, mentioned changes of structure cause the essential variation of frictional wear for these materials. In this aspect plays an important role the existence of a hydrogen bonds polymer-filler.

REFERENCES

[1] Lipatov Yu.S. Polumer reinforcement. *Chem. Techn.* Publishers, Toronto, 1995, 256 p.
[2] Burya A.I., Tchigvintseva O.P., Sutchilina-Sokolenko S.P. *Polyarylates*. Synthesis, Properties, Composite Materials. (Russ.). Dnepropetrovsk, *Nauka i osvita*, 2001, 152 p.
[3] Fomitchev A.I., Burya A.I., Gubenkov M.G. *Electron.obrabotka materialov*, 1978, No. 4, p. 26-27.
[4] Burya A.I. *Trenie i iznos*, 1998, v. 19, No. 5, p. 671-676.
[5] *A methods of calculated estimation of wear-proofing of friction surfaces of mashines detals*. (Russ.). Moscow, Izd. Standartov, 1979, 156 p.
[6] Kozlov G.V., Beloshenko V.A., Kuznetsov E.N., Lipatov Yu.S. *Doklady NAN* Ukrainy, 1994, No. 12, p. 126-128.

[7] Kozlov G.V., Zaikov G.E. *Structure of the Polymer Amorphous State*. Leiden, Brill Academic Publishers, 2004, 465 p.
[8] Kozlov G.V., Novikov V.U. Synergetics and fractal analysis of cross-linked polymers. (Russ.). Moscow, *Klassika*, 1998, 112 p.
[9] Kozlov G.V., Novikov V.U. *Uspekhi Fizich. Nauk*, 2001, v. 171, No. 7, p. 717-764.
[10] Budtov V.P. A Physical Chemistry of Polymer Solutions. Sanct-Peterburg, *Khimiya*, 1992, 384 p.
[11] Aharoni S.M. *Macromolecules*, 1983, v. 16, No. 9, p. 1722-1728.

Chapter 15

A INFLUENCE OF FEEDBACK IN THE STRUCTURE OF CARBON PLASTICS ON THEIR PROPERTIES

G. V. Kozlov, A. I. Burya and G. V. Zaikov

Agrarian State University
Voroshilov st., 25, Dniepropetrovsk, 49 027, Ukraine
Institute of Biochemical Physics Russian Academy of Sciences
Kosygin str. 4, 119991, Moscow, Russia

ABSTRACT

The received in the present paper results are allowed to find a structural sense of feedback effect for carbon plastics based on phenylone and demonstrated its influence on the strength of these materials. The decrease of feedback parameter can be a result to a substantial increase of carbon plastics macroscopic strength. To control the value of this parameter is possible by the change of fibres orientation factor, which is controlling parameter for interfacial regions.

Keywords: Carbon plastics, feedback, interfacial regions, mechanical properties.

INTRODUCTION

As it is shown in paper [1], carbon plastics structure which is produced with the help of the method of components preliminary blending in rotating electromagnetic field, is synergetical system. This is expressed in its main characteristics behavior as a function of blending duration t: at small t ($t \leq 120$ s) is observed periodical (ordered) behavior, closed to sinucoidal with twofold period and then the transition to chaotic behaviour is realized [1]. As it is known [2], one of the main features of synergetical systems is the existence of feedback in them. Structural sense of feedback for considered carbon plastics is expressed by simple relationship [1]:

$$\varphi_{cl} = 0{,}74 - \varphi_{\text{int}}, \tag{1}$$

where φ_{cl} and φ_{int} are relative fractions of local order regions (clusters) and interfacial regions, accordingly, i.e., densely-packed regions of structure.

It should be pointed out, that the constant 0,74 in relationship (1) is equal to maximally possible relative fraction of densely-packed regions of composite according to the conception of thermal cluster [3].

Therefore, the physical sense of feedback effect in considered carbon plastics is very simple: the increase φ_{int} results to decrease φ_{cl} and on the contrary. Therefore the purpose of the present paper is finding of influence degree of feedback in structure of carbon plastics based on phenylone on their mechanical characteristics, particularly, on fracture stress (strength).

EXPERIMENTAL

As the polymer matrix is used an aromatic polyamide – phenylone [4] is used and as filler – carbon fibres (CF) with diameter 7-9 mcm and length 3 mm. The composite is produced by "dry" method including components blending in rotating electromagnetic field. For this in reactor were placed powdery polymer, CF and nonequiaxial ferromagnetic particles with length 40 mm. Then reactor was placed in the end window of the generator of the electromagnetic apparatus. Through the action of rotating electromagnetic field ferromagnetic particles begin to rotate, colliding between them selves, that results to equipartition (chaotic) distribution of CF in polymer matrix. In the result of collisions particles are wear down and products of worn get to composition. For taking away ferromagnetic particles after blending two methods are used: magnetic and mechanical separation [5].

Specimens for studying of mechanical properties are prepared by method of hot pressing at temperature 603 K and pressure 55 MPa. The compression testing is made on the machine FP – 100 at temperature 293 K and strain rate 10^{-3} s^{-1}.

The thermal properties were determined on differential scanning calorimeter of model UT-C-400 at heating rate 10 K/min.

RESULTS AND DISCUSSION

At first we give the brief description of calculation method of parameters φ_{cl} and φ_{int} in the equation (1), which will be necessary at the next estimations. The value of fractal (Hausdorff) dimension d_f of carbon plastics structure is determined from the equation [6]:

$$d_f = 2(1 + \nu), \tag{2}$$

where ν is Poisson's ratio, the value of which can be calculated with the help of mechanical testing results with the usage of relationship [7]:

$$\frac{\sigma_Y}{E} = \frac{1-2\nu}{6(1+\nu)}, \tag{3}$$

where σ_Y is yield stress, E is elasticity modulus.

Then the value φ_{cl} with the usage of the equation [8] is determines:

$$d_f = 3 - 6\left(\frac{\varphi_{cl}}{SC_\infty}\right)^{1/2} \tag{4}$$

In the equation (4) S is the cross-sectional area of a macromolecule, C_∞ is the characteristic ratio, which is an indicator of polymer chain statistical flexibility [9]. For phenylone S=17,6 Å^2 [10], C_∞=3 [11].

The value φ_{int} can be calculated according to the following equation [12]:

$$\varphi_{int} = 1 - \frac{\Delta C_p^c}{\Delta C_p^p}, \tag{5}$$

where ΔC_p^c and ΔC_p^p are the values of heat capacity at constant pressure jump at glass transition temperature for composite and matrix polymer, accordingly.

It wsa shown in paper [13], that the controlling parameter at carbon plastics structure formation (exactly, their interfacial regions) is fibers orientation factor η. Then feedback parameter λ can be calculated according to the Puancare equation [2]:

$$\eta_{n+1} = \lambda(1-\eta_n)\eta_n, \tag{6}$$

where indexes n, $n+1$, ... are called sequential intervals of blending duration t of components in rotating electromagnetic field (t_1=5 s, t_2=10 s, ..., and so on) and values η are accepted according to the data of paper [13].

In Fig. 1 is presented the dependence $\varphi_{cl}(\lambda)$, from which the φ_{cl} increase at amplification of feedback expressed by increase λ follows. Such picture corresponds completely to the relationship (8): the increase λ results to "redistribution" of polymer material from interfacial regions in bulk polymer matrix and, as a consequence, to the increase of local order regions fraction in it. Differently speaking, in most general terms the change of feedback degree results to variation of polymer matrix structure. It is naturally to expect, that the mentioned structure change causes the variation of polymer matrix properties, particularly, its strength σ_f^m, which can be determined according to the equation [14]:

$$\sigma_f^m = 0,14 \left(\frac{\varphi_{cl}}{2N_A S l_0 C_\infty} \right)^{5/6}, \text{MPa}, \qquad (7)$$

where N_A is Avogadro number, l_0 is the length of main chain sceletal bond (for phenylone l_0=1,25 Å [11]).

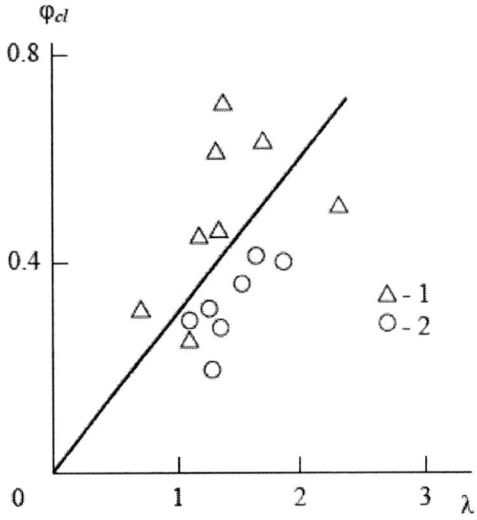

Figure 1. The dependence of clusters relative fraction φ_{cl} on feedback parameter λ. The specimens are prepared with the usage of magnetic (1) and mechanical (2) separation

As follows from the data of Fig. 2, where the dependence $\sigma_f^m(\lambda)$ is presented, the change of feedback level is influenced really substantially on bulk polymer matrix stength: the increase λ from 0,67 up to 2,27, i.e., approximately in three times, results to σ_f^m increase approximately in 2,7 times, from 78 up to 218 MPa. Note the characteristic feature of linear correlation $\sigma_f^m(\lambda)$: it passes through origin and this means, that in case of feedback absence polymer matrix will have zero strength.

On the basis of the reported above observations it is necessary to propose, that the happening because of feedback existence "redistribution" of material from interfacial regions in bulk polymer matrix changes the strength of interfacial regions σ_{int} and determines also the interrelation of stresses σ_f^m and σ_{int}. The value σ_{int} can be calculated according to the equation (7) with replace φ_{cl} on φ_{int} and supposing for interfacial regions C_∞=9 [1]. In Fig. 3 the relationship between strengths of bulk polymer matrix σ_f^m and interfacial regions σ_{int} is presented, from which follows the decrease σ_f^m at increase σ_{int} and vica versa. This interrelation was expected because of feedback effect and its structural expression: the "redistribution" of polymer material from one densely-packed structural component into the other.

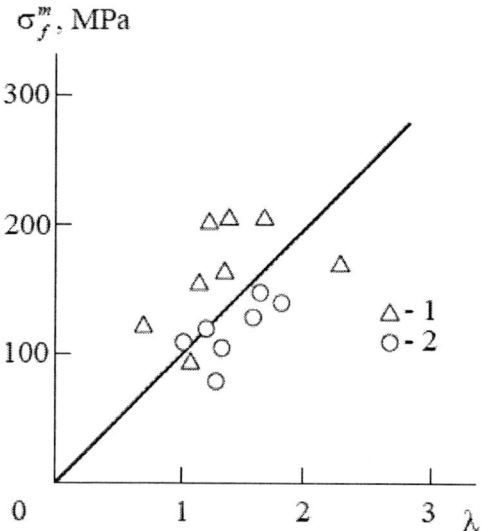

Figure 2. The dependence of bulk polymer matrix strength σ_f^m on feedback parameter λ. Meanings are the same, as in Fig. 1

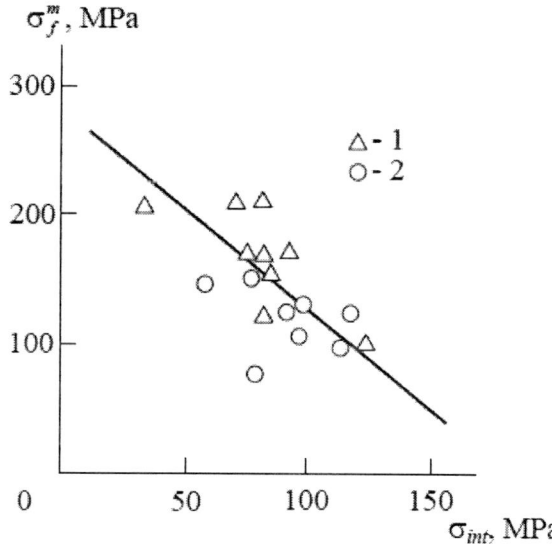

Figure 3. The relationship between the strengths of bulk polymer matrix σ_f^m and interfacial regions σ_{int}. Meanings are the same, as in Fig. 1

As it is well known [15, 16], the properties of interfacial regions determine to a larger extent properties of polymer composites as engineering materials. The data of Fig. 4, on which the dependence of experimentally determined macroscopic strength of carbon plastics

σ_f on σ_{int} is presented, demonstrates this law: the increase σ_{int} from 35 up to 125 MPa results to increase σ_f from ~ 300 up to ~ 406 MPa.

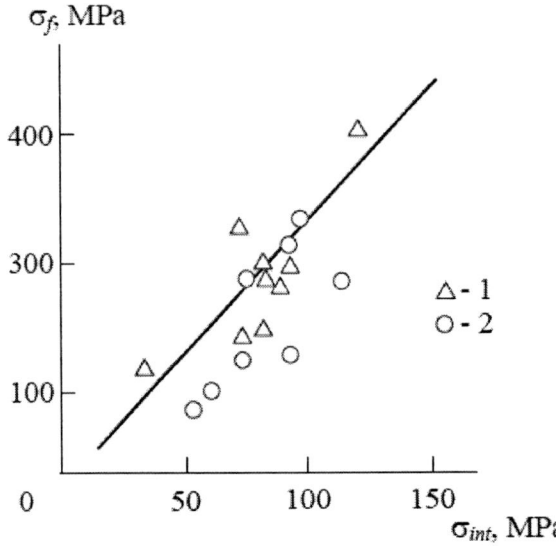

Figure 4. The dependence of microscopic strength σ_f on strength of interfacial regions σ_{int}. Meanings are the same, as in Fig. 1

Then, the joning of Fig. 1-4 allows to follow the influence of carbon plastics structural changes, that take place because of feedback existence, on their mechanical properties: an amplification of feedback (the increase λ) results to increase of clusters relative fraction φ_{cl} in bulk polymer matrix (Fig. 1), growth its strength σ_f^m (Fig. 2), decrease of interfacial regions strength σ_{int} (Fig. 3) and finally decrease of composite macroscopic strength σ_f (Fig. 4). The practical conclusion from the said above is obvious: for improvement of carbon plastics strength it is necessary to decrease the feedback parameter. So, from data of Fig. 2-4 follow two limiting cases: at $\lambda=0$ $\sigma_f^m=0$, $\sigma_{int}=180$ MPa and $\sigma_f=440$ MPa, then at $\lambda=2,40$ $\sigma_f^m=240$ MPa, $\sigma_{int}=28$ MPa and $\sigma_f=290$ MPa, i.e., the decrease σ_f approximately in one and a half times at increase λ from 0 up to 2,40 is observed.

On the basis of the spoken above appears the question: how can a feed back parameter be regulated purposely. The answer on this question gave the plot of Fig. 5, where the dependence λ on controlling parameter of interfacial regions is presented. As follows to wait some retardion (reaction, see the equation (6)) λ in comparison with η, then in Fig. 5 this dependence gave as $\lambda_{n+1}(\eta_n)$. From data of Fig. 5 follows, that decrease value λ can be by increase of fibres orientation factor λ. So, for mentioned above increase σ_f from 290 up to 440 MPa or decrease λ from 2,40 up to 0,55, that generally is reasonable result.

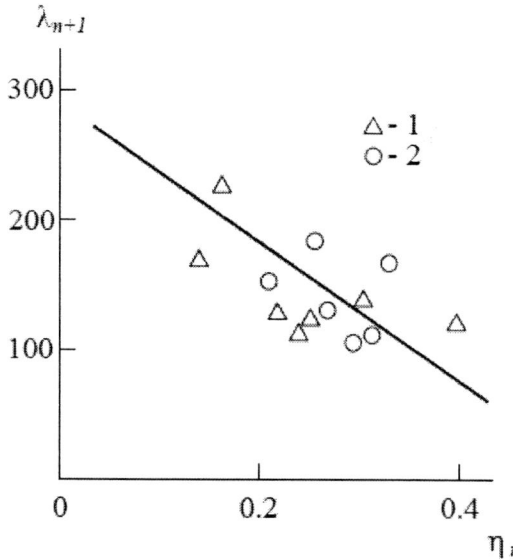

Figure 5. The dependence of feedback parameter λ_{n+1} on fibres orientation factor η_n. Meanings are the same, as in Fig. 1

CONCLUSIONS

Therefore, received in the present paper results allow to find a structural sense of feedback effect for carbon plastics based on phenylone and demonstrated its influence on the strength of these materials. The decrease of feedback parameter can result to substantial increase of carbon plastics macroscopic strength. The value of this parameter can be controlled by the change of fibres orientation factor which is controlling parameter for interfacial regions.

REFERENCES

[1] A.I. Burya., G.V. Kozlov. *Synergetics of polymer composites structure, formed in rotating electromagnetic field.* Proc. conf. "Science, Engeenirig, Higher Education". Rostov-na-Dony, RSU, 2004, p.p. 56-58.

[2] V.S. Ivanova, I.R. Kuzeev, M.M. Zakirnichnaya. *Synergetics and Fractals. An Universality of Mechanical Behaviour of Materials.* Ufa, Publ. USSEU, 1998, 366 p.

[3] G.V. Kozlov, Yu.S. Lipatov. *Mekhanika Kompozitnykh Materialov*, Vol. 39, 2003 № 1, p.p. 89-96.

[4] L.B. Sokolov, G.A. Kuznetsov, V.D. Gerasimov. *Plast. Massy*, 1967 № 9, p.p. 21-23.

[5] A.I. Burya ., G.V. Kozlov. *Trenie i iznos*, Vol. 24, 2003 № 3, p.p. 279-283.

[6] A.S. Balankin. *Synergetics of Deformable Body.* Moscow, Publ. Ministerstva Oborony SSSR, 1991, 404 p.

[7] G.V. Kozlov, D.S. Sanditov. Anharmonic Effects and Physico-Mechanical Properties of Polymers. Novosibirsk, *Nauka*, 1994, 261 p.
[8] G.V. Kozlov, V.U. Novikov. Synergetics and Fractal Analysis of Cross-linked Polymers. Moscow, *Klassika*, 1998, 112 p.
[9] V.P. Budtov. Physical Chemistry of Polymer Solutions. Sankt-Peterburg, *Khimia*, 1992, 384 p.
[10] S.M. Aharoni. *Macromolecules*, Vol. 18, 1985 № 12, p.p. 2624-2630.
[11] S.M. Aharoni. *Macromolecules*, Vol. 16, 1983 № 9, p.p. 1722-1728.
[12] Yu.S. Lipatov. Physical Chemistry of Filled Polymers. Moscow, *Khimia*, 1977, 304 p.
[13] A.I. Burya., G.V. Kozlov, M.E. Kazakov. *Description of formation of carbon plastics structure in terms of solid body synergetics.* Proc. 24 International Conf. "Composite Materials in Industry" ("Slavpolycom"). Yalta-Kiev, 31 May-4 June 2004, p.p. 246-248.
[14] V.U. Novikov, G.V. Kozlov, Yu.S. Lipatov. *Plast. Massy*, 2003 № 10, p.p. 4-8.
[15] V.U. Novikov, G.V. Kozlov, O.Yu. Buryan. *Mekhanika Kompozitnykh Materialov*, Vol. 36, 2000 № 1, p.p. 3-32.
[16] G.V. Kozlov, Yu.G. Yanovskii, Yu.S. Lipatov. *Mekhanika Kompozitnykh Materialov i Konstruktsii,* Vol. 8, 2002 № 1, p.p. 111-149.

FACTORS DEFINING THE SELECTIVITY POLYMER MEMBRANES: A FRACTAL MODEL

G. V. Kozlov, I. V. Dolbin[1] and G. E. Zaikov[2]

[1] Researching Institute of Applied Mathematics and Automatization of Kabardino-Balkarian Scientific Center of Russian Academy of Science, Nal'chik – 360000, Shortanov str., 89 "a", Russian Federation
[2] Institute of Biochemical Physical of Russian Academy of Science Moscow – 119991, Kosygin str., 4, Russian Federation

ABSTRACT

It is shown that the selectivity of the polymer nonporous membranes by the diffusivity depends on the polymer structure by rather complex way. The most important characteristic in this case is the dimension D_f controlling the gas transport processes and the molecular mobility degree influences only at large enough values of diameter of the selected gas molecules, the difference of these diameters and the dimension D_f.

Keywords: Polymer membranes, gas diffusion, selectivity, voids, fractal analysis.

INTRODUCTION

It was repeatedly assumed earlier that the selectivity of polymer membranes by the coefficients of gas permeability [1] and the diffusion [2] depends on the chain kinetic rigidity. In favour of such assumption several confirmations were cited, but the direct correlation between these properties of polymers is absent. The purpose of the present article is to determine the factors influencing on the selectivity of polymer membranes and to give the quantitative description of this influence in the terms of a fractal model.

EXPERIMENTAL

For the calculation of a coefficient of the separation of gases i and k by diffusivities α_{ik}^{D}, the diffusivities D by He and CH_4 for polyvinyltrimethylsilane (PVTMS) and polydimethylsiloxane (PDMS) and the equation of type Arrenius are used [3]:

$$D = D_0 e^{-E_D/RT}, \qquad (1)$$

where D_0 is a constant, E_D is diffusion activation energy, R is a universal gas constant, T is the testing temperature, for the calculation of D temperature dependence in interval $T=293$-420 K in case of the diffusion of C_2H_4 and C_3H_8 in PVTMS [4]. The values of a molecule diameter d_m for the used gases (He, CH_4, C_2H_4 and C_3H_8) are accepted according to the data of papers [3, 5].

RESULTS AND DISCUSSION

As it is known [6], the part of a polymer chain in the amorphous phase between the points of macromolecule topological fixation (entanglement nodes, cross-links, etc.) is a fractal with dimension D_{ch} ($1 < D_{ch} \leq 2$). The value of D_{ch} is a characteristic of the chain rigidity or, more exactly, of molecular mobility degree. The authors [7] received the following analytical relationship between D and D_{ch}:

$$D = K D_{ch}^{\Delta}, \qquad (2)$$

where K and Δ are constants for each gas-penetrant.

It was also found that the values K and Δ for PVTMS are the function of gas-penetrant molecule diameter d_m and are determined so [7]:

$$K = 0{,}39 d_m^{-12,5}, \qquad (3)$$

$$\Delta = 1{,}94 \times 10^{-4} d_m^{6,25}. \qquad (4)$$

If the values of d_m are accepted in Å, then D, according to the equation (2), isreceived in units cm^2/s. Within the framework of a fractal model of gas transport the value of D is determined as follows [8]:

$$D = D_0' f_g (d_h / d_m)^{2(D_f - d_s)/d_s}, \qquad (5)$$

where D_0' is a universal constant, f_g is a relative free volume, d_n is a diameter of the microvoid of this volume, D_t is the dimension controlling the processes of gas transport, d_s is a spectral dimension of the polymer structure (further it is accepted equal to 1,0).

As it is shown before [9], for PVTMS the exponent in the equation (5) is equal to 12,5. Therefore, comparing this value with the exponents in the equations (3) and (4), it can be written:

$$K = 0{,}39 d_m^{-2(D_t-d_s)/d_s}, \qquad (6)$$

$$\Delta = 1{,}94 \times 10^{-4} d_m^{(D_t-d_s)/d_s}. \qquad (7)$$

The checking of this assumption will be made below. Entirely the equation for the determination of a selectivity coefficient of two gases i and k by the diffusivity α_{ik}^D can be written as follows:

$$\alpha_{ik}^D = \left(\frac{d_{m_i}}{d_{m_k}}\right)^{-2(D_t-d_s)/d_s} D_{ch}^{1{,}94\times10^{-4}\left(d_{m_i}^{(D_t-d_s)/d_s} - d_{m_k}^{(D_t-d_s)/d_s}\right)} \qquad (8)$$

for $d_{m_i} < d_{m_k}$.

It is necessary to mention the important feature of the equation (8): the dimension D_{ch} influences on value α_{ik}^D only at large enough absolute values d_{m_i} and d_{m_k}, difference $(\alpha_{m_i} - \alpha_{m_k})$ and dimension D_t. At small values of these parameters, the exponents for D_{ch} in the equation (8) is less than one and the molecular mobility effect in this case is sharply decreased (the second multiplier of the equation is approaching to one).

The influence of the molecular mobility characterized by dimension D_{ch} can be followed most simply on the example of D temperature dependence since the D_{ch} increase at T raising is well known [6]. The value of D_{ch} as function T, can be theoretically calculated as follows. The relative fraction of the local order regions (clusters) φ_{cl} can be determined according to the next percolation relationship [6]:

$$\varphi_{cl} = 0{,}03(T_{cr} - T)^{0{,}55}, \qquad (9)$$

where T_{cr} is a critical temperature, equal to the glass transition temperature T_g for the amorphous polymers and the melting temperature T_m – for the semicrystalline polymers.

Then the following simple relationship [6] can be used:

$$\frac{2}{\varphi_{cl}} = C_\infty^{D_{ch}}, \qquad (10)$$

where C_∞ is a characteristic ratio, which is the indicator of the statistical flexibility of chain [10] and is equal to ~ 4 for PVTMS [9].

The comparison of the experimental (where the values of D were calculated according to the equation (1)) and the estimated according to the equation (8) α_{ik}^D values for PVTMS as function of D_{ch} in Fig. 1 is shown. As a gas i C_2H_4 is supposed and as a gas k – C_3H_8. As one can see, the good correspondence of α_{ik}^D values calculated by both mentioned methods is observed. As it is assumed [1, 2], the increase of D_{ch} results to α_{ik}^D decrease and at $D_{ch} \approx$ 1,50 the inversion of D values is observed: the value of D for C_3H_8 becomes bigger than the corresponding value for C_2H_4. Therefore, the plot of Fig. 1 clearly confirms the thesis about the selectivity decrease at raising the molecular chain mobility.

As it is known [3], PVTMS and PDMS have large enough D values, but sharply differing coefficients of the selectivity. So, the value of α_{ik}^D for He and CH_4 in the PVTMS case is equal to ~ 206 and in the PDMS case – about 4,8 [3]. The equation (8) allows to explain this difference. The parameters of this equation in the PVTMS case are equal to: $2(D_f-d_s)/d_s$=12,5, D_{ch}=1,046. For PDMS the following magnitudes are obtained: $2(D_f-d_s)/d_s$=3,8, D_{ch}=1,8. Then the calculation according to the equation (8) gives the following values: α_{ik}^D=328 for PVTMS and 6 for PDMS, that is corresponded with the experimentally obtained values of this parameter. It is important that in this case the α_{ik}^D value is independent on D_{ch} (for PVTMS because of the small value D_{ch} and for PDMS because of the small value D_f), but is entirely defined by D_f difference: 7,25 for PVTMS and 3,8 for PDMS. It is necessary to mention, that in virtue of the condition $d_{m_i} < d_{m_k}$ the D_{ch} increase will decrease α_{ik}^D.

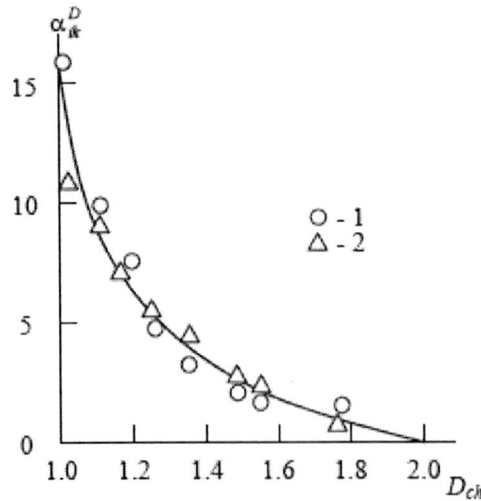

Figure 1. The dependences of selectivity coefficient CH_2 and C_3H_8 by diffusivity α_{ik}^D on fractal dimension of a chain part between the entanglements D_{ch} for PVTMS. The calculation according to the equation (1) (1) and equation (8) (2)

The equation (8) also allows to predict the α_{ik}^{D} for the various polymers and the gas pairs according to the following technique. In the paper [11] the D_t increase at $T_g(T_m)$ raising was shown and the corresponding calibrating plot is adduced, allowing the estimation of exponents in the equation (8). In Fig. 2 the comparison of the experimental [3, 12] and calculated according to the equation (8) α_{ik}^{D} values for 12 polymers by H_2 and CH_4 is shown. As for these gases the d_m and $(d_{m_i} - d_{m_k})$ are relatively small, then D_{ch} influence becomes appreciable only at large enough $D_t(D_t>5,0)$. As one can see even in this simple variation good enough correspondence of the theory and the experiment (the average discrepancy is equal to ~ 27 %) is obtained, though the experimentally determined α_{ik}^{D} values can have the essential scatter. So, for PVTMS the value of α_{ik}^{D} for the pair H_2-CH_2 is equal to 178 according to the data of paper [12] and to 261 according to the data of paper [3], i.e., the discrepancy is about 32 %. Therefore, we can believe that the accuracy of the theoretical prediction according to the equation (8) is about equal to the experiment error. This correspondence also confirms the correctness of the choice of exponents in the equations (6) and (7).

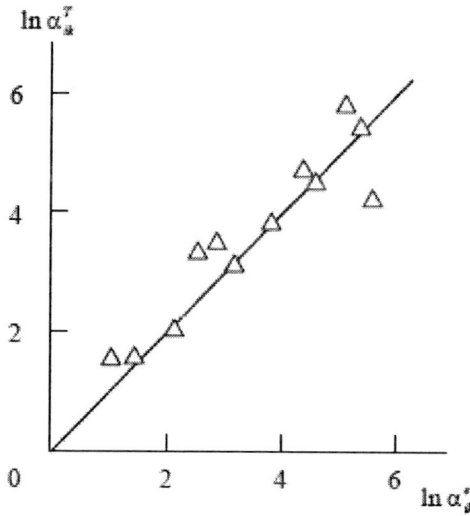

Figure 2. The comparison of the experimental α_{ik}^{e} and calculated according to the equation (8) α_{ik}^{T} selectivity coefficient H_2 and CH_4 for 12 polymers in double logarithmic coordinates

And in the conclusion it is necessary to mention one important theoretical aspect. As follows from the above cited estimations, D_t value influences on the polymer selectivity much more strongly than D_{ch}. In its turn, as D_t either fractal (Hausdorff) dimension of structure d_f or dimension of the areas of localization of the excessive energy D_f can be used depending on the mechanism of gas diffusion [13]. These dimensions are connected between themselves by the relationship [14]:

$$D_f = 1 + \frac{1}{3 - d_f}, \quad (11)$$

and in its turn the value of d_f is determined according to the following equation [6]:

$$d_f = 3 - 6\left(\frac{\varphi_{cl}}{C_\infty S}\right)^{1/2}, \quad (12)$$

where S is the cross-section of a macromolecule in units of Å.

Hence, the dimension $D_t(d_f$ or $D_f)$ is determined by the characteristics both molecular (C_∞ and S) and supermolecular (φ_{cl}) structure of polymer. It can be expected, as the polymer is a thermodynamically nonequilibrium solid and for its description, as minimum, two order parameters [15] are required.

CONCLUSIONS

Therefore, the results of the present paper showed that the selectivity of the polymer nonporous membranes by the diffusivity depends on the polymer structure by rather complex way. The most important characteristic in this case is the dimension D_t and the molecular mobility degree influences only at large enough values of diameter of the selected gas molecules, the difference of these diameters and the dimension D_t.

REFERENCES

[1] Volkov V.V., Durgar'yan S.G. *Vysokomolek. Soed.* A, 1983, v. 25, № 1, p. 30-36.
[2] Volkov V.V., Nametkin N.S., Novitskii E.G., Durgar'yan S.G. *Vysokomolek. Soed.* A, 1979, v. 21, № 4, p. 920-926.
[3] Teplyakov V.V., Durgar'yan S.G. *Vysokomolek. Soed.* A, 1984, v. 26, № 7, p. 1498-1505.
[4] Volkov V.V., Nametkin N.S., Novitskii E.G., Durgar'yan S.G. *Vysokomolek. Soed.* A, 1979, v. 21, № 4, p. 927-931.
[5] Teplyakov V.V., Durgar'yan S.G. *Vysokomolek. Soed.* A, 1986, v. 28, № 3, p. 564-572.
[6] Kozlov G.V., Zaikov G.E. *Structure of the Polymer Amorphous State*. Utrecht-Boston, Brill Academic Publishers, 2004, 465 p.
[7] Khalikov R.M., Kozlov G.V. *Uspekhi sovremen. estestvozn.*, 2004, № 10, p. 63-64.
[8] Kozlov G.V., Zaikov G.E. In book: *Fractal Analysis of Polymers*: From Synthesis to Composites. Ed. Kozlov G., Zaikov G., Novikov V. New York, Nova Science Publishers, Inc., 2003, p. 107-112.
[9] Khalikov R.M., Kozlov G.V. *Uspekhi sovremen. estestvozn.*, 2004, № 10, p. 64-65.

[10] Budtov V.P. *Physical Chemistry of Polymer Solutions* (in Russian). Sankt-Peterburg, Khimiya, 1992, 384 p.
[11] Khalikov R.M., Kozlov G.V., Shetov R.A. *Manuscript deposed to VINITI RAS*, Moscow, 2005 (in press).
[12] Volkov V.V., Gol'danskii A.V., Durgar'yan S.G., Onishchuk V.A., Shantorovich V.P., Yampol'skii Yu.P. *Vysokomolek. Soed.* A, 1987, v. 29, № 1, p. 192-197.
[13] Kozlov G.V., Afaunov V.V., Mashukov N.I., Lipatov Yu.S. *Doklady NAN* Ukraine, 2000, № 10, p. 140-145.
[14] Balankin A.S. *Synergetics of Deformable Body* (in Russian). Moscow, MO SSSR, 1991, 404 p.
[15] Kozlov G.V., Novikov V.U. *Uspekhi Fizich. Nauk*, 2001, v. 171, № 7, p. 717-764.

INDEX

A

acceptance, 152
access, 2, 30
accounting, 211
accumulation, vii, 10, 37, 213, 244
accuracy, 61, 70, 71, 182, 303
acetic acid, 69
acetonitrile, 63
acid, 6, 8, 10, 12, 17, 19, 21, 29, 31, 36, 39, 41, 50, 56, 59, 63, 65, 71, 74, 120, 122, 144
acidity, 61, 64, 66, 67, 70
acrylonitrile, 39
activation, 124, 174, 232, 300
activation energy, 124, 174, 232, 300
active centers, vii, 203, 206, 207, 208, 209, 210, 211
active radicals, 36
active transport, 6
additives, 29, 30, 39, 52, 53, 56, 140, 146, 260, 261, 266, 268, 270, 275
adenovirus, 10, 15
adhesion, 6, 7, 191
adsorption, 67
affect, vii, 25, 34, 52, 53, 104, 139, 140, 141, 160, 194
agent, 4, 12, 29, 31, 59, 65, 182
aggregates, 8, 10, 20, 21, 96, 218
aggregation, 4, 5, 9, 153, 154, 155, 280
aggregation process, 280
aging, vii, 27, 139, 142, 143, 144, 145, 215, 219, 247
AIDS, 23
alcohol, 7, 36, 95, 120, 125, 132, 263, 271, 272
alcohols, 41, 63, 68, 69, 73, 113, 114
aldehydes, 13
aliphatic compounds, 28, 66, 72
allergy, 20
alternative, 179, 184, 185

aluminum, 32, 40, 52
aluminum oxide, 40
amines, 41, 85, 90, 105, 113, 114, 141, 142, 146
amino acids, 2
ammonia, 12, 30, 31, 32, 39
ammonium, 30, 31, 32, 36, 37, 90
ammonium salts, 36
amorphous polymers, 106, 116, 120, 152, 164, 168, 170, 196, 214, 218, 220, 301
amorphous-crystalline polymers, 150, 151, 152, 153, 154, 155, 157, 162, 163, 164, 165, 171, 172, 181, 182, 183, 185, 186, 196
amplitude, 142
aniline, 113
animals, 6
anisotropy, viii, 52, 285, 287, 289
annealing, 178
annihilation, 222, 223, 224, 225
antibody, 16
antigen, 4, 16
antigen-presenting cell, 4
antioxidant, 56, 140
apoptosis, 2
aqueous solutions, 8, 9, 20
arabinogalactan, 13
argon, 50, 51, 90, 104, 120
argument, 191
aromatic compounds, 70, 72
aromatic hydrocarbons, 121
aromatic rings, 39
assessment, 121, 150, 164, 166, 167, 171, 174, 175, 178, 179, 180, 181, 184
association, 153, 214
assumptions, 26, 206, 211, 251
atmospheric pressure, 27
atoms, 32, 84, 90, 91, 93, 104, 105, 118, 171, 225, 241, 262, 271
attention, vii, 1, 2, 11, 16, 36, 49, 86, 113, 150, 182, 224, 231, 278
averaging, 68

Avogadro number, 148, 259, 267, 294

B

backlash, 168, 179, 180
barriers, 3, 14, 17
basicity, 61, 64, 65, 66, 67, 68, 69, 71, 77
behavior, 7, 27, 51, 56, 59, 64, 67, 95, 147, 152, 153, 155, 160, 166, 170, 171, 184, 192, 193, 217, 226, 228, 232, 291
Belarus, 43, 139
benzene, 28, 72, 82, 87, 90, 96, 115, 117
benzoyl peroxide, 203, 204, 205, 207, 208, 209, 210, 211
bile, 10
binary blends, viii, 249
binding, 5, 9, 10, 11, 69, 76
bioavailability, 9
biocompatibility, 9, 21
biodegradability, 9
biological processes, 2
biomaterials, 6
biopolymer, 8
bisphenol, 149, 222
blends, 249, 256
blocks, 56
blood, 5, 9
blood plasma, 5
body, 2, 6, 10, 21, 25, 153, 214, 245, 298
Boltzmann constant, 148
bonds, 13, 29, 30, 38, 39, 56, 90, 91, 93, 94, 111, 140, 285, 287, 288
bone marrow, 4
boric acid, 41
branching, 86, 95, 113
breakdown, 108
bromine, 30, 31, 36, 39
butadiene, 27, 28
butadiene-styrene, 27, 28

C

calorimetry, 68
cancer, 15
cancer cells, 15
capillary, 35, 250
carbides, 27
carbohydrate, 5, 12, 14, 15, 16
carbon, viii, 13, 20, 26, 27, 28, 32, 34, 38, 40, 54, 65, 78, 277, 278, 279, 280, 281, 282, 291, 292, 293, 295, 296, 297, 298
carbon atoms, 27
carbon dioxide, 20

carbonization, 31, 34, 36, 39
carbonyl groups, 193
carrier, 4, 11, 15, 18, 21, 113
cartilage, 6
catalyst, 122, 132, 133
catalytic effect, 29
catalytic system, 203
cation, 64, 77
CEC, 148
cell, vii, 1, 2, 3, 4, 5, 6, 7, 9, 10, 11, 12, 14, 16, 18, 20, 21, 22
cell line, 10, 11, 20
cell membranes, 7
cell surface, 5, 10
CH2-groups, 49
chain mobility, 164, 174, 302
chain propagation, 103, 203, 206, 207, 209, 210, 211
chain rigidity, 101, 104, 160, 300
chain transfer, 81
channels, 215, 241
chaos, 164, 165
characteristic viscosity, 92
chemical kinetics, 71
chemical properties, 15, 64, 65, 66, 76, 81, 115, 116, 119
chemical reactions, 31, 56, 59, 75
chemical reactivity, 77
chitin, 8
chlorine, 30, 84
chlorobenzene, 262, 271, 272
chloroform, 142
cholesterol, 17
chromatograms, 209
chromatography, 34, 109, 204
chromosome, 2
circulation, 9
classes, 36, 75, 76, 113, 114, 140, 152, 162, 182, 215
classification, 228
cleavages, 113
cluster, 147, 148, 149, 150, 152, 153, 154, 155, 156, 158, 160, 162, 163, 165, 168, 169, 170, 171, 174, 175, 176, 177, 178, 179, 180, 181, 182, 183, 184, 185, 188, 190, 191, 194, 196, 214, 215, 218, 220, 221, 223, 226, 227, 228, 229, 231, 279, 285
cluster entanglement network, 148, 163
cluster functionality, 149, 155, 178, 179, 181, 183, 185, 194, 227
cluster model, 147, 149, 153, 155, 156, 158, 162, 169, 171, 175, 180, 214, 215, 218, 223, 226, 231, 285

cluster network, 149, 165, 168, 171, 174, 177, 178, 180, 183, 188, 190, 191, 196, 227, 228, 229
C-N, 13
CO2, 32, 53, 54, 56, 241, 242, 243
coal, 27, 28
cobalt, 37, 40
coherence, 106
coke, 26, 27, 28, 29, 30, 31, 32, 34, 35, 36, 37, 38, 39, 40, 41, 50, 51
coke formation, 26, 28, 29, 31, 39, 40, 41
collagen, 19
collisions, 278, 292
colon, 22
combined effect, 66
combustibility, 25, 26, 30, 34, 35, 36, 37, 39, 40
combustion, 26, 30, 31, 32, 33, 34, 35, 36, 37, 39, 40
combustion decelerators, 26, 40
compatibility, 142, 144, 145, 255, 268
compliance, 96
components, 5, 13, 25, 26, 28, 29, 30, 33, 36, 38, 39, 41, 54, 60, 61, 71, 73, 77, 81, 96, 105, 117, 129, 203, 211, 235, 254, 255, 256, 278, 286, 291, 292, 293
composites, 27, 30, 57, 186, 187, 188, 189, 190, 191, 192, 193, 194, 195, 196, 278, 281, 286, 287, 288, 289
composition, 13, 14, 25, 26, 29, 31, 37, 38, 52, 54, 91, 96, 113, 125, 157, 164, 194, 278, 292
compounds, 22, 26, 27, 28, 29, 30, 31, 32, 36, 37, 38, 40, 41, 62, 72, 73, 75, 76, 81, 82, 86, 96, 103, 104, 109, 113, 114, 116, 118, 120, 121, 122, 124, 127, 130, 131, 132, 133, 157, 210, 255, 256, 257, 261, 262, 264, 268, 269, 271, 273
compressibility, 171
concentration, vii, viii, 9, 31, 33, 34, 37, 38, 63, 96, 103, 116, 117, 120, 125, 126, 133, 139, 140, 141, 142, 143, 144, 145, 169, 180, 186, 187, 188, 189, 190, 192, 193, 195, 204, 229, 237, 253, 254, 256, 259, 260, 261, 262, 263, 264, 265, 266, 267, 269, 270, 271, 273, 274, 275, 288
conception, 140, 292
conceptualization, 101
concordance, 177
condensation, 4, 18, 39, 82, 85, 115, 118, 130
conductivity, 192, 277, 278, 280, 281
conductor, 278
configuration, 104, 229
conflict, 148
conformity, 62, 182
conjugation, 11
connective tissue, 6
constitution, 217

construction, vii, 1, 6, 237
consumption, 140, 211
continuity, 188
control, viii, 2, 4, 9, 61, 196, 291
convergence, 164, 254
conversion, 82, 84, 85, 90, 109, 111, 115, 118, 123, 204, 205, 206, 207, 208, 209, 210, 211
conversion rate, 205, 206, 207, 208, 211
cooling, 27, 41, 249, 262, 263, 264, 271, 274
copolymers, viii, 18, 39, 82, 85, 86, 87, 88, 89, 90, 92, 95, 96, 97, 98, 99, 100, 101, 102, 103, 104, 105, 111, 114, 115, 116, 117, 118, 119, 120, 122, 123, 124, 125, 127, 128, 129, 132, 133, 249
copper, 37, 56, 191
corporations, 224
correlation, vii, 63, 64, 66, 67, 68, 69, 70, 71, 72, 73, 75, 110, 139, 141, 144, 145, 147, 156, 157, 162, 165, 168, 173, 181, 183, 218, 221, 288, 294, 299
cotton, 30
coupling, 10
cristallinity, 286
critical density, 172
critical value, 171
cross-linked polymers, 81, 122, 215, 245, 290
cross-linking reaction, 96
crude oil, 27
crystal polymers, 49, 50
crystallinity, 150, 183, 189, 190, 216, 230, 251
crystallites, 148, 150, 152, 163, 183, 184, 186, 188, 189, 190, 231, 235, 262, 271
crystallization, 106, 130, 157, 163, 174, 182, 183, 188, 189, 190, 196, 262, 263, 266, 271, 272, 275, 286
culture, 4
curing, 182
current limit, 15
cycles, 108, 122, 263, 272
cyclohexanol, 113
cytokines, 15
cytotoxicity, 2, 20

D

damage, 9
data analysis, 251
decay, 149, 169, 170, 177, 178, 182
decomposition, 25, 29, 31, 34, 36, 39, 41, 104, 108, 109, 132, 185, 229, 231
decomposition temperature, 104
defects, 52, 56
definition, 153, 218, 233, 286
deformation, 70, 147, 150, 170

degenerate, 253
degradation, vii, 2, 3, 4, 5, 7, 9, 11, 15, 16, 25, 27, 28, 29, 31, 32, 33, 34, 35, 36, 37, 38, 40, 50, 51, 52, 53, 54, 55, 56, 57, 96, 111, 113, 118, 127, 139, 140, 141, 142, 155, 192
degradation mechanism, 56
degradation process, 57, 97, 118, 127
degradation rate, 33, 34, 40
dehydration, 5, 28, 31, 39
dehydrochlorination, 40
delivery, vii, 1, 2, 3, 4, 6, 7, 8, 10, 12, 13, 14, 15, 16, 17, 18, 19, 20, 21, 22, 23
demand, 66, 69, 70, 71, 72, 73
density, viii, 4, 5, 6, 12, 13, 29, 76, 162, 171, 172, 173, 174, 189, 190, 216, 219, 225, 226, 227, 229, 230, 233, 249, 250, 251, 252, 253, 254, 255, 256, 257, 268, 278, 286
density fluctuations, 171, 172
dephosphorylation, 31
depolymerization, 27, 32, 97
deposition, 238
deposits, 144
derivatives, 2, 7, 9, 10, 14, 18, 19, 36, 38, 69, 76, 113, 114, 140
destruction, 27, 33
detachment, 104
detection, 56, 141, 234, 237, 238
deviation, 62, 159, 162, 172, 277
DGEBA, 215, 217
diaminodiphenylmethane, 182, 215
dielectric constant, 67, 89
differential scanning, 40, 103, 186, 279, 292
differential scanning calorimeter, 292
differential scanning calorimetry, 103, 186
differentiation, 19
diffraction, 100, 103, 105, 106, 116, 120, 286, 287
diffusion, viii, 6, 30, 32, 34, 41, 213, 215, 232, 241, 242, 243, 244, 299, 300
diffusion process, 232
diffusivities, 300
diffusivity, viii, 213, 231, 232, 233, 236, 242, 244, 299, 301, 302, 304
difractogram, 51
digestion, 6
diglycidyl ether of bisphenol, 215
dihydroxy-containing oligomers, 117
dimethylformamide, 6
disclosure, 120
disorder, viii, 160, 166, 167, 168, 169, 170, 171, 173, 259, 260, 262, 264, 266, 267, 268, 271, 273, 274, 275
dispersion, 60, 103, 108, 153, 188, 190, 193, 194
displacement, 225, 286

disposition, 60, 63, 71, 81, 82, 84, 86, 87, 88, 89, 90, 97, 98, 99, 100, 101, 104, 124, 125, 127, 128, 133, 174
dissipative structure, 218
dissipative structures, 218
dissociation, 3, 5
distribution, vii, viii, 2, 5, 7, 10, 12, 13, 21, 35, 39, 59, 71, 75, 109, 113, 186, 190, 203, 204, 205, 206, 207, 208, 209, 211, 213, 214, 217, 218, 232, 234, 236, 244, 259, 260, 266, 267, 269, 275, 277, 278, 292
divergence, 164, 220
diversity, 67
division, 26, 206, 207
DMFA, 74
DNA, vii, 1, 2, 3, 4, 5, 6, 7, 9, 10, 11, 13, 14, 15, 16, 17, 18, 19, 20, 21, 22, 23
domain, 194
dominance, 69
donors, 61, 62, 66
dosage, 9
dosing, 7
double bonds, 41
double logarithmic coordinates, 238, 239, 303
drug delivery, 17, 19, 21, 22, 23
drugs, 9, 11
DSC, 50, 103, 105, 186, 189, 251
DSC method, 50, 186, 189, 251
DTA curve, 50
durability, 40
duration, 27, 109, 291, 293

E

ECM, 6
efficiency criteria, 142
elastic deformation, 152
elasticity, 41, 147, 148, 167, 170, 177, 183, 185, 216, 250, 293
elasticity modulus, 41, 167, 170, 183, 216, 293
elastomers, 27
electron microscopy, 186, 188
electrons, 61, 171, 192
electrophoresis, 7
electroporation, 15
emission, 25, 52
employment, 206
encoding, 7, 16, 20
energy consumption, 27
energy density, 71, 75, 76
entanglement network, 161, 162, 172, 174, 179, 191, 193, 194, 195, 196
entanglement points, 151

entanglements, 148, 150, 194, 195, 228, 229, 243, 302
entropy, 63, 157, 162, 166, 169, 170
environment, viii, 7, 13, 60, 213, 241, 244
enzymatic activity, 12
enzymes, 3
EP-1, 182, 215, 217, 219, 221, 222
EP-2, 182, 215, 217, 221, 222
EP-3, 215, 217, 219, 221, 222
epoxy polymer, 215, 217, 219, 221, 222
epoxy resins, 37
equality, 96, 168, 185, 230, 239
equilibrium, 38, 59, 64, 71, 75, 148, 152, 153, 156, 157, 158, 159, 160, 171, 215, 220, 225, 226, 229, 244, 261, 269
equipment, 52
ESR, 45, 75, 103, 125, 130, 141, 142, 143, 144, 145
ESR spectra, 125, 130, 142
ESR spectroscopy, 145
ester, 12, 23, 55
estimating, 36, 84, 174, 208
ethers, 72, 73, 121
ethylene, vii, viii, 10, 11, 18, 30, 70, 139, 140, 141, 249, 250, 256
Euclidean object, viii, 213, 232, 235, 236, 244
Euclidean space, 216, 217, 228, 239, 240, 279
evidence, 4, 39, 144, 150
evolution, 142, 156, 158
exclusion, 66
exposure, 6, 54, 55, 142
expression, 2, 5, 9, 15, 16, 17, 22, 64, 70, 152, 171, 174
expulsion, 41
extinction, 189, 190
extraction, 82, 84, 96, 97
extrapolation, 92, 243, 251

F

factor analysis, 63, 66
family, 3, 13, 49
feedback, viii, 291, 292, 293, 294, 295, 296, 297
fibers, viii, 52, 277, 285, 287, 288, 289, 293
fibroblast growth factor, 19
fibroblasts, 7, 19
filled polymers, 153, 154
filler particles, 153, 218, 278
fillers, 26, 27, 285
films, 37, 96, 141, 142, 143, 144, 216, 263, 273
filtration, 96
financial support, 146
fire resistance, 25, 39
fixation, 287, 300

flame, 25, 27, 29, 30, 31, 32, 33, 34, 36, 41
flexibility, 5, 92, 151, 159, 164, 218, 288, 293, 302
fluctuation free volume, vii, 213, 214, 216, 218, 219, 221, 222, 223, 224, 225, 226, 227, 228, 229, 230, 231, 238, 241, 242, 244
fluctuations, viii, 171, 172, 174, 184, 185, 196, 213, 214, 225, 241, 244
fluorescence, 75
fluorine, 36
foaming covers, 40
food, 12
food industry, 12
forecasting, 61, 64, 170
formaldehyde, 28, 36, 37
fractal analysis, 213, 215, 225, 226, 228, 231, 243, 245, 278, 290, 299
fractal cluster, 218, 229
fractal dimension, 216, 218, 226, 229, 233, 234, 242, 243, 244, 278, 279, 282, 285, 287, 288, 302
fractal objects, 220
fractal properties, viii, 213, 216, 226, 236, 241, 244
fractal structure, vii, 213, 244
fractal theory, 239
fractality, 218, 220, 226, 235
fracture stress, 292
framing, 86, 88, 97, 104, 113, 118, 120
France, 139
free energy, 59, 60, 78
free radicals, 39, 211
free rotation, 91, 93, 94
free volume, vii, 140, 166, 167, 168, 169, 170, 172, 213, 214, 215, 217, 218, 220, 222, 223, 225, 226, 232, 233, 234, 236, 237, 238, 240, 244, 256, 260, 268, 301
freedom, 184
freezing, 243, 286
friction, 250, 285, 286, 289
fuel, 30, 33
fulfillment, 179

G

gas diffusion, 232, 241, 242, 243, 299, 303
gas-diffusant molecules, 232
gasification, 28, 33, 35
gel, 7, 13, 90, 122
gel formation, 90
gelation, 13
gene, vii, 1, 2, 3, 4, 5, 6, 7, 8, 9, 10, 11, 13, 14, 15, 16, 17, 18, 19, 20, 21, 22, 23
gene expression, vii, 1, 2, 3, 9, 15
gene therapy, vii, 1, 3, 12, 15
gene transfer, vii, 1, 4, 5, 10, 15, 17, 19, 23

generalization, 59, 60, 62, 63, 64, 66, 70, 75
generation, 211
genes, vii, 1, 2, 4, 7, 15
genetic disorders, 2
Georgia, 81, 133, 134, 136, 137
Germany, 139
glass transition, 86, 88, 103, 104, 116, 117, 128, 149, 152, 153, 159, 163, 168, 175, 177, 182, 214, 223, 226, 227, 228, 280, 287, 293, 301
glass transition temperature, 88, 104, 116, 117, 128, 149, 163, 168, 182, 223, 228, 280, 287, 293, 301
glasses, 226
glassy polymers, 147, 150, 152, 153, 167, 168, 170, 182, 214, 216, 221, 226, 228
glycol, 5, 10, 11, 18, 74
graph, 157, 163, 184
graphite, 27, 29, 39, 54
gravity, 105
grouping, 38
groups, 2, 4, 8, 9, 10, 11, 12, 13, 14, 29, 38, 41, 52, 55, 56, 66, 69, 72, 75, 81, 82, 84, 86, 88, 90, 91, 95, 96, 97, 99, 101, 102, 106, 113, 118, 120, 121, 122, 123, 124, 125, 131, 144, 145, 164, 250, 260, 268
growth, 7, 13, 15, 19, 94, 108, 162, 164, 192, 205, 206, 209, 210, 211, 214, 220, 251, 252, 253, 254, 255, 282, 296
growth factor, 7, 15

H

halogen, 30
halogens, 30, 31, 39
halos, 100, 103
HALS, 146
hardener, 215
harm, 67
harmony, 67
Hausdorff dimension, 239, 240, 241
HDPE, 150, 151, 152, 154, 155, 161, 162, 163, 172, 173, 178, 179, 180, 181, 183, 184, 185, 186, 187, 188, 189, 190, 191, 192, 193, 194, 195, 196, 216, 229, 230, 231, 238, 239, 240, 241, 242, 243, 244
heat, viii, 25, 26, 27, 29, 30, 32, 33, 34, 35, 36, 40, 50, 51, 56, 105, 106, 142, 160, 166, 171, 175, 238, 265, 274, 277, 278, 279, 280, 281, 282, 293
heat capacity, 35, 160, 175, 280, 293
heat conductivity, viii, 25, 26, 27, 35, 36, 40, 277, 278, 279, 280, 281, 282
heat loss, 36
heat release, 32, 35
heat shield, 25, 26, 40
heat transfer, 25, 26, 33

heating, 26, 27, 33, 34, 35, 36, 37, 38, 39, 40, 41, 50, 51, 52, 87, 88, 120, 121, 122, 259, 263, 264, 265, 266, 267, 271, 273, 274, 275, 279, 292
heating rate, 34, 41, 50, 52, 87, 88, 279, 292
height, 35, 40, 106, 108
helium, 238
heptane, 262, 263, 272, 273
heterogeneous systems, 152, 254
hexane, 113, 271
high density polyethylene, 150, 271
high-molecular compounds, 39
HIV, 16
HIV-1, 16
homopolycondensation, 56, 129
homopolymers, 39
host, 5
house dust, 20
hydrocarbons, 32, 40, 72, 73, 113
hydrocortisone, 19
hydrogels, 6, 7, 17
hydrogen, viii, 30, 36, 39, 54, 60, 61, 63, 66, 82, 84, 96, 101, 115, 117, 118, 193, 285, 287, 288, 289
hydrogen bonds, viii, 60, 61, 285, 287, 288, 289
hydrogen bromide, 39
hydrogen chloride, 82, 84, 101, 115, 117, 118
hydrolysis, 6, 29, 52, 54, 81, 95, 101, 121
hydroperoxides, 56
hydroxide, 132
hydroxyl, 13, 30, 52, 82, 85, 193
hydroxyl groups, 13, 82, 85

I

ideas, 30, 168
identification, 27
identity, 163, 235
immune response, 4
immunization, 16, 20
immunogenicity, vii, 1, 2
in vitro, 4, 7, 10, 13, 14, 18, 19, 20, 21
inclusion, 108, 236
incubation time, 7
independence, 75
indication, 144
indices, 36, 37, 60, 63, 148, 164, 173, 188
induction, 37
induction period, 37
industry, 27
infancy, vii, 1
infection, 16, 23
inflammation, 6
influence, viii, 5, 9, 13, 27, 28, 32, 34, 37, 40, 41, 59, 60, 61, 63, 64, 65, 66, 69, 70, 72, 75, 77, 90,

91, 96, 118, 140, 141, 142, 144, 145, 254, 255, 256, 257, 285, 289, 291, 292, 296, 297, 299, 301, 303
infrared spectroscopy, 38
inhibition, 5, 36, 56
inhibitor, 39
inhomogeneity, 152
initial reagents, 85, 120, 130
initiation, 37, 41, 50, 52
injections, 2
insects, 8
insertion, 3, 176, 210
insight, 145
instability, 9, 39
insulation, 40
intensity, 34, 38, 40, 55, 60, 61, 96, 107, 108, 142, 188, 191, 286, 288, 289
interaction, 5, 7, 9, 10, 11, 30, 32, 36, 38, 39, 54, 59, 60, 61, 62, 63, 64, 65, 67, 70, 71, 180, 191, 196, 203, 209, 249, 254, 255, 256, 268, 269
interactions, 2, 4, 17, 59, 60, 63, 67, 141
interest, 2, 40, 56, 69, 89, 121, 214, 222, 243, 249
interface, 285
interfacial layer, viii, 285, 287, 289
interlacements, 261, 262, 270, 271
intermolecular interactions, 108, 156
internalization, 5, 19
interphase, 196
interpretation, 62, 64, 71, 147, 165, 166, 168, 171, 193, 244, 249
interval, 175, 184, 209, 220, 226, 227, 228, 236, 237, 238, 239, 279, 286, 300
intravenously, 10
inversion, 302
iodine, 36, 62, 69
ions, 52, 60, 72
IR spectra, 121
iron, 27, 192, 193
irradiation, 9, 25, 35, 36, 142
IR-spectra, 64, 68
IR-spectroscopy, 30, 32, 55, 68, 186, 189
isolation, 12, 13, 96
isomers, 85, 86, 90, 94, 104, 121, 122, 125
isoprene, 27
isotactic polypropylene, 146, 261, 262, 263, 264, 265, 266, 270, 271, 273, 274, 275
isothermal heating, 38
isotherms, 261, 262, 263, 270, 272
Israel, 13
Italy, 46
iteration, 221

J

justification, 72

K

KBr, 63
kidney, 10, 11
kinetic parameters, 206
kinetics, 5, 7, 27, 41, 53, 54, 55, 56, 60, 64, 66, 71
knowledge, 104, 156, 180, 278
KOH, 132

L

Landau theory, 163
lead, 15, 39, 81
lifetime, 57, 140
ligands, 5, 11, 15
light scattering, 96
limitation, 61, 190
linear dependence, 69, 71, 72
linear function, 214, 226
linear polymers, 28, 162, 169
linkage, 12, 33, 140, 141, 227
links, vii, 123, 139, 141, 217, 242, 250, 251, 254, 287, 300
lipids, 5
liposomes, 7, 11, 22
liquid chromatography, 103, 113
liquid phase, 29, 64, 71, 260, 268
liquids, 35, 41, 121, 130, 244, 249, 256
liver, 10, 11
liver cells, 11
local order, 149, 150, 151, 156, 158, 160, 162, 163, 165, 168, 171, 172, 173, 178, 180, 182, 183, 194, 196, 214, 220, 226, 231, 235, 242, 246, 247, 248, 286, 292, 293, 301
localization, 6, 18, 140, 215, 218, 221, 232, 303
location, 2, 104
low density polyethylene, 146, 172
low temperatures, 26, 103, 265, 274
LTD, 282
luciferase, 16
lysosome, 3

M

mAb, 7
macromolecular coil, 90, 92
macromolecular coils, 180, 181
macromolecular entanglements, 147, 150, 171, 174, 183, 184, 185, 188, 194, 227

macromolecules, 11, 21, 38, 52, 90, 91, 92, 94, 95, 101, 104, 105, 148, 150, 152, 159, 164, 165, 166, 171, 177, 178, 181, 206, 214, 215, 226, 241, 244, 254, 255, 257, 260, 268, 285, 287
macroradicals, 210
magnetic field, 142, 191, 194
magnetic moment, 191, 192
magnetization, 191
marker genes, 14
Marx, 249
mass, 30, 32, 34, 38, 39, 51, 52, 85, 88, 96, 97, 110, 113, 116, 118, 119, 123, 125, 127, 128, 129, 133, 142, 175, 195, 205, 206, 207, 208, 209, 210, 211, 250, 259, 262, 267, 271, 278, 286
mass loss, 38, 51, 88, 97, 116, 118, 119, 123, 127, 129, 133
mass spectrometry, 110
material degradation, 33, 35
material surface, 25, 32, 33, 36
matrix, 3, 6, 15, 19, 26, 32, 37, 40, 41, 148, 149, 153, 154, 169, 178, 182, 192, 214, 215, 223, 225, 226, 229, 277, 278, 280, 281, 285, 286, 292, 293, 294
maturation, 12
meanings, 262, 263, 271
measurement, 92, 96, 142, 189, 233, 234, 236, 279
measures, 164
mechanical properties, 153, 161, 162, 291, 292, 296
mechanical testing, 216, 292
media, 7, 9, 28, 60, 61, 186
median, 7, 237
mediation, 11
medium composition, 28
melt, viii, 26, 35, 38, 41, 49, 50, 52, 54, 103, 104, 142, 144, 145, 150, 171, 184, 187, 188, 190, 193, 194, 195, 196, 213, 231, 232, 233, 235, 236, 237, 244, 249, 250, 251, 252, 253, 254, 255, 256, 263, 266, 272, 275
melt flow index, 188, 193, 195
melting, 26, 35, 38, 50, 51, 54, 108, 144, 156, 157, 163, 166, 186, 187, 228, 231, 261, 262, 263, 264, 270, 271, 272, 274, 301
melting temperature, 26, 35, 156, 157, 163, 186, 187, 228, 261, 264, 270, 274, 301
melts, 188, 190, 254
membranes, viii, 245, 299, 304
memory, viii, 259, 265, 267, 274, 286
mercury, 142
metallocenes, 203
metals, 26, 52, 152, 218
methanol, 65, 90, 115
methyl groups, 99
methyl methacrylate, 34, 39, 46, 153, 172

methylene chloride, 216
MFI, 141, 188, 195, 196
mice, 20, 23
microcavity, 169, 172
microheterogeneity, 176
microscope, 96
microspheres, 4, 7, 17, 19, 21, 23
microstructure, 113
microvoid, viii, 213, 214, 215, 216, 219, 221, 222, 223, 224, 225, 229, 230, 232, 234, 236, 237, 238, 239, 240, 241, 244, 301
migration, 32, 140
mixing, 6, 9, 39, 62, 70, 142, 249, 255
mobility, 149, 166, 171, 174, 242, 243, 287
mode, 39, 142, 146, 242
modeling, 90, 215, 241, 280
models, 59, 62, 64, 70, 106, 140, 150, 152, 153, 155, 175, 228, 229, 280
modulus, 147, 148, 152, 154, 225
moisture, 216
molar volume, 157, 166
mold, 51
mole, 53, 241, 251
molecular liquids, 254
molecular mass, vii, 20, 21, 85, 86, 89, 91, 92, 95, 96, 101, 104, 115, 118, 125, 130, 133, 148, 162, 169, 175, 179, 184, 185, 195, 196, 203, 204, 205, 206, 207, 208, 209, 210, 211, 216, 250, 256, 260, 262, 268, 271
molecular mobility, viii, 178, 242, 243, 244, 286, 287, 288, 299, 300, 301, 304
molecular structure, 141, 142, 145, 146, 177, 249
molecular weight, 9, 10, 11, 13, 16, 19, 20, 21, 108, 217
molecules, viii, 5, 8, 13, 29, 64, 77, 91, 111, 144, 209, 215, 232, 238, 239, 244, 249, 255, 259, 260, 262, 267, 268, 271, 299, 304
monograph, 71
monomers, 52
morphology, 56, 150, 186, 188, 189, 190
Moscow, 25, 42, 44, 45, 46, 49, 57, 58, 59, 76, 77, 78, 81, 133, 135, 146, 147, 197, 198, 199, 200, 201, 212, 213, 244, 245, 247, 259, 267, 277, 283, 285, 289, 290, 291, 297, 298, 299, 305
motion, 35, 174
moulding, 216, 225
mucin, 16
multifractal formalism, 165
multifractality, 232, 233, 236
multiplier, 301
muscles, 15

N

nanoparticles, 4, 9, 10, 16, 17, 20
natural polymers, 19
needs, 218, 268
neglect, 260, 269
neon, 238
network, viii, 30, 31, 32, 88, 113, 147, 148, 150, 151, 161, 163, 168, 174, 178, 179, 180, 194, 196, 217, 223, 277, 278, 281, 282
nitrides, 27
nitrobenzene, 72
nitrogen, 30, 33, 34, 38, 41, 90, 186, 238
nitroxyl radicals, 141, 143, 145
NMR, 53, 55, 57, 64, 70, 75, 86, 103, 105, 110, 118, 119, 121, 122, 124, 131
nonequilibrium, viii, 213, 220, 244, 304
nucleus, 3, 4, 6, 18
nutrients, 6

O

obligate, 97
observations, 19, 147, 188, 209, 231, 294
oil, 7, 130
oligomeric structures, 86
oligomers, 5, 20, 55, 56, 95, 109, 111, 113, 117, 119, 132
organic compounds, 25, 76, 96, 113
organic polymers, 104
organic solvents, 9, 63, 85, 89, 90, 103, 118, 123, 125, 133
organization, 6
organocyclosiloxanes, 82, 90, 96, 129
orientation, viii, 49, 52, 60, 188, 281, 291, 293, 296, 297
oscillograph, 216
output, 56
oxidation, 13, 28, 29, 34, 35, 37, 54, 56, 57, 140, 141, 142, 144, 193
oxidation products, 29, 193
oxidation rate, 54
oxides, 27, 31, 193
oxygen, 27, 30, 31, 32, 33, 34, 36, 37, 39, 40, 50, 53, 56, 142, 143
oxygen absorption, 53

P

parameter, viii, 62, 63, 64, 65, 66, 67, 68, 69, 70, 71, 72, 73, 75, 76, 92, 95, 115, 140, 142, 151, 154, 155, 162, 163, 164, 165, 166, 167, 169, 170, 174, 181, 182, 183, 206, 217, 220, 226, 228, 229, 230, 236, 241, 280, 281, 291, 293, 294, 295, 296, 297, 302
parameter estimation, 92
particles, 4, 9, 10, 15, 19, 21, 29, 191, 192, 193, 194, 218, 278, 286, 292
passive, 26
pathogens, 15
percolation, viii, 213, 215, 228, 229, 241, 242, 244, 287, 301
percolation cluster, viii, 213, 228, 229, 244
percolation theory, 228, 241
permeability, 11, 35, 36, 37, 41, 299
permeation, 34, 40, 204
peroxide, 203, 204, 205, 208
peroxide radical, 203
PET, 141, 142, 144, 145
pH, 9, 11, 13, 16, 19, 20, 21
phagocytosis, 4
pharmacology, 5
phase transformation, 33
phase transitions, 33, 51, 160, 163, 177
phenol, 28, 36, 37, 56, 61, 62, 65, 113
PhOH, 68, 69, 76
phosphates, 30, 31, 32, 40
phosphorus, 29, 30, 31, 32, 34, 36, 37
phosphorylation, 31
photooxidation, 142
photooxidative degradation, vii, 139, 144
physical aging, 159, 219
physical properties, 73, 140, 144, 146, 245, 278
physical-mechanical properties, 245
physics, 214, 246, 247, 256
plants, 12
plasma, 11, 16, 52
plasmid, vii, 1, 3, 4, 5, 6, 9, 11, 13, 15, 20, 21, 22
plastics, viii, 26, 247, 277, 278, 279, 280, 281, 282, 291, 292, 293, 295, 296, 297, 298
platinum, 122
PMMA, 164, 167, 168, 169, 170, 176
Poisson ratio, 220
polar media, 60, 63
polarity, 60, 64, 66, 67, 69, 71
polarization, 60, 63, 70
poly(ethylene terephthalate), 30
poly(methyl methacrylate), 33, 34, 35
polyamides, 52
polyarylate, 149, 178, 216, 227, 285, 287
polybutadiene, 27, 28
polycarbonate, 52, 149, 216, 222, 227
polycondensation, 26, 82, 83, 84, 85, 87, 124
polydimethylsiloxane, 86, 300
polydispersity, 205, 206

polyethylenes, 150, 163, 182, 184, 188, 189, 216, 230, 236, 244
polyheteroarylenes, 56
polyisobutylene, 27, 28
polyisoprene, 28
polymer amorphous state, 147, 148, 150, 152, 153, 154, 156, 160, 165, 173, 178
polymer amorphous state structure, 147, 160
polymer chains, 179, 196, 261, 262, 270, 271
polymer clusters, 180
polymer combustibility, 35, 36
polymer combustion, 29, 33, 40
polymer composites, viii, 278, 285, 288, 289, 295, 297
polymer density, 148, 169, 175, 180, 262, 271
polymer materials, 256
polymer matrix, 140, 145, 169, 191, 194, 278, 280, 281, 286, 287, 292, 293, 294, 295, 296
polymer melts, 35
polymer molecule, 40
polymer networks, 157, 158, 160, 169
polymer oxidation, 144, 145
polymer properties, 154, 249
polymer solutions, 246, 254
polymer structure, viii, 108, 110, 122, 153, 158, 160, 164, 165, 166, 170, 171, 181, 213, 214, 216, 217, 219, 220, 221, 226, 233, 236, 244, 263, 272, 299, 301, 304
polymer synthesis, 133
polymeric chains, 29, 32
polymeric composites, 29, 39, 218
polymeric materials, 26, 27, 34, 36, 37, 39, 40, 245, 246, 247, 248, 257
polymeric matrices, 15, 32
polymeric melt, 195
polymeric membranes, 215, 232, 241
polymeric products, 85
polymerization, vii, 26, 81, 84, 108, 122, 124, 129, 131, 132, 133, 203, 204, 205, 206, 207, 208, 209, 210, 211
polymerization process, 206
polymerization temperature, 211
polymers, vii, 2, 4, 5, 9, 13, 16, 18, 25, 27, 28, 30, 31, 33, 34, 35, 36, 39, 49, 52, 53, 56, 57, 81, 82, 88, 91, 96, 101, 105, 108, 109, 119, 120, 121, 129, 133, 140, 141, 145, 146, 149, 150, 151, 152, 153, 154, 155, 157, 158, 159, 160, 161, 162, 164, 166, 170, 171, 175, 178, 180, 182, 183, 188, 193, 194, 196, 204, 213, 214, 215, 216, 217, 218, 219, 220, 223, 226, 227, 228, 229, 231, 232, 242, 243, 244, 245, 246, 247, 249, 250, 254, 255, 256, 257, 266, 267, 285, 287, 299, 303
polyolefins, vii, 139, 140, 145, 146

polypeptide, 9
polyphosphates, 30, 31, 37
polypropylene, 30, 31, 32, 150, 182, 257, 262, 263, 271, 272
polypyromellitimides, 57
polystyrene, 32, 33, 34, 37, 52, 159, 204, 205, 209, 210, 211
population, 5
porosity, 215, 226, 278
positron, 215, 223, 225
positrons, 222, 224, 225
potassium, 37, 132
power, 25, 33, 142, 218
precipitation, 95, 96, 105, 125, 264, 266, 274, 275
prediction, 278, 303
preparation, vii, 1, 2, 18
pressure, 41, 156, 166, 250, 280, 292, 293
prevention, vii, 1, 23
primary products, 113
principle, vii, 1, 28, 40, 59, 61, 63, 71, 168, 176
probability, 32, 62, 73, 144, 152, 153, 174, 206, 211, 234, 236, 237, 238
probe, 17
production, vii, 1, 16, 27, 39, 41, 113, 279
production technology, 279
program, 13, 110, 212
proliferation, 7
propagation, 210
proportionality, 230
propylene, 32
proteins, 15, 22
protons, 118, 125, 130
purification, 89, 90, 96
PVC, 176
PVP, 21
pyrolysis, 27, 28, 29, 30, 31, 33, 34, 35, 36, 37, 38, 40, 41, 96, 108, 109, 110, 111, 112, 113, 117
pyrophosphate, 30, 32

Q

quartz, 142
quasi-equilibrium, viii, 174, 213, 219, 243, 244
quasi-equilibrium state, 174, 219, 243

R

radiation, 286
radical mechanism, 96
radical polymerization, vii, 203, 204, 206, 209, 210, 211
radius, 9, 222, 224, 233
radius of gyration, 9, 233

range, 9, 14, 35, 36, 37, 38, 39, 50, 51, 52, 54, 55, 56, 62, 63, 64, 69, 86, 90, 95, 96, 103, 104, 105, 106, 107, 108, 109, 113, 118, 122, 127, 130, 132, 147, 150, 152, 156, 161, 162, 166, 167, 174, 175, 176, 177, 184, 191, 193, 195, 207, 216, 221, 223, 226, 241, 268
reaction rate, 59, 65, 67, 124
reaction rate constants, 124
reaction temperature, 120, 125, 132, 133
reaction zone, 55
reactive groups, 34, 38, 41, 84
reading, 245, 250
reagents, 7, 10, 117
reality, 160, 168, 215
receptors, 2, 7
recognition, 2, 7
recombination, 30, 39, 210
reconstruction, viii, 259, 267
rectification, 121
redistribution, 38, 293, 294
reduced combustibility, 25
reduction, 13, 26, 28, 33, 39, 162, 193, 252, 288
refining, 27
reflection, 106, 107, 108
reflexes, 105
regeneration, 6, 19, 140
regression, 63, 66, 75
regression equation, 66
regulation, 6, 103, 104, 181
reinforcement, 27, 285, 289
relationship, 4, 9, 62, 140, 216, 217, 218, 219, 220, 221, 222, 228, 229, 230, 233, 238, 239, 243, 251, 281, 282, 286, 287, 291, 292, 293, 294, 295, 300, 301, 303
relationships, 59, 61, 66, 71, 279
relative size, 237
relaxation, 163, 174, 256, 266, 275
reliability, 73
replacement, 66, 230, 235, 241
residues, 22, 28, 36
resins, 28, 36, 108
resistance, 26, 29, 32, 41, 186
resolution, 205
resonator, 142
respiratory, 23
respiratory syncytial virus, 23
restructuring, 175
retention, 109, 110
risk, 2
room temperature, 27, 82, 84, 150, 182, 263, 271
root-mean-square, 206
roughness, 241, 286
RSV infection, 16

rubber, 27, 28, 147, 148, 185, 223
rubbers, 26, 27, 28, 113, 160, 161, 162
rubbery state, 41, 147, 161
Russia, 25, 49, 59, 81, 147, 203, 213, 249, 256, 257, 259, 267, 285, 291

S

safety, vii, 1, 4
salts, 5, 9, 39, 52, 63
sample, 28, 32, 33, 34, 50, 51, 142, 161, 259, 264, 267, 274
satisfaction, 221
saturation, 192, 260, 268
scaling, 174, 237
scanning calorimetry, 246
scatter, 280, 303
scattering, 36, 96, 165, 186, 188, 189, 286
scattering patterns, 188, 189
search, 189
secretion, 15
sedimentation, 266, 269, 275
selectivity, viii, 7, 113, 299, 301, 302, 303, 304
self, 9, 10, 11, 18, 21, 40, 156, 157, 218, 220, 221, 222
self-assembly, 18, 156
self-control, 40
self-similarity, 218, 220, 221, 222
semi-crystalline polyethylene, viii, 213, 232, 233, 234, 236, 237, 238, 244
semicrystalline polymers, 301
semi-crystalline polymers, 216, 228, 229, 231, 232
sensitivity, 60, 65, 144
separation, 60, 67, 81, 108, 113, 114, 169, 278, 280, 282, 292, 294, 300
series, 33, 39, 40, 63, 64, 65, 66, 69, 75, 81, 111, 113, 121, 141, 154, 161, 172, 182, 189, 191, 242
serum, 9, 10, 20
shape, 5, 20, 28, 66, 94, 108, 149, 157, 163, 170, 181, 192, 260, 267, 268
Sharpy method, 216
shear, 142, 148, 153, 154, 190
sign, 28, 94, 217, 230, 233, 239
signal peptide, 18
signals, 6, 105, 144
silicon, 26, 82, 84, 90, 97, 99, 105, 118
similarity, 163, 220, 221, 241
simulation, 141, 146, 160, 219, 224, 225, 241
sites, 2, 281, 287, 288
skeleton, 227
skin, 15, 19
smoke, 32
solid phase, 31, 144, 153, 196, 254

solid polymers, 160
solid state, 41, 214, 231, 249
solubility, 10, 11, 63, 66, 73, 75, 95, 140, 142, 238, 239, 254, 255, 256, 260, 262, 264, 266, 269, 270, 271, 274, 275
solvation, 59, 60, 61, 63, 64, 65, 66, 67, 68, 69, 70, 71, 72, 73, 75, 77
solvent acidity, 63
solvent molecules, 262, 268, 270, 271
solvents, 63, 64, 65, 66, 67, 68, 69, 70, 71, 72, 73, 74, 75, 76, 77, 92, 101, 261, 270
sorption, 238, 260, 261, 262, 263, 264, 268, 269, 270, 271, 273, 274
sorption isotherms, 261, 262, 263, 269, 270, 271, 273
specific surface, 33, 191, 219
specificity, vii, 1, 4, 11, 159, 163, 241
spectral dimension, 301
spectrophotometry, 52
spectroscopy, 38, 52, 68, 70, 110, 121, 141, 215, 225
spectrum, 25, 26, 39, 61, 65, 69, 96, 118, 125, 130, 131, 169
speed, 174, 250
spherulite, 186
spin, 142
spleen, 4
stability, 4, 5, 9, 10, 11, 27, 50, 51, 52, 56, 81, 97, 113, 116, 117, 118, 127, 140, 142, 176, 177, 178, 265, 274, 285
stabilization, vii, 56, 57, 104, 139, 140, 141, 142, 144, 145, 247
stabilization efficiency, 145
stabilizers, vii, 139, 140, 141, 142, 143, 144, 145, 146
stages, 25, 51, 54, 126, 129
standard error, 73
standards, 109
statistical processing, 62
statistics, 148, 179, 182
steel, 286
strain, 147, 161, 216, 292
strategies, vii, 1, 17, 18
strength, viii, 28, 29, 50, 62, 66, 70, 153, 154, 155, 291, 292, 293, 294, 295, 296, 297
stress, 147, 150, 152, 153, 154, 155, 161, 170, 216, 293
stretching, 150
strong interaction, 62, 104
structural changes, 186, 190, 224, 296
structural characteristics, 164, 176, 231
structure formation, 31, 156, 293
structuring, 34, 96

styrene, vii, 39, 203, 204, 207, 208, 211
substitution, 32, 99, 118, 166, 261, 270
subtraction, 65
sugar, 12
sulfonamide, 38
sulfur, 38
Sun, 20
superimposition, 50, 168
suppression, 34, 103, 196
surface area, 7, 219
surface layer, 34
surface modification, 5, 11
surface structure, viii, 33, 285, 289
surface tension, 41
surface treatment, 26
susceptibility, 189, 191
swelling, 7
symbols, 112
symmetry, 142
syndiotactic sequences, 203
synergetic system, 245
synergetics, 247, 298
synthesis, 6, 14, 20, 81, 82, 85, 90, 109, 115, 120, 122, 133
synthesized copolymers, 89, 115, 118, 122, 125
synthetic polymers, 16
systems, vii, 2, 3, 5, 6, 15, 16, 17, 18, 19, 21, 22, 23, 27, 29, 30, 31, 33, 41, 61, 62, 63, 71, 100, 127, 152, 156, 157, 171, 203, 204, 205, 207, 208, 209, 210, 219, 249, 253, 254, 255, 257, 291

T

tacticity, 101, 106, 107, 108
technology, vii, 1, 15, 18, 29, 246, 257
temperature, 9, 13, 25, 26, 27, 28, 29, 32, 33, 35, 36, 37, 38, 39, 40, 41, 44, 46, 50, 51, 53, 55, 56, 59, 86, 87, 89, 90, 93, 94, 95, 96, 103, 104, 105, 106, 107, 108, 109, 110, 113, 117, 118, 122, 123, 127, 132, 148, 149, 152, 154, 156, 157, 159, 161, 163, 167, 168, 169, 170, 172, 174, 175, 176, 177, 178, 179, 180, 181, 182, 190, 194, 211, 214, 216, 220, 223, 224, 225, 226, 227, 228, 229, 231, 232, 238, 240, 241, 242, 243, 244, 250, 251, 252, 253, 254, 255, 261, 263, 269, 270, 271, 287, 292, 300, 301
temperature dependence, 107, 154, 163, 177, 180, 181, 226, 228, 238, 241, 300, 301
tension, 147, 182, 183, 194, 250, 255
tetrahydrofuran, 120
TGA, 50, 51, 52, 53, 119
T-helper cell, 22

theory, 72, 152, 153, 156, 162, 163, 169, 175, 177, 185, 214, 222, 223, 224, 225, 227, 236, 244, 256, 259, 267, 303
therapy, vii, 1, 5, 7, 14, 16, 17, 18, 22
thermal analysis, 105
thermal cluster, 229, 292
thermal decomposition, 39
thermal degradation, 26, 27, 29, 96, 108, 133
thermal oxidation, 53, 54, 55, 56
thermal properties, 279, 292
thermal resistance, 39
thermal stability, 10, 31, 51, 52, 53, 57, 96, 101, 103, 104, 113
thermodynamic equilibrium, 219, 220
thermodynamic nonequilibrium, 219, 244
thermodynamic properties, 162, 278
thermodynamics, 59, 166
thermogravimetric analysis, 39
thermooxidation, 31, 142, 145
thermooxidative degradation, 29, 33, 37
thermooxidative stability, 97
thermostabilization, 142
threshold, 215, 229, 241, 242
Tikhonov regularization method, vii, 203, 206
time, 7, 9, 15, 27, 31, 32, 33, 40, 69, 75, 82, 92, 95, 110, 142, 144, 160, 174, 183, 209, 216, 221, 241, 259, 262, 265, 267, 271, 274
tin, 37
tissue, 4, 15, 22
titanium, 40
TMC, 11
toluene, 82, 84, 85, 89, 92, 93, 94, 95, 102, 106, 115, 116, 118, 120, 121, 123, 125, 128, 129, 204
total energy, 60
toxicity, 2, 4, 9, 12, 15, 16, 34
trade, 286
transcription, 3
transfection, vii, 1, 2, 4, 5, 6, 7, 9, 10, 11, 14, 15, 16, 17, 18, 20, 21, 22
transformation, 117, 142, 144
transformations, 26, 39, 54
transition, 26, 38, 41, 50, 56, 78, 91, 94, 97, 101, 122, 127, 133, 156, 157, 158, 159, 160, 162, 163, 167, 174, 210, 211, 214, 223, 228, 231, 243, 281, 291
transition metal, 56
transition rate, 41
transition temperature, 162, 163, 214
transitions, 51, 91, 94, 105, 106, 157
translation, 3
translocation, 3
transmission, 35
transmits, 25
transplantation, 18
transport, viii, 9, 11, 22, 232, 241, 281, 299, 300, 301
transport processes, viii, 299
transportation, 6
trend, 143, 144
turbulence, 218

U

Ukraine, 59, 213, 277, 285, 291, 305
universal gas constant, 148, 232, 238, 300
unstructured liquid, 243
USSR, 133, 135, 136, 137, 199, 201
UV, vii, 64, 68, 70, 75, 139, 140, 142, 143, 144, 145, 261, 269
UV light, 261, 269

V

vaccines, 15, 16, 17, 22, 23
vacuum, 86, 90, 216, 263, 271
valence, 27, 38, 91, 93, 104, 118
validation, 63, 140, 156
validity, 66, 75, 175, 184
values, viii, 11, 31, 33, 62, 63, 66, 68, 69, 71, 72, 82, 83, 84, 87, 90, 91, 92, 93, 94, 101, 106, 108, 110, 111, 118, 121, 123, 127, 140, 142, 144, 148, 150, 151, 153, 154, 157, 159, 161, 168, 171, 172, 173, 175, 176, 177, 178, 180, 182, 184, 186, 189, 192, 193, 196, 204, 205, 210, 213, 214, 216, 218, 220, 221, 222, 223, 224, 225, 226, 227, 228, 229, 230, 231, 232, 236, 238, 239, 240, 243, 244, 251, 253, 254, 255, 256, 262, 271, 278, 280, 293, 299, 300, 301, 302, 303, 304
variable, 60, 64, 73, 157, 182
variables, 278
variance, 227
variation, viii, 81, 106, 108, 116, 121, 153, 162, 165, 174, 177, 178, 180, 184, 186, 188, 190, 194, 196, 224, 225, 232, 233, 235, 237, 285, 289, 293, 303
vector, vii, 1, 2, 3, 4, 5, 10, 11, 15, 16, 17, 18, 20, 22, 64, 153
vehicles, 17
vessels, 27
vinyl monomers, 210
viral vectors, vii, 1, 2, 17
viscoelastic properties, 41
viscosity, 35, 36, 37, 38, 41, 52, 67, 87, 90, 115, 120, 121, 122, 142, 186, 188, 190, 195, 250, 255

W

water, 2, 7, 10, 11, 17, 30, 31, 32, 54, 262, 271
wear, viii, 278, 285, 286, 288, 289, 292
words, 4, 207, 251
work, 14, 61, 64, 69, 76, 85, 92, 94, 141, 145, 148, 150, 152, 157, 162, 164, 174, 176, 180, 189, 190, 195, 204, 212, 219
wound healing, 6, 15, 19

X

X-ray analysis, 51
X-ray diffraction, 54, 57, 110, 120, 127

Y

yield, 13, 31, 32, 35, 37, 41, 85, 86, 87, 88, 105, 115, 123, 129, 141, 142, 143, 144, 145, 170, 203, 210, 216, 230, 293

Z

zeolites, 31, 32, 36